ADVANCES IN
NUMERICAL HEAT TRANSFER

ADVANCES IN
NUMERICAL HEAT TRANSFER

Editors
W.J. Minkowycz, *University of Illinois, Chicago, Illinois*
E.M. Sparrow, *University of Minnesota, Minneapolis, Minnesota*

ADVANCES IN
NUMERICAL HEAT TRANSFER

Volume 2

Edited By

W.J. Minkowycz

Professor of Mechanical Engineering
University of Illinois
Chicago, Illinois

E.M. Sparrow

Professor of Mechanical Engineering
University of Minnesota
Minneapolis, Minnesota

CRC Press
Taylor & Francis Group
Boca Raton London New York

CRC Press is an imprint of the
Taylor & Francis Group, an **informa** business
A TAYLOR & FRANCIS BOOK

CONTENTS

In putting together the contents of this second volume of *Advances in Numerical Heat Transfer*, the editors faced the daunting task of striking a balance between generic fundamentals, specific fundamentals, generic applications, and specific applications. The final contents of the volume reflects the diversity of this balance.

The volume begins (Chapter 1) with a broad overview of the field of high-performance scientific computing as personified by its role in fluid flow and heat transfer problems. Future expectations are viewed through the window of present capabilities. Generic fundamentals are then treated in a pair of articles (Chapters 2 and 3). The first of these deals with unstructured meshes used in conjunction with finite volume methods. The second sets forth the spectral element method, which combines finite elements and spectral schemes to produce high-order spatial accuracy.

The next group of articles fall into the category of generic applications. Chapter 4 provides a tutorial presentation of the use of the finite volume method for the numerical solution of radiative heat transfer problems. The next chapter (Chapter 5) is focused on heat conduction and on the use of the boundary element method for both steady and unsteady problems. A special feature of the presentation is the consideration of non-Fourier constitutive equations.

Recent interest in processes at the microscale motivated the inclusion of a pair of articles (Chapters 6 and 7) dealing with the special numerical issues related to solving microscale heat transfer problems. The first of these features the molecular dynamics method while the second includes other solution methods such as Monte Carlo. The Monte Carlo method is also featured in Chapter 8, where the current status of the use of parallel computing for turbulent reacting flows is reviewed.

Two highly computation-dependent specific applications are reviewed in Chapters 9-11. The first of these is flow and heat transfer in porous media, respectively discussed for forced convection in Chapter 9 and for natural convection and mixed convection in Chapter 10. The second specific application (Chapter 11) is the thermal management of electronic systems.

The editors would appreciate comments on the choice of the content for Volume 2, with particular focus on the diversity of the balance of the subject matter.

W.J. Minkowycz
E.M. Sparrow

C.H. AMON
Carnegie Mellon University
Department of Mechanical Engineering
Schenley Park
Pittsburgh, PA 15213-3890

J. C. CHAI
Tennessee Technological University
Department of Mechanical Engineering
Box 5014
Cookeville, TN 38505

C.P. GRIGOROPOULOS
University of California
Department of Mechanical Engineering
6189 Etcheverry Hall #1740
Berkeley, CA 94720-1740

H.A. HADIM
Stevens Institute of Technology
Department of Mechanical Engineering
Castle Point on Hudson
Hoboken, NJ 07030

Y. JOSHI
University of Maryland at College Park
Department of Mechanical Engineering
College Park, MD 20742-3035

A.J. KASSAB
University of Central Florida
Department of Mechanical and
Aerospace Engineering
Box 162450
Orlando, FL 32816-2450

J.M. LOMBARDO
University of Nevada
Department of Mechanical Engineering
4505 Maryland Parkway
Las Vegas, NV 89154-4027

S. MARUYAMA
The University of Tokyo
Department of Mechanical Engineering
7-3-1 Hongo, Bunkyo-ku
Tokyo 113-8656
Japan

S.R. MATHUR
Fluent Inc.
10 Cavendish Court
Lebanon, NH 03766

J.Y. MURTHY
Carnegie Mellon University
Department of Mechanical Engineering
Pittsburgh, PA 15213

S.V. PATANKAR
University of Minnesota
Department of Mechanical Engineering
125 MEB
111 Church Street S.E.
Minneapolis, MN 55455

D.W. PEPPER
University of Nevada
Department of Mechanical Engineering
4505 Maryland Parkway
Las Vegas, NV 89154-4027

M.S. RAJU
Dynacs Engineering Co. Inc.
NASA Lewis Research Center
2001 Aerospace Parkway
Brook Park, OH 44142

K. VAFAI
Ohio State University
Department of Mechanical Engineering
206 W. 18th Avenue
Columbus, OH 43210-1107

L.C. WROBEL
Brunel University
Department of Mechanical Engineering
Uxbridge, Middlesex UB8 3PH
United Kingdom

M. YE
University of California
Department of Mechanical Engineering
6189 Etcheverry Hall #1740
Berkeley, CA 94720-1740

HIGH-PERFORMANCE COMPUTING FOR FLUID FLOW AND HEAT TRANSFER

D.W. Pepper
J.M. Lombardo

1 INTRODUCTION

1.1 Overview

The use of computers in heat transfer and fluid flow has become so commonplace today that no one would consider working in either field without some knowledge of computing. Some of the earliest computing efforts were spent on obtaining solutions to specific heat transfer and fluid flow problems. While such problems solved using computers over 30 years ago are now considered trivial exercises, they were run on what were then considered state-of-the-art, or high-performance computers. Solving 3-D problems with up to 640K available memory was extraordinary. When mentioning such problems and computer systems to students today, most laugh and wonder how one lived without a PC and access to the web. When one sees how far we have come in just a few decades, it is truly amazing to have ready access to tremendously powerful machines and software right from our desk. What is even more amazing is what is yet to come – both at the PC level and at the high-performance end of computing. Problems are now being solved on a daily basis that even a few years ago were considered intractable. While we once thought that a problem with a few million nodes was huge a few years ago, researchers are now considering problems with over 100 million nodes. At such levels of detail, one can begin to model processes at the micro level of physics. When researchers are able to quickly analyze these gigantic data sets and can generate insightful graphical displays, the understanding of fundamental processes and governing relations will escalate tremendously.

There is no doubt that a lot of fluid flow and heat transfer problems can be solved on PCs and workstations. The power of these desktop machines continues to increase at an exponential rate – 1000+ MHz PCs are just around the corner.

While many problems can be solved on such machines, it isn't long before one begins to push the envelope of the problem – to add more physics or more nodes to simulate more of the real processes. At this point, the role of high-performance computers comes into play. Once the specialized programming techniques and operating systems are mastered, the user begins to drive the machine to its maximum performance – and eventually saturates the machine. The need to develop even faster high-performance machines with more memory is insatiable.

1.2 What is High-Performance Computing?

In the late 1960's, new and powerful computers began to develop in universities and commercial laboratories. These computers came to be known as supercomputers. Today, such computers are referred to as high-performance computers. High-performance (or supercomputer) delineates the most powerful, state-of-the-art computer, and usually denotes a system with fast logic elements and multiprocessors, or parallel architectures. Such systems consist of a collection of processors that are connected together to work jointly on a single problem.

The speed of logic elements will ultimately approach the limits of performance imposed by the laws of physics; to overcome this limit, it is felt that parallel architectures are the key to future advances in speed and power. Parallel architecture allows many parts of a computation to be done simultaneously and is the major factor that distinguishes such machines from earlier scalar systems.

The use of high-performance computing has become an accepted practice for solving complex problems in many different fields. One needs only a high-speed modem or web access and log-on privileges to connect to a large-scale machine. While there are now fewer large supercomputer centers throughout the country, the current centers contain computers of extraordinary power surpassing the scalar and vector supercomputers of just a few years ago. Providing one has a plausible project that requires some extensive computing, it is neither difficult nor costly to obtain access to any of the sites. The cost of such high-performance machines has now come down to such levels that many universities and laboratories have their own systems. Only a few years ago, it was not uncommon to spend many millions of dollars to acquire a supercomputer, e.g., Y-MP Cray supercomputers were generally the fastest and most expensive of the supercomputers. Today, parallel computer systems not much larger than workstations exist that cost only a few hundred thousand dollars, are air cooled, and can be placed almost anywhere.

A high-performance computer is either a machine capable of storing large sets of numbers in its core memory and performing calculations at a high rate, or the hardware architecture of the machine permits access to many numbers and performs calculations at fast speeds. The Cray class machines of a few years ago provided computing power and permitted massive storage for in-core calculations using scalar and vector operations. Today, high-performance computers use low-power workstations and PC-level chips in specific architectures that permit parallel computing which surpasses previous Cray performance. It is the ability to perform millions or billions of operations per second using a defined set of operations that classify a computer as a high-performance machine.

Over 90% of a high-end computer's calculational effort is spent performing arithmetic operations [1]; performance in this area is rated in floating point operations per second - *Megaflops* (10^6). Workstations and low-end machines spend over 80% of their time executing input/output (I/O) and operating systems calls - referred to as millions of integer operations per second, or *Mips*. To qualify as a high-performance computer over 10 years ago, the sustained rate of performance had to be around 20 Mflops with an average capacity of a million or more words. Today, many workstations perform at such rates and are typically standard for normal computing on the desktop. High-performance computers now deal with *Gigaflop* (10^9) performance - it is not too far away that such machines will routinely reach *Teraflop* (10^{12}) performance. However, one must be careful when evaluating capabilities of a high-performance computer for specific tasks. The ability to solve complex problems involving numerous variables may be more important than how fast the computer processes a set of numbers. For example, aircraft designers principally desire values for drag and lift at various angles of attack whereas detailed viscous solutions of the flow around an airfoil are of secondary interest. Performing such detailed calculations for every possible variation would require an excessive amount of time. However, once a preliminary design is chosen, detailed flow calculations are desired to finalize the design. An example of such an analysis is the Boeing 777, which was principally designed using computers.

A major factor in the improvement in computing performance over the last few years has been a steady decrease in the size of the microelectronic circuits. The number of transistors that is now being placed on a silicon chip exceed several million, which is a change from the dozen or so first produced in the early 1960's. Such chips now permit large, fast memories to be achieved at very reasonable cost and cycle times. The state of a machine is determined at each tick of its clock. The shorter the clock period, the faster the operations. The primary limit on clock rate is the response time of the gates, or switches, which make up the computer. With the advent of the transistor, cycle times were reduced to nearly 100 nsec. Today, integrated circuit technology has reduced clock times to nearly 1 nsec. One of the simplest ways to reduce clock time is to reduce the distance between various parts of the computers. This approach recognizes the fact that no signal can travel faster than the speed of light. The reduction of the distance between processing centers is now apparent in the decreasing size of computers, especially when compared to large, room-sized machines of a decade ago. Increasing the density of the integrated circuitry has created microcomputer chips (or microprocessors) that continue to improve in performance with each new design. However, a tremendous amount of memory and associated logic circuitry must also be provided. This has led to compacting everything into a small package and has become a very challenging problem dealing with heat transfer – the removal of waste heat generated by the chips [2].

1.3 A Brief History

The development of the electronic computer first began in 1939 with the Model I developed by Bell Labs. It operated at around 0.5 flops. A few years later, the Z3 and ABC were constructed at Iowa State University. In 1944, the Mark I was built by Harvard University and IBM. The Bletchley Park Colossus Mark I and Mark II were built in 1943 and 1944 to break the German Enigma code. After the war, the University of Pennsylvania built ENIAC, which was used to calculate the ballistics of cannon shells. The ENIAC proved to be over 800 times faster than the Model I computer. The UNIVAC computer, built in 1951, proved to be five times faster than the ENIAC and is considered the machine that launched the performance race among commercial companies, e.g., IBM, CDC, and Remington-Rand.

IBM began its dominance of the high-end computers with the 700 series introduced in 1952, ending with the 7094 in 1962. IBM had the clever ability to double the performance of its machines every 12 months – a practice now followed by Intel and other chip manufacturers. In 1956, IBM introduced the "stretch" concept for its 700 series computers and delivered one to LANL in 1961. The machine was 25 times faster than its 704. About this time, CDC introduced the 1604, followed quickly by the 3600 to compete against the IBM 7094. While IBM may have had the edge in speed, they began to withdraw from the technical focus and shifted towards the business side of computing, thus allowing CDC to begin domination of the scientific computing side. Seymour Cray introduced the CDC 6600 in 1964 and set a standard for high-end computing in the 1960's. The 6600 had a clock speed of 100 nsec. Many universities and national laboratories obtained the 6600. To compete against CDC, IBM introduced the 360 series of computers. Although originally built for data processing, the 360/91, 360/95, and 360/195 were adapted to more efficiently deal with scientific computations. When CDC introduced the 7600 in 1969, IBM phased out of the high-end scientific computing race.

The CDC 7600 had a clock time of 27.5 nsec (about 4 Mflop) and had a memory capacity of 500K. The CDC 6600 and 7600 machines dominated the scientific computing field from 1964 until 1975. Although other machines were developed during this period that were faster, including the first parallel SIMD (single instruction, multiple data) computer (ILLIAC IV by Burroughs Corporation), none was a commercial success. This was the beginning of the new architecture known as *pipelined vector computers* (the data processor continuously works on current instructions without waiting for the next set).

In 1976, Seymour Cray (Cray Research, Inc.) introduced the first modern supercomputer, the Cray-1. The machine had a clock speed of 12.5 nsec and used vector pipelining, although the operating system was fairly crude. About this time, J. Dongarra at Argonne National Laboratory developed the LINPACK set of benchmark codes to assess computing performance [3]. In peak mode, the Cray-1 was about 3.5 times faster than the CDC 7600. Utilizing clever packaging and innovative hardware in lieu of recent chip advances, the Cray-1 was a huge commercial success. This success prompted IBM and CDC to re-enter the supercomputer race.

CDC produced the CYBER 203 and 205, which had a clock speed of 20 nsec. The CYBER had a 5-fold increase in peak performance rate over the Cray by streaming out its memory (and not staging it through registers for vector operation), but it was slower in scalar performance. However, the CYBER was more difficult to use, especially in dealing with data. IBM introduced the 3090 machine that performed well on scalar operations but didn't compare in vector operations to the performance of the Cray. In the mid-1980s, three Japanese firms produced vector pipeline computers with fast scalar processors that were IBM-compatible.

In 1983, Cray produced the X-MP, which was a multiprocessing supercomputer with a clock speed of 9.5 nsec. The speed was increased to 8.5 nsec in 1985 in an effort to compete against IBM's 3090 which had been improved to include multiple processors and one operating system. The Cray-2 appeared in 1985. It had a clock speed of 4.1 nsec with a main storage of 256 MWords. The memory was soon increased to 512 MWords. However, the Cray-2 did not perform any faster than the X-MP on most jobs. In 1988, Cray introduced the Y-MP, which had a clock time of 6.3 nsec. Both the X-MP and Y-MP were among the most successful machines, at least based on the number of installed machines. Around this same time, CDC introduced the ETA-10 with a clock time of 7 nsec. However, the company dissolved in 1989. Similarly, Seymour Cray left Cray Research, Inc. and founded a new company to develop the Cray-3 and Cray-4, but the company dissolved in the mid-1990s.

In 1989, Fujitsu and NEC both announced the achievement of exceeding 1 Gflop performance on their machines. The NEC SX-3 reached 5.5 Gflops and the Fujitsu VP-2600 obtained 4 Gflops. The SX-3 ultimately reached 22 Gflops using 4 CPUs. The Cray-3 achieved 16 GFLOPS using 16 CPUs and had a clock time of 2 nsec with 2 GWord memory. Although a number of computer companies sprung up and developed interesting and affordable parallel-type machines (Alliant, Thinking Machines, Convex, etc.) in the late 1980's and early 1990's, every one of these companies eventually succumbed to declining market shares and saturation. A collapse and falling-out of high-end computer makers was inevitable.

In 1996, SGI, well known for their high-end visualization workstations, bought Cray Research, Inc. and merged the companies into SGI-Cray. A multiple CPU machine using the R10000 chips used in the SGI workstations was developed and introduced into the market as the SGI-Cray ORIGIN 2000. The machine eliminated the need to use specialized cooling unique to the Cray Y-MP and could simply be plugged into an electrical outlet within any room. Utilizing room air for cooling, the machine was significantly smaller and considerably more portable than any supercomputer built to date. This machine has become the de facto commercial standard computer for high-end computing and has begun to appear in supercomputer centers around the world.

1.4 Grand Challenge Problems

In the late 1980's, a five-year strategy for federally supported research and development on high-performance computing was developed. This plan subsequently became the Federal High-Performance Computing Program (HPCP); approximately $800 million was proposed for this program in 1993. The plan provided a list of "grand challenge" problems: fundamental problems in science and engineering with potentially broad economic, political, or scientific impact, which could be advanced by applying high-performance computing resources. These grand challenge problems are now often cited as prototypes for the kinds of problems that demand the power of a supercomputer. An abbreviated list of the grand challenge problems are listed as follows [1, 4]:

1 *Prediction of weather, climate, and global change*: the aim is to understand the coupled atmosphere-ocean biosphere system in enough detail to be able to make long-range predictions about its behavior.
2 *Materials science*: high-performance computing provides invaluable assistance towards improving our understanding of the atomic nature of material; examples include semiconductors and high-temperature superconductors.
3 *Semiconductor design*: a fundamental understanding is required of faster materials, such as gallium arsenide used for electronic switches, as to how they operate and how to change their characteristics.
4 *Superconductivity*: the discovery of high-temperature superconductivity in 1986 has provided the potential for spectacular energy-efficient power transmission technologies, ultra-sensitive instrumentation, and new devices; massive computing is needed to obtain a deeper understanding of high-temperature superconductivity.
5 *Structural biology*: the aim of this work is to understand the mechanism of enzymatic catalysis, recognition of nucleic acids by proteins, anti body/antigen binding, and other phenomena central to cell biology.
6 *Drug design*: predictions of the folded confirmation of proteins and of RNA molecules by computer simulation.
7 *Human genome*: comparison of normal and pathological molecular sequences is the most powerful method for understanding genomes and the molecular basis for disease.
8 *Quantum chromodynamics (QCD)*: computer simulations of QCD are needed to examine the properties of strongly interacting elementary particles, including new phases of matter, and computations of properties in the cores of stars.
9 *Astronomy*: greater computational power is needed to examine the volumes of data generated by radio telescopes.
10 *Transportation*: substantial computations are needed to examine vehicle performance, including modeling of fluid dynamic behavior about automobiles and complete aircraft geometries, flow inside turbines, and flow around ship hulls.

11 *Turbulence*: turbulence in fluid flow affects the stability and control, thermal characteristics, and fuel needs of virtually all vehicles, especially aerospace vehicles; understanding the fundamental physics of turbulence is necessary to reliably assess the performance of vehicle configurations.

12 *Combustion systems*: an understanding is needed of the interplay between flows of various substances involved in combustion processes and the quantum chemistry that causes those substances to react; the quantum chemistry required to understand the reactions is beyond the reach of current supercomputers.

13 *Oil and gas recovery*: improved seismic analysis techniques and understanding of fluid flow through geological structures are needed in order to devise new and economic ways of extracting oil from the earth.

14 *Computational ocean sciences*: the objective is to develop a global ocean production model that incorporates temperature, chemical composition, circulation, and coupling to the atmosphere along with other oceanographic features.

15 *Speech*: automatic speech understanding by computer is a large modeling and search problem in which millions of computations are required to evaluate the many possibilities of what a person might have said.

16 *Vision*: a challenge exists to develop human-level visual capabilities for computers and robots. Machine vision requires image signal processing, texture and color modeling, geometric processing and reasoning, and object modeling.

In the not too distant future we will have *petaflop* computers (10^{15}) [1, 4]. The power of such computers will allow us to delve deeper and in more detail into the fundamental physics and bases for many processes currently unknown. It is interesting that solutions to long-standing problems tend to create new challenges; this is the nature of science, its challenge, and its mystery. Most of the grand challenge problems involve modeling a physical system in a computer and using the model to create a simulation of its behavior. Others involve a reduction analysis of experimental data on a very large scale. Hence, the need for high-performance computing will continue to be with us and will become an ever-increasing part of our research.

The National Science Foundation (NSF) supports several supercomputer centers within the U.S. that are available to scientists and students for research and education. These centers include:

1   Cornell Theory Center, Cornell University, Ithaca, NY
2   National Center for Atmospheric Research (NCAR), Boulder, CO
3   National Center for Supercomputer Applications (NCSA), University of Illinois, Champaign, IL
4   Pittsburgh Supercomputering Center, Carnegie Mellon University and the University of Pittsburgh, Pittsburgh, PA
5   San Diego Supercomputer Center, University of California at San Diego, San Diego, CA

These facilities can be accessed via worldwide networks. In addition to these five centers, there are other supercomputer centers at universities and national laboratories that can provide network access to their facilities. A few of these facilities are:

1   Arctic Region Supercomputering Center, University of Alaska, Fairbanks, AK
2   Army High Performance Computing Research Center, University of Minnesota, Minneapolis, MN
3   Center for Computational Science, Oak Ridge National Laboratory, Oak Ridge, TN
4   Research Institute for Advanced Computer Science, NASA Ames Research Center, Moffett Field, CA
5   National Supercomputer Center for Energy and the Environment (NSCEE), University of Nevada Las Vegas, Las Vegas, NV

2   ARCHITECTURE

2.1 Vector and Parallel Computers – Configurations

In a typical computer, there is only one processor. In order to improve the speed further, we would like to have more processors in a computer. Such a computer is called a parallel computer. Computers with just one processor are called sequential computers, or scalar machines. In order to solve a problem using a parallel computer, one must decompose the problem into small subproblems, which are then solved in parallel. These must be efficiently combined to get the final result of the main problem. Since it is generally not easy to decompose a large problem into subproblems, a *data dependency* usually exists among the subproblems. Because of this data dependency, the processors must communicate with each other. An important point here is the time taken for this communication; usually the time for communication between two processors is very high when compared with the processing time. Hence, the communication scheme must be very well planned to get a good parallel algorithm [5].

When a series of steps is solved in order, each step can begin only when the previous steps are complete – a process known as being inherently *sequential*. To perform the work more efficiently, the work can be divided into subworks, i.e., divided into a series of parallel operations occurring at the same time. This procedure of parallelizing is called *pipelining*.

Another concept that is important in parallel processing is called *multiprocessing*. Multiprocessing refers to simultaneous processing of more than one task by different processors. There are generally two ways to handle multiprocessing: multiprocessors and multicomputers. In a multiprocessor system, many processors work simultaneously using a *common shared memory*. In a multicomputer, there is a group of processors in which each of the processors has a sufficient amount of local memory. The communication between the processors is through messages. This is known as *distributed processing*.

Parallel computers with individual processors that execute instructions asynchronously and send messages to each other are labeled MIMD, which stands for *multiple instruction, multiple data* streams. Each processor executes its own private set of instructions. The messages are passed between the processors by send-and-receive commands. There are mechanisms for causing the processor to enter a wait state in which it waits until receiving data from another processor. Programming these computers at a low level, i.e., specifying the individual send-and-receive commands is difficult, and various languages are being developed to simplify them. MIMD computers in which individual processors execute instructions asynchronously but share common address spaces are referred to as *shared memory* computers. Examples of such machines include the Cray Y-MP, the C-90, the Silicon Graphics Challenge, and the Convex Exemplar. They tend to be easier to program than the MIMD message-passing computers, but they are not without their own set of difficulties, particularly when dealing with synchronization of reads and writes from two processors into the same memory location.

The other class of parallel computers in which processors operate synchronously is known as *single-instruction, multiple-data* streams, or SIMD. A single sequence of instructions is obeyed by all of the processors, each acting on its own data. The Illiac IV, built around 1970, was in this class of parallel computers. SIMD machines do not share memory but rather have a distributed memory—one memory module for each processor. Communication of data is by message-passing. The MasPar line of computers built in the late 1980's and early 1990s were SIMD machines.

Two factors limit the number of processors in a multiprocessor [4, 5]. First, the performance of a message-passing computer is determined in part by the distance a message must travel from a given processor to any other processor in the machine. This suggests that, for efficient message-passing, a multiprocessor should have each of its processors connected to every other processor in the machine. In such a machine, a message will always travel from the sending processor directly to the receiving processor without being transferred through intermediate processors. As more processors are used, this completely connected machine encounters the second limiting factor - the maximum number of processors possible is limited by the number of interprocessor connections (physical wires) required. A completely connected machine with n processors requires n-1 connections per processor or a total of n(n-1)/2 connecting wires.

There are several types of interconnection patterns. In a ring interconnection, every processor is connected to only two others. While the number of ring-connected processors can be large, processors diametrically opposite each other in an n-processor ring would have to pass messages through about n/2 processors to communicate. Another interconnection scheme is known as a *hypercube*: n processors each have $\log_2 n$ interconnection wires requiring that a message pass through no more than $\log_2 n - 1$ intermediate processors between its source and destination. This scheme has the advantage of a relatively short path length, measured in terms of the number of processor-to-processor connections the message must traverse to reach its destination. However, a major disadvantage is that the number of wires connected to each processor increases with n. Some multiprocessors connect processors in a two-dimensional mesh: with n processors, assuming $n = m^2$, the mesh has m rows and m columns. In this configuration, the number of wires connected to a processor is 4, regardless of the value of n. Thus, the number of wires connected to a processor does not grow with n as it does in a hypercube. However, the messages in the mesh have longer paths than in a hypercube. The issue of scalability deals with the number of processors and the increase in speed of a computer. Doubling the number of processors doesn't truly double the speed of the computer; however, if the computer is designed such that its speed increases approximately in proportion to the number of processors, and its complexity in terms of the number of interconnecting wires also increases proportionately, the computer is *scalable*.

2.2 MIMD and SIMD Systems

The MIMD model refers to a system that has multiple processors capable of working independently and producing results for a global system. Each processor is capable of executing a separate instruction with a separate set of data. In a MIMD system, the processors can run independently. Each processor can execute different portions of the same program or completely different programs. MIMD is seen as a more general design capable of performing well over a broader range of applications. However, in a MIMD machine, each processor needs enough memory to store its own copy of at least part of the program and enough logic to decode instructions and manage its program counter; this makes designs with more than a few thousand processors difficult to achieve.

Parallel processor designs tend to cluster into three general configurations [2]: systems containing 1-10 processing elements (PE), systems with 10-5000 PEs, and those with PEs $\geq$ 5000. The bulk of the earlier high-end commercial computers belonged to the first group, i.e., supercomputers developed by Cray and IBM. Such systems were normally pipeline vector machines that used high-power bipolar transistor technology and required sophisticated cooling techniques. Parallelism of the systems was applied not to speed up a particular job, but to enhance throughput, i.e., to handle more jobs. Also in this category are bus-based machines that are designed to decrease cost/performance. The hardware is slower but smaller, lower-powered, and more easily cooled. While the bus is economical and permits direct communication to all the PEs, the fixed bandwidth limits the number of PEs.

Systems with 10-5000 PEs tend to use microprocessors. More powerful networks are employed which allow the bandwidth to be increased. Examples of this kind of system include the earlier MasPar and similar parallel hypercubes. Massively parallel systems use the maximum number of PEs, i.e., bit-serial microprocessors, implemented in very large system-integrated (VLSI) chips with sparse interconnections. The trade-off between the number and power of PEs is still questionable, and active research continues in the design of such massively parallel computers.

If a PE stores its own program and has exclusive access to its part of the data, the architecture is MIMD with private memory; otherwise, it is a SIMD machine. If a PE stores its own program but does not have exclusive access to its part of the data, it is a MIMD with shared memory; the opposite produces a SIMD with shared memory (which is difficult to design because of memory conflicts). The node of a MIMD machine may consist of either a collection of smaller subunits or a computer with SIMD elements. The most powerful processors are either small-grain SIMD or large-grain MIMD. Large grain machines are typically divided into medium-grain computers, such as hypercubes, or large-grain computers like the Cray X-MP.

In a MIMD computer, each node executes its own program and stores its own data. In situations where the same code is given to each node, the program at any one node is separate from all the other nodes. Obviously, it is important to determine the optimum way to decompose the problem into pieces for distribution to various nodes. For example, a single program can be divided into different parts, with the nodes executing different parts of the program. Several separate programs can be executed simultaneously on different notes.

The SIMD system is a system in which the same instruction is carried out for different sets of data in parallel. The number of datasets is the number of processors working simultaneously. An example of a SIMD model is shown in Fig. 1.

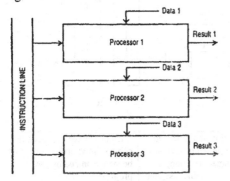

Figure 1 SIMD model (from [4])

Each processor contains data. Notice that the data for different processors are different, whereas the instruction is the same. In a SIMD model there are two types of architecture: shared-memory model and direct-connection networks. In the shared-memory model, there is a common memory that is shared by all the processors. Communication between the two processors takes place only through the shared memory. This is illustrated in Fig. 2. In the direct-connection network, independent processors are connected using wires, and they may be connected according to any desired topology such as rings, hypercubes, etc. A comparison of MIMD and SIMD architecture is shown in Table 1 [5].

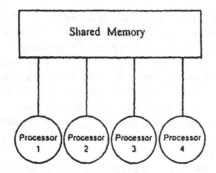

Figure 2 Shared-memory model (from [4])

Table 1 MIMD versus SIMD architecture (from [5])

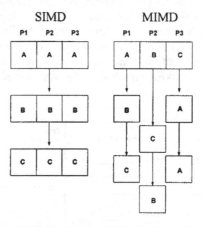

Flow of control (architect's view)	One instruction at a time von Neumann style, all processors doing the same operation; mask bits may disable a subset of processors	Each processor has its own control flow; processors must be synchronized by the program
Programming requirements (programmer's view)	Code must be expressed as regular operations on aggregates and must match inherent regularity of hardware	Code must be broken into relatively independent tasks that minimize communications and synchronization
Interprocessor communications	Subvectors must be moved in patterns fitting the regularity of the interconnections and algorithm	Hardware scheduling of data path resources, data packets with destination addresses, and buffering of data in network; lots of hardware for this runtime partitioning
Memory bandwidth	Vectorized reference allows full access to all memory banks; unvectorized references like a uniprocessor	When possible, local memory is used; global memory references involve interconnection network delay

2.3 Performance Measures

Two terms that are used to describe performance of a computer are *speedup* and *efficiency* [1-5]. The speedup associated with parallel performance can be assessed using the simple relation

$$S_p = T_s / T_p$$

where S_p is the speedup on p nodes, T_s is the time obtained from the best sequential algorithm, and T_p is the time required for the parallel algorithm. Efficiency is defined as the speedup per node,

$$E_p = S_p / S_1$$

where S_1 is the linear speedup. Performance advantage depends upon the type of problem and the way the problem is decomposed. For example, linear speedup is expected when the processes running on different nodes are simply different cases of a single program. Likewise, a complex single program using a set of processes might be more difficult to achieve linear speedup, although performance is significantly enhanced. For a parallel machine, speedup and efficiency are defined as

$$Speedup = T_s/T_p$$

$$Efficiency = T_s/P$$

where T_p is the parallel-time computation and P is a measure of the processor complexity [6]. The speedup is at most equal to the number of processors, and we always attempt to develop parallel algorithms with the speedup nearly equal to the number of processors. In reality, we can achieve this speedup only for a limited number of problems.

Performance measurements in the parallel domain are made more complex by our desire to know how much faster we're running our applications on a parallel computer. That is, what benefit are we deriving from the use of parallelism? This is measured in terms of speed-up. There are various methods of defining serial and parallel execution times. This diversity has resulted in several different definitions of speedup, i.e., relative speedup, real speedup, absolute speedup, and asymptotic and relative speedups. More information on the definitions of speedup can be found in Sahni and Thanvantri [6].

Parallel efficiency can be expressed as a function of three main parameters: set-up time for data transfer (or *latency* time); data transfer rate (usually expressed in MB); and computing time per floating point operation (usually expressed in Mflops). For a given algorithm and communication pattern, a model equation can be developed to express parallel efficiency. Schreck and Peric [7] discuss such a model and show that parallel efficiency can be predicted.

CPU speed indicates how fast the computer works; performance tells how fast the CPU can execute given tasks. The speed of the CPU clock is defined by two indicators: (1) clock rate – the number of clock cycles executed per second (usually in MHz) and (2) clock speed – length of time it takes to execute a single CPU clock cycle (msec or nsec). These two definitions are the reciprocals of each other. If the rate of the CPU clock is 50 MHz, then the length of a single clock cycle is 20 nsec. If one cycle occurs every 12.5 nsec, the clock rate is 80 MHz.

There are various times related to computer execution. *User time* is the amount of time the CPU is active for a given process excluding operating system overhead. *System time* is the amount of time the operating system is active for the given process. *Elapsed time (or wall clock time)* is the difference between the time when the time process begins on a given processor and the time that it ends. This is the total time the process runs.

Mips measurements are highly dependent on the instruction mix used to produce the execution times. For example, a program executing a large number of simple integer operations is much faster than one dealing with a lot of complex floating-point operations. Hence, two different programs with the same number of instructions can produce different mips measurements on the same machine.

Flops is the common performance unit used to define the speed of a computer in terms of the number of floating-point operations it can perform in one second. However, not all floating-point operations require the same number of clock cycles – a floating-point division may take 4-20 times as many cycles as a floating-point addition. The flop is an especially popular measure of machine performance on scientific and mathematical programs, but it is not a reasonable measure for programs or benchmarks using few or no floating-point operations. Peak (theoretical) performance is the maximum number of flops – but is almost never achieved. Table 2 lists several supercomputers and their theoretical peak performances in Mflops, including their maximum number of processors [1, 8].

Table 2 Theoretical peak performance rates of several supercomputers [1, 8]

Machine	Manufacturer	Number of Processors	Theor. Peak Mflops
CM-2	Thinking Machines	2,048	20,000
CM-5	Thinking Machines	16,384	2,000,000
Cray-2	Cray Research	8	4,000
Cray Y-MP	Cray Research	16	15,000
Cray T3D	Cray Research	2,048	307,000
Cray-3	Cray Computer Corp	16	16,000
IBM ES/9000	IBM	6	2,700
IBM 9076 SP1	IBM	64	8,000
Intel iPSC/2	Intel	128	250
Intel iPSC/860	Intel	128	5,120
Intel Delta	Intel	512	20,480
Intel Paragon	Intel	4,000	300,000
KSR1	Kendall Square Rs	32	1,300
KSR2	Kendall Square Rs	5,000	400,000
MP-1	MasPar	16,384	550
MP-2	MasPar	16,384	2,400
NEC SX-A	Nippon Electric Co	4	22,000

Benchmark refers to a set of programs or program segments that are used to measure performance. The time required for a computer to execute a benchmark provides the performance measure. Benchmarks are usually written by a user needing information, such as CPU performance, file server performance, I/O, communications speed, etc. The codes for many commonly used benchmarks are available on the Internet [1]. Many of the more common benchmark codes can be found in *netlib*, which is a file maintained by the Oak Ridge National Laboratory in Tennessee. The listing of the contents of this library can be accessed at netlib@netlib.att.com or netlib@ornl.gov with a single line message - *send index*. To reach *netlib* by anonymous ftp, type *ftp netlib.att.com* and change to the directory called */netlib* [1]. Some of the more common benchmark routines include LINPACK [3], the Livermore Loops (or Livermore Fortran Kernels) [9], and Whetstones, originally used to measure floating-point performance (see *netlib* website). A set of benchmarks was also developed by the NASA Ames Research Center for the purpose of comparing the performance of parallel machines. These benchmark kernels are known as the NAS kernels or the NAS Parallel Benchmarks.

3 PROGRAMMING

In order to achieve greater speeds, parallel computers are needed. The advantage of parallel computers over vector supercomputers is scalability. Parallel computers use standard chips and are cheap to produce. Commercially available parallel computers now employ thousands of processors, gigabytes of memory, and computer power measured in gigaflops. However, algorithms designed for serial machines may not run efficiently on parallel computers.

It makes sense to run the most numerically intensive codes on parallel machines, and frequently these are very large programs assembled over many years. The programming languages for parallel machines are primarily FORTRAN with explicit parallel extensions (FORTRAN 90) and to a lesser extent C. While the language may be familiar, parallel programming becomes more demanding as the number of processors increases. It is usually best to rewrite a code written for one CPU than to try to convert it to parallel.

The operating system (or the user) organizes the work into units called *tasks*, and the tasks assign work to each processor. There is a main task to control the overall execution as well as subtasks to run independent parts of the program (called parallel subroutines, slaves, or subtasks). These parallel subroutines can be distinctive subprograms or multiple copies of the same subprogram. The main-task program does its own computations as well as calling and scheduling the parallel subroutines.

The programmer helps speed up run time by keeping as many processors as possible busy and by avoiding storage conflicts from different parallel subprograms. This load balancing is achieved by dividing the program into subtasks of approximately equal numerical intensity that will run simultaneously on different processors. The rule of thumb is to make the task with the largest granularity (workload) dominant by forcing it to execute first, and to keep all the processors busy by having the number of tasks be an integer multiple of the number of processors.

While vector processors usually work on the innermost loop of a program, parallel subroutines work best on the outermost loops. For example, vectorizing an outer matrix of dimension three would not produce much speedup. But if many floating-point operations are needed to calculate the elements of this matrix, then this may be a good choice for three-way parallel.

To avoid storage conflicts, programs should be designed so that parallel subtasks use data that are independent of the data in the main task and in other parallel tasks. This means that the data should not be modified or even examined by different tasks simultaneously. In organizing these multiple paths, some concern about overhead costs is appropriate. These costs tend to be high for fine-grain programming and to vary for different scheduling commands.

An approach to concurrent processing that has gained wide acceptance for coarse- and medium-grain systems is distributed memory. In it, each processor has its own memory and the processors exchange data among themselves over a high-speed network with a fast switch. The data exchanged or passed among processors have encoded forward and return addresses and are called *messages*.

For a messages-passing program to be successful, the data must be divided among nodes so that, at least for awhile, each node has all the data it needs to run an independent subtask. When a program begins execution, data are sent to all nodes. When all nodes have completed their subtasks, they exchange data again in order for each node to have the complete new set of data to perform the next subtask. This repeated cycle of data exchange followed by processing continues until the full task is completed. In message-passing MIMD programs, the programmer writes a single program that is executed on all the nodes. Often a separate host program, which starts the programs on the nodes, reads the input files and organizes the output.

Although MIMD systems are popular, a number of standards, such as a standard programming language, have yet to be adopted. Some popular systems are Express, a commercial product from ParaSoft Corporation, PVM, and MPI. All hide the messy details of passing messages and synchronization, and are available for use on MIMD machines or clusters of workstations. The existence and utility of these packages means that parallel systems do not have to consist of only a set of dedicated processors; there can also be a number of workstations from various manufacturers connected by some network. While the top priority of these workstations may be the work of their owners, when there are no local demands on them they automatically switch over to helping someone else's big problem get done concurrently. This integrated sum produces tremendous computing power that would otherwise go wasted. This is discussed in more detail in the next section.

When parallelization is performed at the local level (e.g., as in auto-parallelizing compilers), Amdahl's law, which says that speed is basically determined by the least efficient part of the code, becomes important. To achieve high efficiency, the portion of the code that cannot be parallelized has to be very small.

It is best to subdivide the solution domain into subdomains and assign each subdomain to one processor. The same code runs on all processors, i.e., on its own set of data. Since each processor needs data that resides in other subdomains, exchange of data among processors and/or storage overlap is necessary.

In fluid flow and heat transfer problems, the governing equations are typically discretized using a mesh consisting of many thousands (or millions) of nodes. These discretized equations are solved in either implicit or explicit mode.

Explicit schemes are fairly easy to parallelize, since all operations are performed on data from preceding time steps. It is only necessary to exchange the data at the interface regions between neighboring subdomains after each step is completed. The sequence of operations and the results are identical on all processors. The most difficult part of the problem is usually the solution of the elliptic Poisson equation for pressure when solving fluid flow problems, which typically requires an implicit solver. The steady state conduction equation for heat transfer is also elliptic.

Implicit methods are more difficult to parallelize. Calculation of the coefficient matrix and the source vector using 'known' (or previously calculated) values can be efficiently performed in parallel. However, solution of the overall linear system of equations is not easy to parallelize. For example, Gauss elimination, in which each computation requires the result of the previous calculation, is very difficult to perform on parallel machines. Considerable effort has been spent on developing efficient matrix solvers for use on parallel machines. Two popular and successful solvers are a sparse Cholesky matrix solver developed at ORNL [10], and GPS, developed at NASA Langley Research Center [11].

4   CLUSTERS AND NETWORKS OF WORKSTATIONS

In recent years, high-speed networking and improved microprocessor performance are making networks of workstations an appealing vehicle for cost-effective parallel computing [12]. Clusters of workstations and personal computers have become increasingly popular. The incremental scalability of processors, memories, and mass storage systems together with the high-performance interconnection networks have made clusters a cost-effective platform for distributed and parallel computing. Clusters and networks of computers built using commodity hardware or software are playing a major role in redefining the concept of high-performance computing. Since 1995 we have seen an explosive growth in the use of high-performance communications for information access, research at the frontiers of science, and commercial endeavors.

Traditionally, collections of complete computers with dedicated interconnects, called *clusters*, have been used to serve multiprogramming workloads and to improve computer resource availability [13, 14]. A single front-end machine usually acts as an intermediary between a collection of computer servers and a large number of remote machines. The front-end machine tracks the load on the cluster nodes and schedules tasks onto the most lightly loaded nodes. One of the keys to the development of clusters is the availability of affordable, high-performance networks to serve as support for the cluster.

Increasingly, clusters are being used as parallel machines, often called networks of workstations (NOWs). In an ARPA-funded effort at the University of California at Berkeley, the NOW project has developed hardware and software support for using a network of workstations as a distributed computing system on a building-wide scale. A major influence on clusters has been the rise of popular public domain software, such as PVM [15], MPI and Condor [16]. NOW-enabled software allows users to farm out jobs over a collection of machines or to run a parallel algorithm on a number of machines connected by a local- or even wide-area network.

4.1 Technology

The technology breakthrough that presents the potential of clusters taking on an important role in large-scale parallel computing is a scalable, low-latency interconnect [12]. The association of different hardware systems and autonomous operating system kernels requires high bandwidth, low-latency communication and efficient coordination within LAN and WAN clusters. Several potential candidate networks have evolved for both local- and wide-area networks. Local area networks (LANS) have traditionally been either a shared bus Ethernet or a ring topology such as Token Ring or Fiber Distributed Data Interface (FDDI). In recent years a strong push to switch-based LANS utilizing HPPI, FDDI, ATM switches, and Fibre Channels [17] provides scalable bandwidth that can support large networks of high-performance machines.

Fibre Channel (FC) was developed to be a practical, inexpensive, yet expendable means of quickly transferring data between workstations, mainframes, supercomputers, desktop computers, storage devices, displays and other peripherals [12]. Fibre Channel is the general name of an integrated set of standards [X3T9.3 Task Group of ANSI: Fibre Channel Physical and Signaling Interface (FC-PH), Rev. 4.2 October 8, 1993] being developed by the American National Standards Institute (ANSI). There are two basic types communication: data communication between processors (networks) and data communication between processors and peripherals (channels). A channel provides a direct or switched point-to-point connection between the communicating devices. In contrast, a network is an aggregation of distributed nodes (like workstations, file servers or peripherals) with a distinct protocol that support interaction among the nodes. A network has relatively high overhead, since it is software-intensive and consequently is slower than a channel.

A recent and significant development is the widespread adoption of the ATM (asynchronous transfer mode) standard. ATM is emerging as the primary networking technology for next-generation, multi-media communications. Several vendors offer switches with 16 ports that support LAN bandwidth up to 155-Mb/s (19.4-MB/s) which can be cascaded to form larger networks. Over the long run, ATM may be seen as having the greatest flexibility in serving voice, data, and video traffic. ATM carries the additional benefit of being able to be used in both LAN and WAN applications. ATM protocols are designed to handle isochronous (time-critical) data such as telephony (audio) and video, in addition to more conventional inter-computer data communications. ATM protocols are designed to be scalable in bandwidth, with the ability to support real multi-media applications. There are standards in place today to implement ATM over OC-1 (51 megabits-per-second) to OC-48 (2.488 gigabits-per-second).

4.2 Increasing Capacity to 10Gb/s and Beyond

Synchronous Digital Hierarchy (SDH) is a bit-rate, synchronous transmission standard derived from the synchronous optical network (SONET) format devised in the USA. One of the principal features of the SONET/SDH systems is the extensive provision of signal overheads intended for flexible management and control. Another benefit of the SONET/SDH is that it is backwards compatible with the existing plesiochronous digital hierarchy (PDH) networks and is also expected to provide the physical layer support for broadband integrated services digital networks (B-ISDN) based on the asynchronous transfer mode (ATM).

Two approaches to achieving transmission capacities of 10 Gb/s that are currently being supported in various network testbeds are: (1) expanding the SONET/SDH transmission hierarchy to the OC-192 (9.953 Gb/s) data rate and (2) using wavelength division multiplexing (WDM) techniques. The WDM approach offers the potential to upgrade existing optical fiber facilities and to use transmission protocols that are already familiar to the data communications community. One of the most significant advantages to OC-192 systems is the potential to reduce network implementation and operations costs.

4.3 A Description of Four Gigabit Networks

4.3.1 CASA. CASA is a 2.4 Gb/s SONET testbed connecting the NSF San Diego Supercomputer Center (SDSC), the DOE Los Alamos National Laboratory (LANL), NASA's Jet Propulsion Laboratory (JPL), and California Institute of Technology (Caltech). Each of these sites has significant computing resources, and CASA focused on interconnecting these supercomputers. While LANL and the other sites are about 1,600 km apart, the link is 2,000 km long. The CASA testbed allows two groups of climate researchers, one in California and the other in New Mexico, each working on different parts of a problem, to combine their software models into a single metacomputer-based execution using heterogeneous computing systems.

4.3.2 MAGIC. The MAGIC (Multidimensional Applications and Gigabit Internetwork Consortium) project was established to develop a very high-speed, wide-area networking testbed to address challenges in heterogeneous computing, distributed storage, coordination of multiple data streams, and techniques for managing the effects of network delays in a defense applications context. Sustained rates of several hundred Mb/s have been demonstrated on MAGIC.

4.3.3 ATDnet. The Advanced Technology Demonstration Network (ATDnet) is a recently inaugurated high-performance networking testbed in the Washington, DC area. It is intended to be representative of possible future Metropolitan Area Networks. Initially architected by NSA and the Naval Research Laboratory and funded by ARPA to enable collaboration among DOD and other Federal agencies, ATDnet has a primary goal to serve as an experimental platform for diverse network research and demonstration initiatives.

4.3.4 ACTS. In FY 1994, NASA launched its Advanced Communications Technology Satellite (ACTS). ACTS provides a "network in the sky" and is capable of multiple low-speed access, four 155 Mb/ s full-duplex connections, or a single 622 Mb/s full-duplex connection.

4.3.5 Summary. Many challenges must be addressed in order to develop a successful cluster or NOW which would enable organizations on those networks to participate in application experiments. Some of these include:

1. ensuring that heterogeneous computing and networking devices interoperate
2. coordinating multiple data streams destined for a single location
3. accommodating bursty and steady traffic
4. compensating for effects of network delays on high-throughput data transmissions
5. supporting very high (1 Gb/s) network throughput
6. enabling a wide range of end-user access speeds (10s of kb/s to 1 Gb/s) and host capabilities (laptops to high-end workstations)

An important success factor of the Network of Workstation projects is the high level of cooperation among the academic, industrial, and governmental participants leading to new capabilities in the science community, new markets for the industrial community, and new visions for high-performance communications.

5 APPLICATIONS

5.1 Governing Equations

The equations of fluid motion are not difficult to formulate for a homogeneous medium; however, once the equations and boundary conditions are established, one must numerically solve these complex equations. The types of numerical methods used by the majority of researchers and applications-oriented engineers to solve the partial differential equations fall into three categories: 1) finite difference (or finite volume) methods (FDM), 2) finite element methods (FEM), and 3) other approaches (boundary integrals, hybrids, analytical, spectral, etc.). The FDM has been used for a wide variety of problems; nearly all of the early numerical simulations dealing with heat transfer and fluid flow revolved around various solution strategies for the FDM.

As structural analyses using FEM became more routine, researchers began to apply the method to more difficult problem areas, particularly those fields dealing with fluid flow. Some of the earliest work in fluid simulation with finite elements can be traced to the mid-1960s; a comprehensive review is given in Zienkiewicz and Taylor [18]. Based on this early pioneering work, the numerical simulation of fluid flow with finite elements began to proliferate by the 1970s. Today, the FEM is a strong contender for simulating all modes of fluid flow processes, rivaling performance standards associated with FDM. While the FDM is conceptually much simpler than the FEM, difficulties in dealing with irregular boundaries requires the FDM to use boundary fitted coordinates or other means to transform the equations.

Once a solution to a problem is obtained, error estimation and accuracy must be assessed. Sources of error can generally be attributed to the way in which the problem domain and the solution to the governing equations have been approximated. If the domain is discretized with a coarse mesh, one may never get close enough to the right answer; likewise, the mesh may be considered suitable, but the approximation functions may be of too low an order. Furthermore, there are the inherent computational errors associated with round-off, numerical differentiation, and numerical integration.

In general, the following steps are needed in any numerical approximation to the solution of a differential equation: 1) the equation (or system of equations) and its boundary and initial conditions must be defined to ensure that a well-posed problem (i.e., physical system) is formulated, 2) a discretization must be created to define the approximation functions to be used in the solution, 3) a mesh must be created that adequately refines regions where large changes in the solution are expected, and that allows the boundary conditions to be properly imposed, 4) the numerical algorithm must be formulated and used to solve the system of algebraic equations, and 5) the error in the approximation must be calculated to determine if the solution is converged or if a more refined solution is needed.

5.1.1 Mass conservation. The most general form of the mass conservation or continuity *equation is*

$$\frac{\partial \rho}{\partial t} + \frac{\partial \rho u_j}{\partial x_j} = 0 \tag{1}$$

where ρ is the fluid density, u_j are the velocity components ($j = 1,2,3$) in the x_j direction, and t denotes time.

5.1.2 Navier-Stokes equation. The Navier-Stokes equation for a Newtonian, laminar viscous fluid can be written in indicial notation as

$$\rho\left(\frac{\partial u_i}{\partial t} + u_j \frac{\partial u_i}{\partial x_j}\right) = -\frac{\partial p}{\partial x_i} + \frac{\partial}{\partial x_j}\left[\mu\left(\frac{\partial u_i}{\partial x_j} + \frac{\partial u_j}{\partial x_i}\right) + \lambda \frac{\partial u_k}{\partial x_k}\delta_{ij}\right] + B_i \tag{2}$$

where subscripts i,j $\equiv$ 1,2,3 (denoting x, y, or z), p is pressure, μ is viscosity, and B_i is the body force term. Replacing λ by $- 2/3\mu$ yields the usual form for compressible flow. For compressible flows the equation of state is also required.

5.1.3 Energy conservation. The conservation of energy can be written in terms of total energy as

$$\rho\left(\frac{\partial e}{\partial t}+u_j\frac{\partial e}{\partial x_j}\right)=\frac{\partial}{\partial x_j}(k\frac{\partial T}{\partial x_j})-\frac{\partial pu_i}{\partial x_j}+u_j\frac{\partial \sigma_{ij}}{\partial x_i}+\phi+q^{'''} \qquad (3)$$

where e is total energy (per unit mass), T is temperature, k is thermal conductivity, p is pressure, σ_{ij} is the shear tensor, ϕ is the dissipation function ($\phi = \sigma_{ij}\partial u_i/\partial x_j$), and $q^{'''}$ is internal heat generation.

5.1.4 Species conservation. The equation for mass transport is

$$\frac{\partial c}{\partial t}+u_j\frac{\partial c}{\partial x_j}=\frac{\partial}{\partial x_j}\left(D\frac{\partial c}{\partial x_j}\right)+S \qquad (4)$$

where c is the species concentration, D is the mass diffusion coefficient and S represents sources and/or sinks. In general, an equation of the form of Eq. (4) is needed for each species component in the fluid. The coupling between the equations can be quite complicated, particularly in the case of flows with chemical reactions.

Note that Eqs. (2-4) are transient advection-diffusion equations and are essentially similar, except for the nonlinearity of the Navier-Stokes equations. Many numerical schemes exist to solve this general class of equations ranging from simple, centered FDM schemes to elegant, highly accurate hybrid methods [19]. The user typically defines a mesh, choosing whether to use a structured or unstructured grid, then selects a numerical scheme that is either implicit or explicit. Since explicit schemes are easier to parallelize, it is a straightforward step to rearrange Eqs. (2-4) such that

$$\phi_i^{n+1} = \phi_i^n + M_{ij}^{-1}\Delta t[(K_{ij}+A_{ij})\phi_i - f_i] \qquad (5)$$

where $\phi_i = \{u_j, e, c\}$, Δt is the time step (n), M_{ij} is the set of matrix coefficients associated with the transient term, K_{ij} denotes the terms associated with viscosity or diffusion, A_{ij} contains the advection terms (which likely includes some form of upwinding), and f_i is the vector of remaining known right-hand terms of the equations (sources, sinks, fluxes, etc.). Euler and two-step Runge-Kutta schemes are particularly popular for time marching.

For steady state conditions, the form of Eq. (5) becomes an elliptic equation of the form

$$[K_{ij} + A_{ij}]\phi_i^{n+1} = f_i \tag{6}$$

where superscript n+1 now denotes an iteration counter. Equation (6) is elliptic and must be solved using a Poisson solver (see following section on pressure).

5.1.5 Boundary conditions. The boundary conditions must be physically realistic and are dependent on the particular geometry, the materials involved and the values of the pertinent parameters. At a solid boundary, it is appropriate to require a fluid to have the same velocity as the solid - or no-slip boundary condition. However, a no-slip boundary is only valid when the continuum hypothesis is justified; it is not a realistic boundary condition at a solid boundary if the fluid under consideration is a gas with a moderate or large mean free path. In this case, there is a slip velocity of the gas relative to the solid boundary. Boundary conditions for the scalar equations of heat transfer and species transport are straightforward Dirichlet or Neumann values as prescribed by the constraints of the problem.

5.1.6 Pressure. Problems dealing with pressure stem from attempts to obtain solutions to the equations of motion for an incompressible fluid (as opposed to compressible flows where the equation of state can be used). The equation for pressure (assuming constant density and viscosity) is typically written in the form

$$\frac{\partial}{\partial x_j}\left(\frac{\partial p}{\partial x_j}\right) = -\frac{\partial}{\partial x_j}\left(\frac{\partial \rho u_i u_j}{\partial x_j}\right) \tag{7}$$

which yields an elliptic Poisson equation for which there are numerous solvers, the more popular being Gauss-Seidel iterative schemes, alternating direction implicit (ADI), strongly implicit procedure (SIP), and LU decomposition techniques. It is important that one maintain consistency of the operators, i.e., the pressure equation is the product of the divergence operator stemming from the continuity equation and the gradient operator for the momentum equations. The approximation of the Poisson equation must be defined as the product of the divergence and gradient approximations used in the governing equations. Hence, it is best to derive the equation for pressure using the discretized momentum and continuity equations instead of discretizing the Poisson equation directly [19, 20].

In general, the pressure is not required to satisfy boundary conditions, but one must prescribe a reference value for it to be determined uniquely. In some flow problems, the known pressure at a free surface provides the equilibrium condition along that interface, or the prescribed pressure at an inflow or outflow boundary is the driving force in the system. In these cases, the pressure is prescribed along these boundaries.

A good parallel iterative scheme is the red-black Gauss-Seidel method, which consists of performing Jacobi iterations on two sets of points in an alternating manner [19]. In 2D, the nodes are colored as on a checkerboard, and new red values are calculated using data only from black neighbor nodes, and vice versa.

Computation of new values on either set of nodes can be performed in parallel. Communication between processors working on neighbor subdomains takes place twice per iteration, once each set of data is updated. The method is rather inefficient and is best used with a multigrid method.

Implicit schemes require the solution of an equation of the form

$$A_{ij}\phi_i = f_i \tag{8}$$

where A_{ij} is the global coefficient matrix that is usually split into a system of diagonal blocks A_{ii}, which contain the elements connecting the nodes that belong to the ith subdomain, and off-diagonal blocks or coupling matrices A_{ij} $(i \neq j)$. For example, the SIP method is recursive and does not lend itself to easy parallelization. In addition, the mesh must be structured, and a lot of fine-grain communication is required with idle times before and after each iteration. The SIP method employs a form of LU decomposition in which all the variables are unknown, e.g., a 2-D centered difference molecule yields a five-diagonal matrix (as opposed to the tridiagonal ADI method), which gets altered to a seven-diagonal matrix. This matrix is subsequently reduced to a set of LU tri-diagonal matrices. For efficiency, there should be little data dependency (data from neighbors) when dealing with the inner iterations; data dependency results in long communication and idle times. Hence, it is better to decouple the blocks such that the off-diagonal terms of the global matrix, $M = LU$, are equal to zero. Thus, the scheme on subdomain i is

$$M_{ii}\phi_i^m = f_i^{m-1} - (A_{ii} - M_{ii})\phi_i^{m-1} - \sum_j A_{ij}\phi_j^{m-1} \quad (i \neq j) \tag{9}$$

After one iteration is performed on each subdomain, the updated values of the unknown ϕ^m must be replaced so that the residual ρ^m can be calculated at nodes near subdomain boundaries.

The conjugate gradient method (CG), without preconditioning, can be parallelized, although some global communication occurs. However, to be really efficient, the method needs a good preconditioner (for which the SIP method is good, but then suffers from the same problems as mentioned above). A simple pseudo-code for a preconditioned CG-solver is listed as follows [19]:

1. initialize by setting: $k=0$, $\phi^0=\phi_{in}$, $\rho^0=f-A\phi_{in}$, $p^0=0$, $s^0=10^{30}$
2. advance the counter: $k=k+1$
3. on each subdomain, solve the system: $Mz^k = \rho^{k-1}$
   local communication: exchange z^k along interfaces
4. calculate: $s^k=\rho^{k-1}\cdot z^k$
   global communication: gather and scatter s^k
   $\beta^k=s^k/s^{k-1}$
   $p^k=z^k+\beta^k p^{k-1}$
   local communication: exchange p^k along interfaces
   $\alpha^k=s^k/(p^k\cdot Ap^k)$

global communication: gather and scatter α^k

$$\phi^k = \phi^{k-1} + \alpha^k p^k$$
$$\rho^k = \rho^{k-1} - \alpha^k A p^k$$

5. repeat until convergence

In order to update the right-hand-side of Eq. (8), data from neighbor blocks are necessary. On parallel computers with shared memory, these data are directly accessible by the processor. On distributed memory machines, communication between processors is necessary. Each processor needs to store data from one or more layers of cells on the other side of the interface. Note that local communication takes place between processors operating on neighboring blocks; global communication is the gathering of some information from all blocks in a master processor and broadcasting some information back to the other processors. PVM [15] is a common software library used for such communications.

If the number of nodes allocated to each processor remains the same as the grid is refined, the ratio of local communication time to computing time remains the same, i.e., the local communication is fully scalable. However, the global communication time increases when the number of processors increases, independent of the load per processor. The global communication time will eventually become larger than the computing time as the number of processors is increased. Global communication becomes the limiting factor in massive parallelism.

A fast and accurate Cholesky method for the solution of symmetric systems of equations is discussed by Storaasli et al. [11] This direct method is based on a variable-band storage scheme and takes advantage of column heights to reduce the number of operations in the Cholesky factorization. The method uses parallel computation in the outermost DO-loop and vector computation using loop unrolling in the innermost DO-loop. The method avoids computations with zeros outside the column heights, and as an option, zeros inside the band. This program is called GPS (General-Purpose Equation Solver) and can be downloaded from the web, as noted earlier. A similar fast solver is discussed by Ng et al. [10], who developed a parallel version of their Cholesky direct solver at Oak Ridge National Laboratory. Both codes are particularly effective in finite element schemes requiring the solution of large global stiffness matrices.

5.2 Mesh Generation and Adaptive Methods

Significant advances in CFD are expected when computer power increases to about one Tflop. This power will permit an order-of-magnitude increase in the number of grid points in each coordinate direction. Achieving such power is possible with parallel computers and using multigrid techniques. Current research into multigrids is concentrated in finding efficient methods for the Navier-Stokes equations applied to general geometries, utilizing advanced discretization techniques, and constructing highly parallel multigrid algorithms.

Recent work on mesh-generation schemes suggests that adaptive mesh refinement and unstructured grids are considered to be better than block refinement in CFD computations [20]. Adaptive methods are now being used routinely in many commercial codes dealing with CFD and transport problems. Solving equations using finite volume or finite difference methods on rectangular grid leads to relatively straightforward parallel processing. However, when one deals with unstructured grids, efforts in parallelizing become significantly more difficult. By their nature, finite volume methods are ideally suited to parallel processing. However, local mesh refinement is almost always needed to achieve acceptable accuracy. Effective ways to produce local refinement using triangularization in both two- and three-dimensional problems is discussed by Bausch [21].

The implementation of multigrid processes on various parallel machines can be found in the literature. Numerous papers exist on the use of performance characteristics of ADI, Gauss elimination, and multigrid schemes running on multiprocessor machines. In nearly all cases, significant improvements in accuracy and execution speeds are reported using parallel methods [1, 2, 4, 5].

The effects of distorted grids on solution accuracy are investigated by Vichnevetsky and Turner [22]. If the mesh has discontinuous first derivatives, spurious waves are generated. When the values of these derivatives are small, the waves are still present, but the destructive interference of successive nodes eliminates the numerical noise. When spectral methods are used with distorted grids, waves tend to disappear from some regions but appear in other regions.

Carey [23] describes the role of grid generation and accuracy for structured and unstructured meshes using finite volume and finite element techniques. The use of local mesh adaptation has been shown to yield more accuracy and less overall nodal calculations than more conventional methods using globally fine grids. However, the bookkeeping needed to deal with mesh refinement and unrefinement, and the associated parameters used to define the refinement criteria, can be burdensome. Parallelization of adaptive schemes is relatively difficult and can lead to extensive modification of a code that runs relatively well on scalar machines. An example of a complex, three-dimensional unstructured grid using local refinement is shown in Fig. 3 (from [19]).

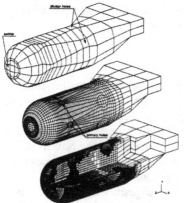

Figure 3 Unstructured, adapted grid (from [19])

5.3 Fluid Flow

5.3.1 Incompressible flow. Numerous works exist on modeling incompressible flows that can be found in the literature and now obtained from the web. Some of the more recent and interesting works dealing with large computations and use of high-performance computing are mentioned.

Rao et al. [24] describe the simulation of vortex shedding from a longitudinally oscillating circular cylinder at Reynolds numbers of 1000 and 4000 (based on cylinder diameter). A 2D moving grid system, based upon time-dependent coordinate transformations, the marker and cell method, and a third-order upwinding were employed using a 200 X 100 grid on a FACOM VP400 supercomputer. The computations predicted the experimentally observed lock-in effect when the oscillating frequency was double the shedding frequency. Simulation of three-dimensional flow with secondary flow and vortex motion in an annular compressor cascade is discussed by Gallus et al. [25]. They successfully modeled corner flow, vortex motion, and radial mixing using a Reynolds-averaged Navier-Stokes solver (RANS). The problem of simulating the transport of air pollutants over Europe is discussed by Zlatev [26], who used a three-dimensional grid of 32 X 32 X 9 to solve for 29 chemical species. More than 10^6 equations had to be solved at each time step. The performance of five different supercomputers was examined and showed differences in computational rates varying by a factor of 20.

Sharp [27] used a finite volume Runge-Kutta time integration scheme to solve the three-dimensional Navier-Stokes on a massively parallel machine using both structured and unstructured grids. Results of this study showed that parallel machines were ideally suited for unstructured grids. This work emphasized the importance of grid generation as a major stumbling block in efficient CFD calculations. A similar computation was reported by Catherasoo and Ramesh [28], who used a message-passing multicomputer with a large memory per processor.

Spectral methods are used to solve some fluid dynamics problems, especially those related to meteorological forecasting. In contrast to the usual FEM and FVM approaches, the spectral method requires large bandwidths and substantial computing power. Most reported simulations have utilized Cray machines. Amon and Mikic [29] studied internal flows with steps and grooves and interconnected flow passages. Lombard et al. [30] used the method to simulate internal gravity waves. Carlenzoli and Gervasio [31] used spectral methods and vectorization techniques and discuss the differences between weak and strong forms of collocation using various vector machines. A commercial code based on spectral elements was initially developed by Patera [32]; this code eventually became known as NEKTON and was ultimately obtained by Fluent, Inc., which it now sells commercially. Liakopoulos [33] used NEKTON to investigate convective heat transfer in tall cavities associated with flow over electronic components.

Andersson and Kristoffersen [34] simulated a steady, wall-driven square cavity and laminar start-up flows in a pipe using a time-dependent Navier-Stokes algorithm. The SIMPLE algorithm, with its staggered grids and sparse matrix associated with the implicit pressure correction, was used in a parallel architecture. Three-dimensional natural convection in a cavity was examined by Schafer [35] using a polynomial preconditioned conjugate gradient method on a SIMD computer with 1024 single-bit processors. Ways to exploit the cyclic connections are described; the best efficiency occurred when each node point in the grid could be paired with a processor.

Bonhaus and Wornom [36] compared implicit upwind and explicit central difference methods for solving the Navier-Stokes equations in transonic flow. Using a grid of 920,000 nodes, the implicit method was more accurate but the explicit method was more efficient. However, for equal degrees of computational effort, the explicit method was preferred. The implicit method used 2-3 hours more computer time to achieve the same level of accuracy.

5.3.2 Compressible flow. CFD is widely appreciated as a tool for analyzing internal flows and external aerodynamics over lifting bodies. Although CFD has been effectively used in engine design, most of the codes are coupled, inviscid boundary layer methods. Stow [37] examined simulations of internal, unsteady engine flow and suggested that future simulations would be based on direct Navier-Stokes simulations. Emphasis would be placed on finer grids, Reynolds stress transport models, unsteady flow, and interacting combustion. Stow's [37] insight has been shown to be true today. While these calculations require improved mathematical models, practical use depends upon implementing the codes on multiprocessor and parallel machines.

A survey of the application of CFD to jet propulsion systems is given by Tindell [38]. The survey discusses the use of CFD Euler codes to explain the differences found between wind tunnel and in-flight tests on the effects of forebody downwash and canopy shocks. ARC3D, a popular NASA-developed code for solving the Navier-Stokes equations, was used to compute the flow through an offset diffuser. Another NASA code, PARC, was used to calculate inlet flow. The inlet calculation normally required 1.5 million grid points; however, the PARC code was merged with a subsonic panel method and required only 51,000 grid points to calculate inlet performance at high angles of attack – angles too high to permit wind tunnel testing.

A transonic unsteady code, based upon a time-accurate approximate factorization algorithm, was used by Bennett et al. [39] to model unsteady flow with small disturbances. The code is capable of treating complete aircraft geometries with multiple lifting services and bodies.

5.3.3 Turbulence. Most turbulence modeling is done using the k-ε equations for closure, along with the continuity, Navier-Stokes, and energy equations. The k-ε equations are also nonlinear and highly coupled. When used in an implicit finite element method, the solution is often difficult to obtain without significant underrelaxation and lengthy computational times. Murthy [40] used a finite element method based on the q-f equations ($q = \sqrt{k}$; $f = \varepsilon/k$) rather than the usual k-ε equations. The q-f equations are more linear and only weakly coupled, but sometimes suffer from weak convergence history. More recent work by Wilcox

[41] shows the effectiveness and improved accuracy of the k-ω equations, where ω is the vorticity.

An important application of the Cray supercomputer is described by Jones et al. [42], who modeled the King's Cross Station fire (London), in which a fire erupted in an escalator and spread rapidly through a large ticketing gallery. A finite volume model (which later became the commercial code CFX), incorporating the k-ε equations, was used to model the fire. Using tight convergence criteria and a time step of 0.1 seconds, approximately 48 hours of computing time was required. The results were displayed on an SGI IRIS workstation and clearly demonstrated an unexpected phenomenon: the flames spread upward along the trench of the escalator. Subsequent 1/10- and 1/3-size experiments confirmed the effect. It was concluded that high-performance computing was critical in arriving at this conclusion in a timely fashion.

Cummings et al. [43] used the Baldwin-Lomax turbulence model with flux-splitting to study the flow at the wing-fuselage region of the F-18 for angles of incident up to 30°. The Chimera overset grid method, which permits holes in the grid, was used with a coarse far-field grid. Using 534,000 grid points, they obtained excellent agreement for the local pressures and the primary/secondary vortices on the fuselage and leading wing extension, as well as accurate predictions of the vortex breakdown.

Kida and Orszag [44] used a Navier-Stokes code to study compressible turbulence for Mach numbers up to 0.9. Flows were randomly forced with energy supplied to either the rotational or compressive components of kinetic energy.

Simulation of high Reynolds number turbulent flow over a backward-facing step was conducted by Friedrich and Arnold [45] using large eddy simulation technique (LES). The LES is generally considered to be better than other current turbulence closure schemes because it permits the investigation of instantaneous flow structure and the prediction of statistical qualities not amenable to measurement. Small-scale effects are eliminated by low-pass filtering. Central differencing and explicit time integration were used. Free shear layer and reattachment information were obtained, but the results showed the need for increased use of locally refined grids.

Ferziger [46] gives an excellent comparison of Reynolds-averaged Navier-Stokes (RANS) approach and the LES method for architectural or wind engineering problems which require very large grids. RANS works well but is not sufficiently accurate for three-dimensional or separating and reattaching flows; generally, adaptive or locally refined grids are required. LES appears to have the ability to treat these problems if appropriate wall-turbulence models can be developed. An excellent collection of numerical simulations of turbulent wind and vortex-shedding problems - all of which require supercomputers because of the large number of grids required - can be found in Volume 35 of the *Journal of Wind Engineering and Industrial Aerodynamics* (1990). A more recent review of CFD techniques and modeling of turbulence is given by Ferziger and Peric [19].

5.3.4 Visualization. Numerical solutions are not useful until specific information can be extracted from the results. The goal of visualization is to provide a readily understood effective visual representation of raw data. Such visual representations convey new insights and improved understanding of physical processes, mathematical concepts, and other important phenomena contained in the data.

The field of computer graphics provides tools to obtain pictures from symbolic or numeric descriptions and to interact with these pictures. Computer graphics is concerned with the development of the algorithms to create pictures on a computer display. It is not concerned with pictures and displays once their appearance is satisfactory. The extraction of meaning from the picture is in the human mind and is not of concern to this field.

Data availability rates began to increase significantly from both measuring devices (experimental data acquisition) and as a result of computations on fast computers, particularly supercomputers, in the late 1980s. In addition, the resolution was multiplied significantly. National supercomputer centers allowed access to virtually any scientist or engineer in the United States for large calculations. Improvements in computer size and performance, and experimental measuring devices, have lead to the generation of huge amounts of data on a daily basis. Dealing with such large amounts of numbers has become a major problem for handling computer output and visualization. In the mid 1980s, computer graphics received a significant boost in the development of improved and faster graphics hardware. New *raster* graphics techniques replaced the previous technology of limited, slower vector graphics. As more powerful and affordable processors were developed, personal graphics workstations began to emerge.

In 1986, the National Science Foundation sponsored an advisory panel to make recommendations in response to the needs developed by high data rates and the opportunity of using graphics workstations. This widely published report produced by the panel [47] called for new tools in a new field termed *Scientific Visualization*. Since 1987, many new applications have confirmed the necessity and power of this new methodology.

There are numerous graphical ways to effectively visualize scientific processes, especially in the fluid flow and heat transfer areas. Two of the more commonly used methods for displaying fluid movement are to plot velocity vectors at grid points and/or to animate the motion of Lagrangian particles suspended in the fluid. The following figures illustrate different ways of depicting fluid motion within a cavity [1]:

Figure 4a shows lines depicting information about strong and weak motion at a given time step. The long rods indicate strong movement in the area surrounding the center but not in the center or at the borders. Figure 4b uses arrows to signify the direction of the velocities with the links of the arrows denoting the magnitudes of the velocities. Figure 4c is a combination of the two previous figures: the arrows denote the direction of the velocity; both the size of their heads and the length of the arrows indicate the magnitude of the velocity. Figure 4d shows only arrowheads; while the directions of the velocity vectors are indicated by the arrowheads, color depicts their magnitude (short vectors are blue, medium vectors are green, and long vectors are orange and yellow). Figure 4e shows one frame of an animation of the data and illustrates the positions of particles at the same time as Figs. 4(a-c).

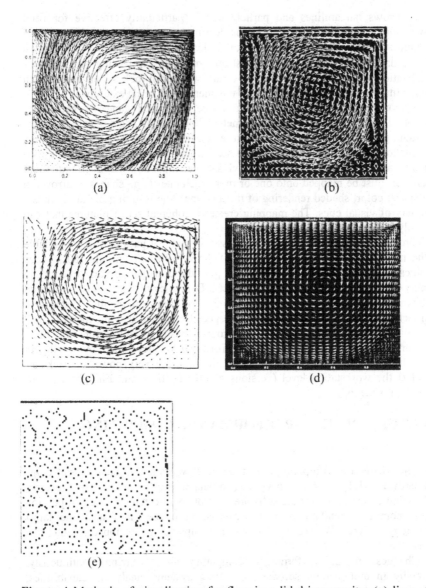

Figure 4 Methods of visualization for flow in a lid-driven cavity: (a) lines, (b) arrows – Application Visualization System (AVS), (c) arrows - MATLAB, (d) arrows – Interactive Data Language (IDL), and (e) particles - AVS (from [1])

Some of the more popular visualization techniques now in use today include the following [1]: scatter plots, glyphs (a glyph is a picture composed of individual parts or segments – visual cues), line graphs, histograms and pie charts, contour plots (isolines), image displays (mapping of a data value into a pixel), surface views, colored transformations, isosurfaces, ray tracing of volumes, data slices, arrows, streamlines and particle tracks (particularly effective for fluid flow), and animation. For most engineering purposes, a combination of techniques is needed to take advantage of all the information that can be extracted from a data set. This fact bears a strong message on writing software for visualization purposes; it is easy to create wonderful-looking pictures which provide little if any physical insight into the phenomena being examined.

In most cases, the numerical data to be visualized generally consist of complex data structures and many parameters. For example, data in aerodynamics research generally consists of pressure (scalar values at 3-D locations), deformation (vectors distributed in 3-D space), and shape (e.g., an airplane). To visualize all the available data, various visual cues must be used. Each individual parameter must be mapped onto one or more such cues (e.g., shapes of arrows in 3-D space; color; shaded rendering of the airplane): the resulting picture is thus a summary of visual cues. The mapping creates a coherent set of pictures that can be easily interpreted and understood.

Three of the more popular post-processing graphical packages now available in the market are TECPLOT [48], FIELDVIEW [49], and EnSight [50]. These graphics programs are quite powerful and give the user the ease and freedom to quickly generate very sophisticated 2- and 3-D images of complex flow-related problems. However, they cannot be run in real time on parallel systems in conjunction with a solver. TECPLOT is especially easy to use, inexpensive, and runs on PCs and SGI workstations; this program has been around for some time, and currently has over 10,000 users. FIELDVIEW, developed by Intelligent Light, Inc., and EnSight, developed by CEI, Inc., are very powerful packages aimed at the workstation level (versions of FIELDVIEW and EnSight are now available for the PC).

6 FUTURE OF HIGH-PERFORMANCE COMPUTING

The general trend of computing over the next few years is to create faster chips and hardware [1,2]. Such advances tend to happen much quicker than expected, not leaving enough time for the software developers to catch up. The competition among computer vendors to be the first to introduce the fastest and latest technology is fierce. We see this today in the chips being developed by Intel and others.

Changes in high-performance computing are occurring continuously, especially in architecture, software, operating systems, analytical capabilities, power, size, and cost. Many of the revolutionary advances in computing have been achieved in the aerospace and defense industries; advancements in these fields are primarily responsible for the continued quest for faster and more powerful machines.

A steady increase in parallelism will occur, especially as operating systems become more standard and easier to use [1,4,5]. Parallel processing is more powerful than sequential calculations, allowing many steps in a problem to be computed simultaneously. In addition, parallel computers are less costly to manufacture and maintain. While miniaturization may set the pace for technological advances, the rate at which software is developed and deployed to the end user will ultimately determine which computing systems penetrate and standardize industrial and engineering disciplines.

Improvements in semiconductor technology will result in new hardware designs. Hybrid machines will arise which employ optical interconnections to link high-speed processors. It is not inconceivable that computing may become totally optical, thus allowing faster analog computations of numerous algorithms which are relatively slow by today's standards.

Computer architects will continue to pursue fine-grain and coarse-grain parallelism. Workstations and mainframes will have multiple CPUs on a single chip. On the fastest computers, multiple VLSI chips will be placed on a single processor. As the number of processors increases, faster processor-to-memory interconnects will be needed. In addition, main memory will continue to increase and will put a heavier demand on data management, volatility, and cost, along with overcoming memory cycle time bottlenecks as the gap between logic, speed, and RAM cycle time grows.

The differences between parallel and vector computers clearly have a dramatic effect on the choice of algorithms afforded the end user. Computational complexity is particularly relevant for vector computers, but not important for parallel computers. In a vector computer, each computation costs time. In parallel computation, extra computation can be supported at no extra cost; likewise, overhead costs associated with communication and synchronization are not affected by computational complexity. Early researchers recognized these factors. Even with the availability of parallel systems today, research still appears to be focused on modifying existing algorithms rather than developing new ones for parallel computing.

High-performance computing of the future will be parallel. Precisely what form of architecture or operating system is still questionable. However, the number of computer systems now available makes it difficult to develop an efficient numerical algorithm that can be easily separated into more manageable parts. Assuming a single standard parallel system were to exist today, considerable work would still be needed to develop efficient software for the machine.

7 REFERENCES

1. L. D. Fosdick, E. R. Jessup, C. J. C. Schauble, and G. Domik, *An Introduction to High-Performance Scientific Computing*, The MIT Press, Cambridge, MA, 1996.
2. D. W. Pepper and A. F. Emery, Chapter 1 - Supercomputing in Heat Transfer, *Annual Review of Heat Transfer, Vol. V*, (ed. C. L. Tien), pp. 1-75, 1994.
3. J. J. Dongarra, C. B. Moler, J. R. Bunch, and G. W. Stewart, LINPACK User's Guide, *SIAM*, Philadelphia, PA, 1979.
4. C. Xavier and S. S. Iyengar, *Introduction to Parallel Algorithms*, J. Wiley & Sons, New York, 1998.

5. G. S. Almasi and A. Gottlieb, *Highly Parallel Computing*, Benjamin Cummings, Menlo Park, CA, 1988.
6. S. Sahni and V. Thanvantri, Performance Metrices: Keeping the Focus on Runtime, *IEEE-PDT*, pp. 43-46, 1996.
7. E. Schreck and M. Peric, Computation of Fluid Flow with a Parallel Multigrid Solver, *Int. J. Num. Meth. Fluids*, vol. 16, pp. 303-327, 1993.
8. J. J. Dongarra, Performance of Various Computers using Standard Linear Equations Software, Tech. Report CS-89-85, Oak Ridge National Laboratory, Oak Ridge, TN 37831; *netlib* version also available, 1994.
9. F. H. McMahon, The Livermore Fortran Kernels: A Computer Test of the Numerical Performance Range, Tech. Report UCRL-53745, Lawrence Livermore National Laboratory, Livermore, CA, 1986.
10. E. Ng and B. W. Peyton, Block Sparse Cholesky Algorithms on Advanced Uniprocessor Computers, *SIAM J. Sci. Comput.*, vol. 14, pp. 1034-1056, 1993.
11. O. O. Storaasli, D. T. Nguyen, M. A. Baddourah, and J. Qin, Computational Mechanics Analysis Tools for Parallel-Vector Supercomputers, *Int. J. Comput. Sys. Eng.*, vol. 4, no. 4-6, pp. 1-10, 1993.
12. D. E. Culler and J. P. Singh, *Parallel Computer Architecture, A Hardware/Software Approach*, Morgan Kaufmann Publishers, Inc., pp. 513-516, 1995.
13. N. P. Kronenberg, H. Levey, and W. D. Strecker, Vax Clusters: A Closely-Coupled Distributed System, *ACM Transactions on Computer Systems*, vol. 4, no. 2, pp. 130-146, 1998.
14. G. F. Pfister, In Search of Clusters-The Computing Battle for Lowly Parallel Computing, Englewood Cliffs, NJ: Prentice Hall, 1995.
15. A. Giest, A. Beguelin, J. Dongarra, W. Jiang, R. Manchek, and V. Sunderam, *PVM – Parallel Virtual Machine: A User's Guide and Tutorial for Network Parallel Computing*, Sci. and Eng. Comput., The MIT Press, Cambridge, MA, 1995.
16. M. Litzkow, M. Livny, and M. W. Mutka, Condor-A Hunter of Idle Workstations. Proc. *Eighth Int. Conference of Distributed Computing Systems* (June), pp. 104-111, 1998.
17. J. Lukowsky and S. Polit, IPPacket Switching on the GIGAswitch/FDDI System, http://www.networks.digital.com:80/dr/techart/gsfip-mm.html, 1997.
18. O. C. Zienkiewicz and R. L. Taylor, *The Finite Element Method*, 4th Ed., vol. 1, McGraw-Hill, London, 1989.
19. J. H. Ferziger and M. Peric, *Computational Methods for Fluid Dynamics*, Springer-Verlag, Berlin, 1996.
20. J. C. Heinrich and D. W. Pepper, *Intermediate Finite Element Method: Applications to Fluid Flow and Heat Transfer*, Taylor and Francis, Philadelphia, PA, 1999.
21. E. Bausch, Local Mesh Refinement in 2 and 3 Dimensions, *Impact Comput. Sci. Eng.*, vol. 3, pp. 181-191, 1991.
22. R. Vichnevetsky and L. H. Turner, Spurious Scattering from Discontinuously Stretching Grids in CFD, *Appl. Num. Math.*, vol. 8, no. 3, pp. 289-299, 1991.

23. G. F. Carey, *Computational Grids, Generation, Adaptation, and Solution Strategies*, Taylor & Francis, Washington, DC, 1997.
24. P. M. Rao, K. Kuwahara, and K. Tsuboi, Simulation of Unsteady Viscous Flow Around a Longitudinally Oscillating Circular Cylinder in a Uniform Flow, *Appl. Math. Modelling*, vol. 16, no. 1, pp. 26-35, 1992.
25. H. E. Gallus, C. Hah, and H. D. Schulz, Experimental and Numerical Investigation of Three-Dimensional Viscous Flows and Vortex Motion Inside an Annular Compressor Blade Row, *ASME J. Eng. Gas Turbines Power*, vol. 113, pp. 198-202, 1991.
26. Z. Zlatev, Running Large Air Pollution Models on High Speed Computers, *Math. Comput. Modelling*, vol. 14, pp. 737-740, 1990.
27. T. Sharp, Computational Fluid Dynamics on the Connection Machine, *Math. Comput. Modelling*, vol. 14, pp. 714-719, 1990.
28. C. J. Catherasoo and A. K. Ramesh, Implementation of a Three-Dimensional Navier-Stokes Code on the Symult Series 2010, *Math. Comput. Modelling*, vol. 14, pp. 710-713, 1990.
29. C. H. Amon and B. B. Mikic, Spectral Element Simulations of Forced Convective Heat Transfer: Application to Supercritical Slotted Channel Flow, *Num. Heat Transfer with Personal Computers and Supercomputers* (ed. R. K. Shah), ASME HTD-Vol. 110, pp. 175-184, 1989.
30. P. N. Lombard, J. J. Riley, and D. D. Stretch, Energetics of a Stably Stratified Mixing Layer, Proc. 9th Symp. on Turb. and Diff., Amer. Meteor. Soc., Roskilde, Denmark, 1990.
31. C. Carlenzoli and P. Gervasio, Effective Numerical Algorithms for the Solution of Algebraic Systems Arising in Spectral Methods, *Appl. Num. Math.*, vol. 10, no. 2, pp. 87-113, 1992.
32. A T. Patera, A Spectral Element Method for Fluid Dynamics; Laminar Flow in a Channel Expansion, *J. Comput. Phys.*, vol. 54, pp. 468, 1984.
33. A. Liakopoulos, P. A. Blythe, and P. G. Simpkins, Convective Flows in Tall Cavities, *Simulation and Numerical Methods in Heat Transfer* (eds. A. F. Emery and D. W. Pepper), ASME HTD-Vol. 157, New York, NY, 1990.
34. H. I. Andersson and R. Kristoffersen, Numerical Simulation of Unsteady Viscous Flow, *Arch. Mech.*, vol. 41, pp. 207-223, 1989.
35. M. Schafer, Parallel Algorithms for the Numerical Simulation of Three-Dimensional Natural Convection, *Appl. Num. Math.*, vol. 7, no. 4, pp. 347-365, 1991.
36. D. L. Bonhaus and S. F. Wornom, Comparison of Two Navier Stokes Codes for Attached Transonic Wing Flows, *J. Aircraft*, vol. 29, no. 1, pp. 101-107, 1992.
37. P. Stow, Computational Fluid Dynamics – The Way of the Future?, *Inst. Math. Appl.*, vol. 26, no. 8/9, pp. 182-189, 1990.
38. R. H. Tindell, Computational Fluid Dynamic Applications for Jet Propulsion System Integration, *ASME J. Eng. Gas Turbines Power*, vol. 113, no. 1, pp. 40-50, 1991.
39. R. M. Bennett, S. R. Bland, J. T. Batina, M. D. Gibbons, and D. G. Mabey, Calculation of Steady and Unsteady Pressures on Wings at Supersonic Speeds

with a Transonic Small-Disturbance Code, *J. Aircraft*, vol. 28, no. 3, pp. 175-180, 1991.

40. C. S. R. Murthy and F. Rajaraman, Analytical and Simulation Studies of a Multiprocessor System for High-Speed Numerical Computations, *Math. Comput. in Simulation*, vol. 32, no. 4, pp. 393-401, 1990.

41. D. C. Wilcox, *Turbulence Modeling for CFD*, 2nd Ed., DCW Industries, La Canada, CA, 1998.

42. I. P. Jones, S. Simcox, and N. S. Wilkes, Modelling the King's Cross Station Fire, *Inst. Math. Appl.*, vol. 26, no. 4, pp. 90-92, 1990.

43. R. M. Cummings, Y. M. Rizk, L. B. Schiff, and N. M. Chaderjian, Navier-Stokes Predictions for the F-18 Wing and Fuselage at Large Incidence, *J. Aircraft*, vol. 29, no. 4, pp. 565-574, 1992.

44. S. Kida and S. A. Orszag, Energy and Spectral Dynamics in Forced Compressible Turbulence, *J. Sci. Comput.*, vol. 5, no. 2, pp. 85-125, 1990.

45. R. Friedrich and M. Arnal, Analysing Turbulent Backward-facing Step Flow with the Lowpass-filtered Navier-Stokes Equations, *J. Wind. Eng. Indus. Aerodyn.*, vol. 35, pp. 101-128, 1990.

46. J. H. Ferziger, Approaches to Turbulent Flow Computation: Applications to Flow Over Obstacles, *J. Wind Eng. Indus. Aerodyn.*, vol. 35, pp. 1-19, 1990.

47. B. H. McCormick, T. A. Defanti, and M. D. Brown, Visualization in Scientific Computing, *Computer Graphics*, vol. 21, no.6, 1987.

48. TECPLOT Reference Manual, Version 7.5, Amtec Engineering, Inc., Bellevue, WA, 98009-3633, 1998.

49. FIELDVIEW, Intelligent Light, Lyndhurst, NJ, 07071, 1998.

50. EnSight, CEI, Inc., Research Triangle Park, NC, 27709, 1998.

TWO

UNSTRUCTURED FINITE VOLUME METHODS FOR MULTI-MODE HEAT TRANSFER

S.R. Mathur
J.Y. Murthy

1 INTRODUCTION

In recent years, there has been considerable interest in the computational fluid dynamics (CFD) community in the development of unstructured mesh methods for fluid flow problems. Unstructured meshes greatly reduce mesh generation time for complex industrial geometries, frequently by an order of magnitude, and allow advances such as dynamic mesh adaption. Aided by the exponential increase in computational speed, unstructured mesh methods are beginning to establish CFD as an integral part of the industrial design cycle.

As unstructured mesh methods make industrial CFD calculations more accessible, there is, increasingly, an interest in addressing multiple physical phenomena in the same problem, and frequently, widely differing scales. Many industrial problems require the coupled computation of conduction, convection and radiation heat transfer in complex domains with laminar and turbulent flows, chemical reactions and other complexities. In automotive underhood cooling, for example, the objective is to minimize the surface-to-surface transfer of radiant heat from the exhaust system to other components under the hood while providing adequate convective cooling of all components. It is not uncommon to couple the underhood calculation with one for external aerodynamics since the partitioning of the air flow between the external and internal flow paths depends on their competing resistances. problem reduces to coupling the conjugate heat transfer problem with a methodology for surface-to-surface radiation in extremely complex geometries, along with a calculation for turbulent convective heat transfer. In glass furnaces, on the other hand, the source of heat is the combustion of oils or gaseous hydrocarbons. Here, the flows in the combustion space are turbulent; combustion products such as carbon dioxide and water vapor, along with particulates such as soot, make wavelength-dependent absorption, emission and scattering important. The glass melt itself is usually laminar and radiatively participating in the near infra red, and is coupled to the thermal and radiative field in the combustion space through a Fresnel interface. Problems of equal complexity occur in materials processing, ceramics processing, optical fiber manufacture, and other areas.

In this chapter, we address the issues associated with the solution of multi-mode heat transfer problems of industrial interest . We focus exclusively on heat transfer issues; the reader is referred to [1, 2] for details about the solution of fluid flow. In the next section, we review available finite volume methods for unstructured meshes. We then present a cell-based finite volume scheme for heat transfer, addressing issues specific to the conservative discretization of the energy equation and its solution using an efficient algebraic multigrid scheme. We then turn our attention to the solution of surface-to-surface and participating radiation using extensions of the finite volume method. We do not attempt an exhaustive tour of the literature; instead, we present only those methods and techniques that we have found useful over the years in solving large-scale industrial problems.

2 REVIEW

Though a variety of finite element (FEM), finite volume (FV) and control-volume-based finite element methods (CVFEM) have emerged over the years for use with unstructured meshes [1, 3, 4, 5, 6, 7, 8], we focus here on the finite volume method. Unstructured finite volume methods can broadly be classified as either node-based or cell-based. In either case, the calculation domain is divided into cells (elements), typically tetrahedra, hexahedra, pyramids and prisms. Node-based schemes store solution unknowns at the cell vertices and impose conservation on the cell-duals, which are non-overlapping polyhedra centered about the cell vertices. Examples include the work by Mavriplis et al. [8] and CVFEM's [9, 10]. Cell-based schemes, on the other hand, store solution unknowns at cell centroids and impose conservation on the cell itself, rather than on its dual [1, 4, 5].

The equations for fluid flow and heat transfer require the evaluation of first gradients for the evaluation of diffusion fluxes and for second-order interpolation schemes. In the absence of line structure, CVFEM's employ finite-element-like shape functions to interpolate solution variables on the element and also to deduce gradients [9, 11]. Both node and cell-based schemes have employed reconstruction, both linear and higher order [1, 12]. Reconstruction has the advantage that it can be made cell-shape independent, facilitating the use of truly arbitrary polyhedral cells for the spatial discretization. Like FEM's, CVFEM's also use shape functions for quadrature; with the use of higher-order elements, numerical schemes of arbitrarily high order may, in principle, be devised. In the cell-based context, it is possible to compute reconstruction gradients and interpolates of arbitrarily high order [12], though spatial quadrature schemes of arbitrary order have not, to our knowledge, been published for arbitrary polyhedra.

As with structured meshes, co-located storage of pressure and velocity leads to checkerboarding. Unlike structured meshes, however, staggered meshes for pressure and velocity are not conveniently devised for unstructured polyhedra. CVFEM's, like FEM's, have employed unequal-order interpolation for pressure and velocity, with pressure being interpolated to lower order than velocity [9]. Within the cell-based context, however, unequal-order interpolation does not follow naturally from the geometry. Instead, most cell-based schemes (and many node-based schemes) have used added-dissipation ideas, storing pressure and velocity at the same location but adding a fourth-order dissipation to damp spurious pressure modes [1, 4, 6, 8].

Regardless of whether cell-based or node-based schemes are used, the same iterative procedures may be used for solving the set of discrete equations once they are derived. Early work in the aerospace community involved the computation of Euler flows over complete aircraft configurations [13], using explicit density-based time-marching schemes; these were extended to the Reynolds-averaged Navier-Stokes equations by a number of researchers ([8], for example). The low-subsonic and incompressible flow regimes were addressed within this framework through the use of pseudo-compressibility and preconditioning ideas [14, 15]. However, the viscous limit poses a strong time-stepping restriction on explicit schemes, necessitating multigrid acceleration, implicit residual smoothing and other remedies [8]. Coupled implicit schemes, on the other hand, impose large memory requirements [16] since practices such as approximate factorization cannot be used on unstructured meshes.

In the incompressible flow community, segregated solution strategies based on the SIMPLE algorithm and its variants have found wide usage with unstructured meshes because of their low memory requirements and their demonstrated success in dealing with industrial flows [1, 4, 6]. Efficient linear solvers based on conjugate gradients or algebraic multigrid schemes are used for solving each equation set. Coupled solution strategies for unstructured meshes have also received much attention recently. In the unstructured mesh context, geometric multigrid methods are difficult to devise since the creation of nested coarse levels by cell agglomeration may lead to non-convex polyhedra and attendant numerical difficulties, especially in the computation of gradients. A variety of coupled algebraic multigrid methods have appeared [17, 18] which sidestep this problem by creating coarse-mesh equations through algebraic manipulations of the fine-mesh equations; no discretization is ever done on the coarse mesh directly. In an alternative attempt to avoid unstructured mesh agglomeration, Jyotsna and Vanka [19] started with an unstructured mesh at the coarsest level and created progressively fine meshes by cell subdivision; they used the SIMPLER algorithm as a relaxation sweep at each level. Substantial increases in solution speed have been obtained with only modest memory penalty. With this type of method, the newly-created boundary nodes must be projected on to the geometry to preserve geometric definition at fine levels, where the solution is really desired. For complex industrial geometries, this necessitates carrying a full CAD description of the geometry in the solver, with attendant difficulties associated with CAD compatibility and connectivity. Also, subdivision of certain cell shapes,

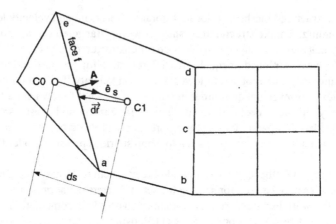

Figure 1 Control volume

such as tetrahedra, leads to a progressive degeneration of mesh quality.

3   FINITE VOLUME FORMULATION

The finite volume discretization is best illustrated for a generic scalar transport equation:

$$\frac{\partial}{\partial t}(\rho\phi) + \nabla\cdot(\rho\mathbf{V}\phi) = \nabla\cdot(\Gamma\nabla\phi) + S_\phi \tag{1}$$

Here, ϕ is the transport variable, t is time, ρ is the density, $\mathbf{V}$ is the velocity vector, Γ is the diffusion coefficient and S_ϕ is the source term. The equations governing the transport of mass, momentum and energy can be cast in the above form by the appropriate choice of ϕ, Γ and S_ϕ.

3.1   Discretization of Convection-Diffusion Equation

The domain is discretized into arbitrary unstructured convex polyhedra called cells. The boundaries surrounding the cells are called faces, and the vertices of the polyhedra are referred to as nodes. Each internal face has two cells on either side; these are referred to as the face neighbors. The neighbors of a cell are defined to be those cells with which it shares a common face.

All transport variables are stored at cell centroids. This arrangement is preferred over node based storage for several reasons. With cell-based storage, conservation can be ensured for arbitrary control volumes with non-conformal interfaces without special interpolation techniques. Consider the mesh shown in Fig. 1, for example. Cell C1 can be considered to have five faces, a-b-c-d-e, and no special treatment is required. On triangular and tetrahedral meshes, cell-based storage also enjoys better resolution (since the ratio of number of cells to nodes is between 3 and 5) for roughly the same amount of work, which is typically proportional to the number of

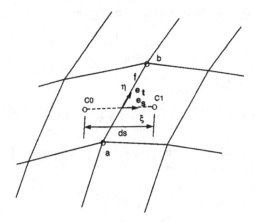

Figure 2 Structured grid control volume

faces. In addition to cell centroids, transport variables are also stored at the centroids of all boundary faces to aid in imposition of non-linear boundary conditions. Integration and discretization of Eq. (1) about the control volume C0 yields:

$$\left[\frac{\partial}{\partial t}(\rho\phi)\right]_0 \Delta V_0 + \sum_f J_f \phi_f = \sum_f D_f + (S_\phi \Delta V)_0 \tag{2}$$

where J_f is the mass flow rate, ΔV_0 is the volume of the cell C0, D_f is the transport due to diffusion through the face f, and the summations are over the faces of the control volume. For the purposes of scalar transport, the mass flow rate J_f is assumed to be known. To obtain a set of linear equations, all other face quantities as well as volume integrals in Eq. (2) must be written in terms of the unknowns, i.e., values of ϕ at cell and boundary face centroids.

3.1.1   Diffusion Term. The diffusion term at a face is given by

$$D_f = \Gamma_f \nabla\phi \cdot \mathbf{A} \tag{3}$$

$\mathbf{A}$ is the area vector associated with the face f. The discretization of this term is derived by a transformation from physical coordinates (x, y) to computational coordinates (ξ, η) at the face f in Fig. 2. The ξ direction is aligned with the vector joining cell centroids; the η direction lies in the direction joining the vertices of the face. We may write

$$\nabla\phi \cdot \mathbf{A} = \phi_\xi \left(\xi_x A_x + \xi_y A_y\right) + \phi_\eta \left(\eta_x A_x + \eta_y A_y\right) \tag{4}$$

A_x and A_y are the components of $\mathbf{A}$. The transformation metrics can be written in terms of derivatives of (x, y) as

$$\begin{aligned}
\xi_x &= \mathcal{J} y_\eta \\
\xi_y &= -\mathcal{J} x_\eta \\
\eta_x &= -\mathcal{J} y_\xi \\
\eta_y &= \mathcal{J} x_\xi
\end{aligned} \tag{5}$$

where $\mathcal{J}$, the Jacobian of the transformation, is given by

$$\mathcal{J} = \frac{1}{x_\xi y_\eta - x_\eta y_\xi} \tag{6}$$

Substituting in Eq. (4) yields

$$\nabla\phi \cdot \mathbf{A} = \phi_\xi \left(\frac{y_\eta A_x - x_\eta A_y}{x_\xi y_\eta - x_\eta y_\xi} \right) + \phi_\eta \left(\frac{-y_\xi A_x - x_\xi A_y}{x_\xi y_\eta - x_\eta y_\xi} \right) \tag{7}$$

For the mesh shown in Fig. 2, writing consistent approximations for the derivatives, we have the following expressions:

$$
\begin{aligned}
\phi_\xi &= \phi_1 - \phi_0 \\
x_\xi &= x_1 - x_0 \\
y_\xi &= y_1 - y_0 \\
\phi_\eta &= \phi_b - \phi_a \\
x_\eta &= x_b - x_a \\
y_\eta &= y_b - y_a
\end{aligned} \tag{8}
$$

From the geometry in Fig. 2, it is seen that

$$
\begin{aligned}
A_x &= (y_b - y_a) \\
A_y &= -(x_b - x_a)
\end{aligned} \tag{9}
$$

We also note that the unit vector in the direction joining the two cell centroids is

$$\hat{e}_s = \frac{(x_1 - x_0)\hat{\imath} + (y_1 - y_0)\hat{\jmath}}{ds} \tag{10}$$

where ds is the distance between the centroids. The tangent vector along the face has the components

$$\hat{e}_t = \frac{(x_b - x_a)\hat{\imath} + (y_b - y_a)\hat{\jmath}}{|\mathbf{A}|} \tag{11}$$

Using Eqs. (8)-(11) in Eq. (7), we obtain

$$\nabla\phi \cdot \mathbf{A} = \frac{\phi_1 - \phi_0}{ds} \frac{\mathbf{A} \cdot \mathbf{A}}{\mathbf{A} \cdot \hat{e}_s} + \frac{\phi_b - \phi_a}{|\mathbf{A}|} \frac{\mathbf{A} \cdot \mathbf{A}}{\mathbf{A} \cdot \hat{e}_s} \hat{e}_t \cdot \hat{e}_s \tag{12}$$

We note that in arriving at Eq. (12), we have not made any assumptions about mesh structure. The diffusion term has been split into two parts, a primary component representing the diffusion component in the direction of the line joining the two cell centroids and a secondary component which arises because of grid non-orthogonality. The primary component is purely in terms of the difference of two cell values on either side of the face f, (i.e., ϕ_0 and ϕ_1). It has the same form as that obtained with a central difference scheme on a structured mesh and is treated implicitly in the discrete equation for the two cells.

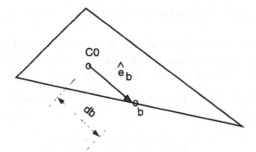

Figure 3 Boundary control volume

In Eq. (12), the secondary component is written in terms of ϕ_a and ϕ_b, the face vertices. In three dimensions, the secondary term would involve gradients along two directions perpendicular to the face normal. This representation is convenient for structured meshes. For unstructured meshes in three dimensions, however, the face f is a polyhedron with an arbitrary number of vertices, and these directions cannot uniquely be identified. We, therefore, write the secondary component as the difference between the total diffusion term and the primary component. Thus,

$$D_f = \Gamma_f \frac{(\phi_1 - \phi_0)}{ds} \frac{\mathbf{A} \cdot \mathbf{A}}{\mathbf{A} \cdot \hat{\mathbf{e}}_s} + \Gamma_f \left(\overline{\nabla \phi} \cdot \mathbf{A} - \overline{\nabla \phi} \cdot \hat{\mathbf{e}}_s \frac{\mathbf{A} \cdot \mathbf{A}}{\mathbf{A} \cdot \hat{\mathbf{e}}_s} \right) \tag{13}$$

where $\overline{\nabla \phi}$ at the face is taken to be the average of the derivatives at the two adjacent cells, determined as discussed in Section 3.1.4.

The discretization of the diffusion term at boundary faces is analogous, with the boundary centroid replacing cell C1. Thus, for face b in Fig. 3 :

$$D_b = \Gamma_b \frac{(\phi_b - \phi_0)}{db} \frac{\mathbf{A} \cdot \mathbf{A}}{\mathbf{A} \cdot \hat{\mathbf{e}}_b} + \Gamma_b \left(\nabla \phi_0 \cdot \mathbf{A} - \nabla \phi_0 \cdot \hat{\mathbf{e}}_b \frac{\mathbf{A} \cdot \mathbf{A}}{\mathbf{A} \cdot \hat{\mathbf{e}}_b} \right) \tag{14}$$

where ϕ_b is the value at the boundary and $\hat{\mathbf{e}}_b$ is the vector from the cell centroid to the boundary face centroid.

3.1.2   Convection Term. As a first order-approximation, the value of the scalar at the face f can be taken to be that of the upwind cell

$$\phi_f = \phi_{\text{upwind}} \tag{15}$$

Higher order schemes on structured grids typically use a bigger stencil along the appropriate grid line to interpolate ϕ_f. On unstructured grids this is not possible, and the variable must be reconstructed using a more general stencil of neighboring cells. In keeping with our philosophy of admitting arbitrary control volumes, we write a linearly reconstructed value of ϕ_f using the *reconstruction* gradient of ϕ in the upwind cell:

$$\phi_f = \phi_{\text{upwind}} + \nabla \phi_{r_{\text{upwind}}} \cdot \vec{dr} \tag{16}$$

A cell-shape independent procedure for computing the reconstruction gradient is discussed below.

At flow boundaries, the value of ϕ is taken to be the specified boundary value at inflow faces. For faces where flow exits the domain, the boundary value is upwinded from the cell and no boundary conditions are required.

3.1.3 Unsteady term. The unsteady term is discretized using backward differences. The first-order scheme is:

$$\left[\frac{\partial}{\partial t}(\rho\phi)\right]_0 = \frac{(\rho\phi)_0^{n+1} - (\rho\phi)_0^n}{\Delta t} \tag{17}$$

Higher order representations of the unsteady term can be written using more levels of storage.

3.1.4 Gradient Calculation. Computation of secondary diffusion terms as well as linear reconstruction for convection terms requires the knowledge of gradients of ϕ at the cell centers. Unlike structured grids, these cannot be obtained by finite differences. CVFEM methods [9] overcome this by employing shape functions; another approach is to use interpolation methods such as least squares [12]. However, many of these techniques are cell-shape specific. A more general approach is to use the discrete form of the divergence theorem. We define the *reconstruction gradient* as:

$$\nabla\phi_r = \frac{\alpha}{\Delta\mathcal{V}}\sum_f(\bar{\phi}_f\mathbf{A}) \tag{18}$$

where the summation is over all the faces of the cell. The face value of ϕ is obtained by averaging the values at the neighboring cells, so that for the face f in Fig. 1

$$\bar{\phi}_f = \frac{\phi_0 + \phi_1}{2} \tag{19}$$

α is a factor used to ensure that the reconstruction does not introduce local extrema [1]. It is easy to see that this procedure can be used for arbitrary control volume shapes, including non-conformal grids.

Using the reconstruction gradient, the value at any face of the cell can be reconstructed as

$$\phi_{f,c} = \phi_c + \nabla\phi_r\cdot\vec{dr} \tag{20}$$

The cell derivatives used for the secondary diffusion terms in Eq. (13) are computed by again applying the divergence theorem over the control volume and using the averaged reconstructed values at the faces:

$$\nabla\phi = \frac{1}{\Delta\mathcal{V}}\sum_f(\tilde{\phi}_f\mathbf{A}) \tag{21}$$

where, for face f in Fig. 1, $\tilde{\phi}_f$ is given by

$$\tilde{\phi}_f = \frac{\phi_{f,0} + \phi_{f,1}}{2} \tag{22}$$

3.1.5   Boundary Conditions. Diffusion and convection terms are evaluated at boundaries as discussed above. In the case of Dirichlet boundary conditions ϕ_b in the diffusion term , (Eq. (14)), is a known quantity. For Neumann conditions, the entire boundary diffusion term D_b is a known quantity, and Eq. (14) may used to determine ϕ_b for post-processing, if desired. For more complex boundary specifications, such as convection or radiation boundary conditions for the energy equation, the boundary value ϕ_b is determined by a local balance. The procedure is illustrated in detail below in conjunction with the energy equation.

3.1.6   Final Form Of Discrete Equations. Combining the discretized diffusion, convection, source and unsteady terms (if any) yields linear equations of the following form at each cell i :

$$M_{ij}\phi_j + S_i = 0 \qquad (23)$$

This system of equations is solved using an algebraic multigrid scheme (Section 6) to obtain the values of ϕ at all cells. In the case of non-linearities (due to either secondary diffusion terms, second-order interpolation or non-linear sources), the coefficients and source terms are updated using the current values of ϕ, and the whole process is repeated until convergence. In the case of transient problems, the converged solution represents the values at the next time step, and the iterations are repeated for the desired duration.

3.2   Computation Of Fluid Flow

The momentum equations are discretized in the same manner as the scalar transport equations. The pressure field is obtained from the continuity using the SIMPLE scheme. Details can be found in [1].

3.3   Solution Adaptivity

Because the mesh is unstructured, it may be adapted conveniently to evolving flow features. Two types of adaption are possible: conformal and non-conformal. In case of conformal adaption, the region marked for adaption is refined by splitting the longest edges. Additional non-marked cells may also need refinement, so as to not create "hanging" nodes, such as node c in Fig. 1. With non-conformal adaption, each marked cell is isotropically subdivided, and "hanging" nodes are allowed. Non-conformal adaption facilitates highly local adaption and preserves mesh quality for cell shapes such as quadrilaterals, hexahedra and triangles. Since the discretization supports arbitrary polyhedra, cells such as $C1$ in Fig. 1, created by non-conformal adaption, are automatically admitted.

Figure 4 shows the mesh used to compute natural convection due to a heated cylinder in a square cavity [1]. A hybrid mesh is set up with quadrilaterals near the walls, and triangles elsewhere, and an initial solution obtained. Non-conformal mesh adaption to the velocity magnitude results in the 16,113-cell mesh shown in Fig. 4; it captures the hot plume rising around the cylinder and the downward flowing stream along the cold wall as seen in the right half of the figure. Figure 5 shows a comparison of the local Nusselt number along the cold wall, and along

Figure 4 Natural convection in a cavity: adapted hybrid mesh and streamlines

the hot cylinder with the benchmark results of [20] using a 256×128 quadrilateral mesh. The comparison is quite good, even with a mesh only half as dense.

3.4 Discussion

The unstructured mesh finite volume scheme presented here is closely related to the cell-based scheme for Cartesian meshes in [21] and the body-fitted-mesh scheme in [22]. Like these methods, the formulation is conservative; unlike these methods, the control volume is an arbitrary polyhedron rather than a rectangle or a quadrilateral. The scheme exhibits the appropriate defaults when a first-order upwind scheme is used for convection. On Cartesian meshes, the discrete equations obtained are identical to those in [21]; similarly, for orthogonal body-fitted meshes, the decomposition of the diffusion terms shown here results in the same discrete equations used in standard structured mesh schemes [22]. The primary difference lies in the evaluation of gradients; secondary diffusion terms and second-order upwind terms, which use cell gradients, result in a different S_i term in Eq. 23 from those in [21, 22]. For unstructured meshes, the matrix M_{ij} in Eq. 23 is sparse, but unstructured; consequently, familiar line-iterative solvers cannot be used, and more general methods, such as conjugate gradient or multigrid solvers, are required. Nevertheless, the similarity with structured mesh solvers means that much of the experience of the last few decades is directly applicable to this class of finite volume schemes.

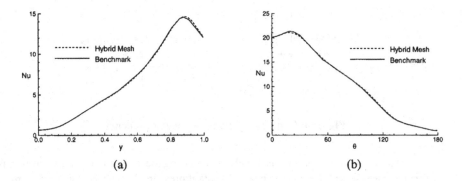

Figure 5 Nusselt number for natural convection in a cavity: (a) cold wall and (b) hot wall

4   ENERGY EQUATION

Heat transfer is governed by the energy conservation equation:

$$\frac{\partial}{\partial t}(\rho E) + \nabla\cdot(\rho \mathbf{V} E) = \nabla\cdot(k\nabla T) + \nabla\cdot(\tau\cdot\mathbf{V}) - \nabla\cdot(p\mathbf{V}) + S_r + S_h \tag{24}$$

Here, k is the thermal conductivity, τ is the stress tensor, p is the pressure and E is the total energy defined as

$$E = e(T) + \frac{\mathbf{V}\cdot\mathbf{V}}{2} \tag{25}$$

and e is the internal energy per unit mass. The terms on the LHS of Eq. (24) describe the temporal evolution and the convective transfer of total energy, respectively. The first three terms on the RHS represent the conductive transfer, viscous dissipation and pressure work, respectively. S_r is the radiative heat transfer (see below), and S_h represents all other volumetric heat sources.

4.1   Iterative Scheme

Equation (24) is similar to the scalar transport equation except for the fact that the convective transfer is described in terms of total energy E while diffusion is governed by gradients of temperature T. To make it amenable to the discretization procedure used for the scalar transport equation, the energy equation is usually cast into the temperature form:

$$\frac{\partial}{\partial t}(\rho C_p T) \;+\; \nabla\cdot(\rho \mathbf{V} C_p T) = \nabla\cdot(k\nabla T) + \tau{:}\mathbf{V} + \rho T\left(\frac{\partial C_p}{\partial t} + \mathbf{V}\cdot\nabla C_p\right)$$
$$+\; S_r + S_h + \left(\frac{\partial \ln(1/\rho)}{\partial \ln T}\right)_p\left(\frac{\partial p}{\partial t} + \mathbf{V}\cdot\nabla p\right) \tag{26}$$

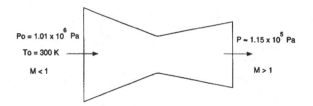

Figure 6 Converging-diverging nozzle: schematic

However, this has the serious drawback of casting the equation in a non-conservative form in cases where the specific heat at constant pressure, C_p, is variable, or if compressibility or viscous dissipation effects are important. Consequently, even the conservative finite volume scheme outlined above cannot guarantee perfect energy conservation on coarse meshes. Another alternative is to write the diffusion terms in terms of energy — but this is not desirable for conjugate heat transfer applications where step changes in C_p make energy double valued at fluid-solid boundaries. Also, boundary conditions are most naturally written in terms of temperature; conversion to the energy form of complex boundary conditions, such as for radiation or surface reaction, is not convenient.

We prefer to retain temperature as the dependent variable while ensuring that the discrete analogue of Eq. (24) is satisfied at convergence. To this end, we rewrite Eq. (24) as

$$\frac{\partial}{\partial t}(\rho C_p T) \;+\; \nabla \cdot (\rho \mathbf{V} C_p T) = \nabla \cdot (k \nabla T) + \nabla \cdot (\tau \cdot \mathbf{V}) - \nabla \cdot (p \mathbf{V}) + S_r + S_h$$

$$+\; \frac{\partial}{\partial t}\left(\rho\left[C_p T^* - E^*\right]\right) + \nabla \cdot \left(\rho \mathbf{V}\left[C_p T^* - E^*\right]\right) \tag{27}$$

Here, T^* and E^* are the values of T and E at the current iteration. In the discretization process, we wish to treat the unsteady, convective and diffusive terms in T implicitly, while absorbing the T^* and E^* terms explicitly. It is easy to see that the procedure described above for the scalar transport equation can now be applied. The coefficients of the algebraic set as well as the non-linear source terms are updated iteratively using prevailing values. At convergence, $T = T^*$; the unsteady and convective terms in T and T^* in Eq. (27) cancel, and Eq. (24) is satisfied identically.

Even though temperature is used as the dependent variable, this approach ensures energy conservation regardless of the mesh size. Since temperature is continuous at fluid-solid boundaries, no special treatment is required for the diffusion terms. Another advantage of this approach over formulations using energy as the dependent variable is that in the latter, temperature must be recovered from energy. For $C_p = C_p(T)$, this requires finding the roots of a non-linear equation for each computational cell in the domain, and can be quite expensive. In the present approach, T is computed directly from Eq. (27); an evaluation of E is required using Eq. (25), but no root-finding.

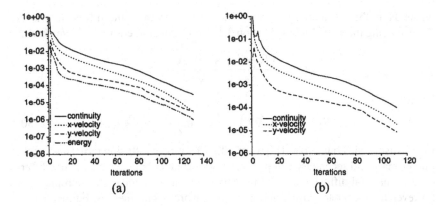

Figure 7 Converging-diverging nozzle: convergence behavior (a) with present scheme, and (b) assuming constant total energy

Since the pressure work and kinetic energy terms appear explicitly as source terms, the form proposed above might be expected to have poorer convergence characteristics compared to energy/enthalpy forms which have no source terms. However, in the latter forms, these terms appear explicitly in the recovery of T from E, and therefore, in practice, the overall convergence behavior of the two approaches is not significantly different. As an illustration, consider the inviscid, compressible flow of a non-conducting perfect gas in a converging-diverging nozzle shown in Fig. 6. For this case, no energy equation need be solved since E everywhere is equal to the inlet value; the temperature field may be deduced from the inlet E and the computed velocity field. The objective is to compare the convergence rates of the iterative scheme outlined here with a calculation assuming constant E. Plots of the scaled continuity, momentum and energy residuals (Fig. 7) for the two approaches show the convergence behavior to be quite similar. The present form requires 128 iterations while the constant E form converges in 114 iterations. Thus, the explicit inclusion of the kinetic energy terms in the present formulation imposes only a small overhead, even for this extreme case. In more general problems involving viscous flow and heat transfer, the difference is even less pronounced. Details of the problem and comparison of the numerical solution with experimental data may be found in [23].

4.2 Anisotropic Conductivity

Another difference between the energy equation and the scalar transport equation is that in some applications the conductivity may not be isotropic, especially in solid regions. The diffusion term then takes the form

$$\nabla \cdot (\mathbf{K} \cdot \nabla T) \tag{28}$$

where $\mathbf{K}$ is the conductivity tensor. The discrete form of the diffusion term is derived along the same lines as Section 3.1.1 [24]. At an interior face f we obtain:

$$D_f = \frac{(T_1 - T_0)}{ds} \frac{\mathbf{A} \cdot (\mathbf{K} \cdot \mathbf{A})}{\mathbf{A} \cdot \hat{\mathbf{e}}_s} + \mathcal{S}_f \tag{29}$$

where

$$\mathcal{S}_f = \mathbf{A} \cdot \left(\mathbf{K} \cdot \overline{\nabla T} \right) - \left(\overline{\nabla T} \cdot \hat{\mathbf{e}}_s \right) \frac{\mathbf{A} \cdot (\mathbf{K} \cdot \mathbf{A})}{\mathbf{A} \cdot \hat{\mathbf{e}}_s} \tag{30}$$

Like its isotropic counterpart (Eq. (13)), the anisotropic diffusion term is also split into primary and secondary components, with the latter being written as the difference of the total diffusion and the primary component. Unlike the isotropic case, however, the secondary term is non-zero even on orthogonal meshes. Further details are given in [24].

4.3 Boundary Conditions

To complete the energy balance in a near-boundary cell, the boundary conduction heat transfer D_b in Eq. (14) must be evaluated. The conduction heat flux from a wall boundary face b (Fig. 3) to the interior is given by an expression analogous to Eq. (14)

$$
\begin{aligned}
q_{b,\text{diff}} &= k_b \frac{(T_b - T_0)}{db} \frac{|\mathbf{A}|}{\mathbf{A} \cdot \hat{\mathbf{e}}_b} + k_b \left(\frac{\nabla T_0 \cdot \mathbf{A}}{|\mathbf{A}|} - \nabla T_0 \cdot \hat{\mathbf{e}}_b \frac{|\mathbf{A}|}{\mathbf{A} \cdot \hat{\mathbf{e}}_b} \right) \\
&= \mathcal{A}_{b,\text{diff}} + \mathcal{B}_{b,\text{diff}} (T_b - T_0)
\end{aligned} \tag{31}
$$

where k_b is the boundary conductivity, and $\mathcal{A}_{b,\text{diff}}$ and $\mathcal{B}_{b,\text{diff}}$ are computed from prevailing values of T.

For given-temperature boundary conditions, the evaluation of the boundary conduction flux is straightforward since T_b is known. For other types of boundary conditions, a boundary heat balance is required to find T_b:

$$q_{b,\text{diff}} + q_{b,\text{rad}} = q_e \tag{32}$$

Here $q_{b,\text{diff}}$ and $q_{b,\text{rad}}$ are the conduction and radiation heat fluxes transferred from the wall to the interior, and q_e is the externally supplied heat flux. The radiative flux $q_{b,\text{rad}}$ has the general form

$$q_{b,\text{rad}} = \mathcal{A}_{b,\text{rad}} + \mathcal{B}_{b,\text{rad}} T_b \tag{33}$$

where $\mathcal{A}_{b,\text{rad}}$ and $\mathcal{B}_{b,\text{rad}}$ are functions of T_b and are evaluated using prevailing values. The specific form of $\mathcal{A}_{b,\text{rad}}$ and $\mathcal{B}_{b,\text{rad}}$ depends on the radiation model used (see Sections 5.1 and 5.2).

The form of the externally-specified heat flux q_e depends on the application. Consider the form

$$q_e = h_e(T_{c,\infty} - T_b) + \epsilon_e \sigma(T_{r,\infty}^4 - T_b^4) \tag{34}$$

where σ is the Stefan-Boltzmann constant, h_e and ϵ_e are the specified external heat transfer coefficient and emissivity respectively and $T_{c,\infty}$ and $T_{r,\infty}$ are the corresponding external convection and radiation temperatures. Linearizing the T_b^4 term, Eq. (34) takes the form:

$$q_e = \mathcal{A}_e + \mathcal{B}_e T_b \tag{35}$$

where $\mathcal{A}_e$ and $\mathcal{B}_e$ are functions of T_b and are evaluated using prevailing values.

Substitution of Eqs. (31), (33) and (35) into the boundary heat balance, Eq. (32), yields

$$\mathcal{A}_b + \mathcal{B}_b T_b + \mathcal{C}_b T_0 = \mathcal{A}_e + \mathcal{B}_e T_b \tag{36}$$

where $\mathcal{A}_b (\equiv \mathcal{A}_{b,\mathrm{diff}} + \mathcal{A}_{b,\mathrm{rad}})$, $\mathcal{B}_b (\equiv \mathcal{B}_{b,\mathrm{diff}} + \mathcal{B}_{b,\mathrm{rad}})$, $\mathcal{C}_b (\equiv -\mathcal{B}_{b,\mathrm{diff}})$ and $\mathcal{A}_e$ and $\mathcal{B}_e$ are known from the specified boundary condition parameters. Equation (36) may be considered an equation for the boundary temperature T_b. We eliminate T_b from Eq. (31) using Eq. (36) to obtain the desired implicit form for the boundary conduction flux.

4.4 Conjugate Heat Transfer

Treatment of conjugate boundaries is along the same lines as the boundary condition treatment discussed above. Consider a face b that lies at a fluid-solid interface between the fluid cell C0 and the solid cell C1. The conduction heat flux from the interface to the fluid side is expressed as in the previous subsection in terms of the interface temperature T_b and the fluid cell temperature T_0 :

$$q_0 = \mathcal{A}_0 + \mathcal{B}_0 T_b + \mathcal{C}_0 T_0 \tag{37}$$

Similarly, the heat transfer from the interface to the solid side can be written as:

$$q_1 = \mathcal{A}_1 + \mathcal{B}_1 T_b + \mathcal{C}_1 T_1 \tag{38}$$

By setting

$$q_0 + q_{b,rad} = -q_1 \tag{39}$$

an expression for T_b may be written in terms of T_0 and T_1. The interface temperature may then be eliminated from either Eqs. (37) or (38) to obtain the interface conduction flux to either cell C0 or C1 in terms of T_0 and T_1.

5 RADIATION HEAT TRANSFER

A variety of methods for computing radiative heat transfer in absorbing, emitting and scattering media has appeared in the literature [25]. Among the most widely used are zonal methods [26], Monte Carlo methods [26], the class of P_N approximations [27], ray tracing methods such as the discrete transfer model [28] and the discrete ordinates and finite volume methods [29, 30, 31, 32]. Recently, the finite volume method [30, 31, 32] has been extended to unstructured meshes [33]. Though the methods for participating radiation may, in principle, be used to compute surface-to-surface radiation, they are not always efficient, and it is sometimes necessary to deal with the two categories separately.

5.1   Surface-to-Surface Radiation

Calculation methods for enclosure radiative transfer have long been available and involve the use of view factors and the solution of a system of equations for the radiosity [34]. Our interest here is in coupling an enclosure radiosity calculation with the calculation procedure for conjugate heat transfer outlined above. In a typical automotive underhood cooling problem, the fluid flow mesh consists of several million cells, with tens of thousands of boundary faces which may be radiating. For an enclosure with N surfaces, the storage required for view factor calculations is $N(N + 1)/2$ words; for 10^4 surfaces, this translates to approximately 200 Mb of storage for view factors alone. The storage and computational requirements of the radiation calculation threaten to overwhelm the overall heat transfer computation in many practical problems. Despite the advent of efficient view factor solvers such as CHAPARRAL [35], it is frequently necessary to compute radiative heat transfer using an agglomerated surface mesh; usually, contiguous boundary faces are agglomerated into groups according to some relevant criterion, such as the similarity of temperature. Since most view factor solvers only admit planar triangular and quadrilateral surfaces, view factor algebra must be employed to find the composite view factors corresponding to the arbitrary faceted surfaces resulting from agglomeration. Furthermore, it is necessary, in formulating a coupling between the enclosure radiation sub-problem and the conjugate heat transfer sub-problem, to preserve the conservative property of the scheme. Also, care must be taken not to create sub-problems with all-Neumann boundary conditions.

Because of space limitations, we do not review view factor calculation methods here. A good summary has been provided by Emery et al. [36]. Among the more recent innovations to be imported into the heat transfer community is the hemicube method for computing 3-D view factors [37]; this method, now commonly used in computer graphics, is substantially faster than traditional double-integration methods. Glass [35] has provided a detailed description of an efficient implementation, as well as related speed and storage issues.

5.1.1   Equations for Enclosure Radiative Transfer. We define the average temperature $\overline{T}_k$ and the average emissivity $\overline{\epsilon}_k$ of an agglomerated boundary surface k consisting of N_k faces of the original mesh as

$$\overline{T}_k = \frac{\sum_{i=1}^{N_k} T_{bi} |\mathbf{A}_i|}{\sum_{i=1}^{N_k} |\mathbf{A}_i|} \tag{40}$$

$$\overline{\epsilon}_k = \frac{\sum_{i=1}^{N_k} \epsilon_{bi} |\mathbf{A}_i|}{\sum_{i=1}^{N_k} |\mathbf{A}_i|} \tag{41}$$

where T_{bi} is the temperature of the ith boundary face, and ϵ_{bi} is its emissivity.

The equations governing radiative heat transfer in an enclosure of M agglomerated surfaces with radiosity J_k and blackbody emissive power $E_{B,k}$ ($= \sigma \overline{T}_k^4$)

are [34]

$$J_k = \frac{\bar{\epsilon}_k E_{B,k} + (1 - \bar{\epsilon}_k) \sum_{\substack{j=1 \\ j \neq k}}^{M} F_{kj} J_j}{\bar{\epsilon}_k + (1 - \bar{\epsilon}_k) \sum_{\substack{j=1 \\ j \neq k}}^{M} F_{kj}} \qquad k = 1, M \qquad (42)$$

F_{kj} is the view factor of surface j from surface k. If the average temperatures $\overline{T}_k$ of the enclosure surfaces are (nominally) known, the set of algebraic equations (42) is easily solved using an iterative Gauss-Seidel scheme. The equation set is diagonally dominant and convergence is guaranteed. Glass [35] has described an alternative method called progressive refinement, which, in some cases, allows a near-exact answer to be computed with only a fraction of the necessary view factors. Once J_k is known, the average radiative heat flux $\bar{q}_{k,\text{rad}}$ is computed from

$$\bar{q}_{k,\text{rad}} = \left(\sum_{\substack{j=1 \\ j \neq k}}^{M} F_{kj} \right) \bar{\epsilon}_k C_k E_{B,k} - \bar{\epsilon}_k C_k \sum_{\substack{j=1 \\ j \neq k}}^{M} F_{kj} J_j \qquad (43)$$

where

$$C_k = \frac{1}{\bar{\epsilon}_k + (1 - \bar{\epsilon}_k) \sum_{\substack{j=1 \\ j \neq k}}^{M} F_{kj}} \qquad (44)$$

5.1.2 Coupling with Conjugate Heat Transfer Calculation. To ensure conservation, we assume that

$$q_{b,\text{rad}} = \bar{q}_{k,\text{rad}} \qquad (45)$$

for any face b belonging to the agglomerate k. To aid in convergence of the iterative solution, we wish to linearize $q_{b,\text{rad}}$ in terms of the boundary face temperature T_b. However, Eq. (43) has been derived in terms of $\overline{T}_k$, the mean temperature of the agglomerated surface. To facilitate linearization, we write the net radiative heat flux transferred from the face b as

$$q_{b,\text{rad}} = \left(\sum_{\substack{j=1 \\ j \neq k}}^{M} F_{kj} \right) \bar{\epsilon}_k C_k (\sigma T_b^4) + \left[\bar{q}_{k,\text{rad}} - \left(\sum_{\substack{j=1 \\ j \neq k}}^{M} F_{kj} \right) \bar{\epsilon}_k C_k (\sigma T_b^4) \right]^* \qquad (46)$$

Here * denotes prevailing values. Equation (46) is linearized with respect to T_b assuming $\bar{q}_{k,\text{rad}}$ constant

$$q_{b,\text{rad}} = q_{b,\text{rad}}^* + \left. \frac{\partial q_{b,\text{rad}}}{\partial T_b} \right|_{T_b^*} (T_b - T_b^*) \qquad (47)$$

to yield

$$q_{b,\text{rad}} = 4 \left(\sum_{\substack{j=1 \\ j \neq k}}^{M} F_{kj} \right) \bar{\epsilon}_k C_k (\sigma T_b^{3*}) T_b + \left[\bar{q}_{k,\text{rad}} - 4 \left(\sum_{\substack{j=1 \\ j \neq k}}^{M} F_{kj} \right) \bar{\epsilon}_k C_k (\sigma T_b^4) \right]^*$$

$$= A_{b,\text{rad}} + B_{b,\text{rad}} T_b \qquad (48)$$

This linearized expression for the net radiative heat flux is included in the wall heat balance in Eq. (36). Both the enclosure radiation sub-problem and the conjugate heat transfer sub-problem are completely conservative, as is the overall formulation; each sub-problem is also well-posed. The overall solution procedure takes the form:

1. From prevailing values of T_b, $\overline{T}_k$ and J_k on the boundaries, find $\mathcal{A}_{b,\text{rad}}$ and $\mathcal{B}_{b,\text{rad}}$.

2. Solve the conjugate heat transfer sub-problem using the methods described in Section 4. Update the boundary temperatures T_b.

3. Check for convergence. If converged, stop; else continue.

4. Find the temperatures $\overline{T}_k$ of the agglomerated surfaces k.

5. Solve the enclosure radiation sub-problem by solving the equation set (42) for J_k using a Gauss-Seidel scheme.

6. Go to 1.

5.2 Participating Radiation

The discrete ordinates method has been used for many years for the computation of radiation in absorbing, emitting and scattering media [38]. More recently, Raithby and Chui [30, 31] developed the finite volume scheme for radiative heat transfer by imposing energy conservation principles over discrete solid angles. A similar approach has been taken by Chai and co-workers [32, 39]. The discrete ordinates and finite volume methods are similar; however, the latter is a natural extension of conservative finite volume schemes for fluid flow and heat transfer and may be implemented relatively easily within the framework of this chapter. Murthy and Mathur [33] have developed an implementation for unstructured meshes. Extensions for axisymmetric and periodic geometries [40, 41] as well as for semi-transparent media with Fresnel boundaries [42] have been published.

5.2.1 Radiative Transfer Equation. The radiative transfer equation for a gray absorbing, emitting and scattering medium in the direction s may be written as [26]:

$$\nabla \cdot (I(\mathbf{s})\mathbf{s}) = -(\kappa + \sigma_s)\, I\,(\mathbf{s}) + B(\mathbf{s}) \qquad (49)$$

where

$$B\,(\mathbf{s}) = \kappa I_B + \frac{\sigma_s}{4\pi} \int_{4\pi} I\,(\mathbf{s}')\, \Phi\,(\mathbf{s}', \mathbf{s})\, d\Omega' \qquad (50)$$

Here κ is the absorption coefficient, σ_s is the scattering coefficient, I_B is the black-body intensity and Φ is the scattering phase function.

For brevity, only gray-diffuse boundaries are considered here. The boundary intensity I_b for all directions outgoing from the boundary ($\mathbf{s} \cdot \hat{\mathbf{e}}_n < 0$) is given by

$$I_b = \frac{(1 - \epsilon_b)}{\pi} \int_{\mathbf{s} \cdot \hat{\mathbf{e}}_n > 0} I\,(\mathbf{s})\, \mathbf{s} \cdot \hat{\mathbf{e}}_n\, d\Omega + \frac{\epsilon_b \sigma T_b^4}{\pi} \qquad (51)$$

Here, T_b is the boundary temperature, and ϵ_b is its emissivity. The unit vector $\hat{e}_n$ is the surface normal pointing out of the domain.

The radiative source term S_r in the energy equation (Eq. 24) is given by

$$S_r = \kappa \int_{4\pi} [I(\mathbf{s}) - I_B] \, d\Omega \tag{52}$$

5.2.2 Discretization. The spatial discretization is done as in Section 3. In addition, the angular space 4π at any spatial location is discretized into discrete non-overlapping control angles ω_i, the centroids of which are denoted by the direction vector $\mathbf{s}_i$ and the polar and azimuthal angles θ_i and ϕ_i. Each octant is discretized into $N_\theta \times N_\phi$ solid angles. The angles θ and ϕ are measured with respect to the global Cartesian system (x, y, z); θ is measured from the z axis, and ϕ is measured from the y axis. The angular discretization is uniform; the control angle extents are given by $\Delta\theta$ and $\Delta\phi$. For each discrete direction i, Eq. (49) is integrated over the the control volume $C0$ in Fig. 8 and the solid angle ω_i to yield

$$\sum_f J_f I_{if} |\mathbf{A}| = [-(\kappa + \sigma_s) I_{i0} + B_i] \omega_i \Delta V_0 \tag{53}$$

Here, I_{if} is the intensity associated with the direction i at the face f of the control volume, and I_{i0} is the intensity at the cell $C0$ in the direction i. J_f is a geometric factor defined below. The solid angle ω_i is given by

$$\omega_i = \int_{\Delta\phi} \int_{\Delta\theta} \sin\theta \, d\theta \, d\phi \tag{54}$$

The integration may be performed exactly so that $\sum_i \omega_i = 4\pi$ is identically satisfied. The source term B_i is given by

$$B_i = \kappa I_{B0} + \frac{\sigma_s}{4\pi} \sum_j I_{j0} \gamma_{ij} \tag{55}$$

where

$$\gamma_{ij} = \frac{1}{\omega_i} \int_{\omega_i} \int_{\omega_j} \Phi(\mathbf{s}_{i'} \cdot \mathbf{s}_{j'}) \, d\omega_{j'} \, d\omega_{i'} \tag{56}$$

The black body intensity I_{B0} is based on the temperature of the cell $C0$, and I_{j0} is the cell intensity in the directions j.

5.2.3 Control Angle Overhang. It would seem possible to "upwind" I_{if} using any of the schemes outlined in Section 3.1.2 for the convection term; however, for unstructured and body-fitted meshes, the computation of I_{if} and J_f are complicated by the possibility of control angle overhang. The situation in 2-D is shown in Fig. 9. Since the directions $\mathbf{s}_i$ are defined with respect to a global coordinate system (x, y, z), the boundaries of the discrete solid angles ω_i do not necessarily align with control volume faces. In 3-D, the intersection of the control volume face with the angular discretization is more complex, involving the intersection of an arbitrary great circle with the control angle. For directions with no overhang on the face f, we write

$$J_f = \hat{e}_n \cdot \int_{\Delta\theta} \int_{\Delta\phi} \mathbf{s} \sin\theta \, d\theta \, d\phi \tag{57}$$

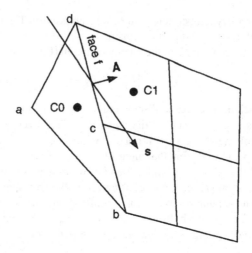

Figure 8 Control volume for radiant energy conservation in direction s

Using a standard "step" approximation for I_{if} [26]

$$I_{if} = I_{i,\text{upwind}} \tag{58}$$

Here $I_{i,\text{upwind}}$ is the value of I_i in the "upwind" cell. A second-order approximation may also be written. The angular integrations are done analytically.

When there is control angle overhang, it is necessary to account for the degree of overhang in writing I_{if} and J_f. Chui and Raithby [31] proposed a treatment for control angle overhang at boundaries for two dimensions. Chai et al. [32] proposed a treatment based on the *mean* direction vector s_i. They assumed that the *entire* solid angle associated with the direction i was outgoing to the cell if $s_i \cdot \hat{e}_n > 0$ and incoming otherwise. The treatment of I_{if} in either event was identical to the treatment for non-overhanging directions. Using the step scheme, the cell intensity at the appropriate "upwind" cell was used for I_{if}. Murthy and Mathur [33] proposed an alternative treatment. If the direction i exhibits overhang at the face f, the incoming and outgoing portions of the solid angle are differenced differently. Equation (57) is written as:

$$J_f I_{if} = I_{if,\text{out}} \alpha_{i,\text{out}} + I_{if,\text{in}} \alpha_{i,\text{in}} \tag{59}$$

where

$$\alpha_{i,\text{out}} = \hat{e}_n \cdot \int_{\Delta\theta} \int_{\Delta\phi} s \sin\theta \, d\theta \, d\phi, \quad s \cdot \hat{e}_n > 0$$

$$\alpha_{i,\text{in}} = \hat{e}_n \cdot \int_{\Delta\theta} \int_{\Delta\phi} s \sin\theta \, d\theta \, d\phi, \quad s \cdot \hat{e}_n \leq 0 \tag{60}$$

The incoming and outgoing face intensities are then written as in Eq. (58). Thus, referring to Fig. 1,

$$J_f I_{if} = I_{i0} \alpha_{i,\text{out}} + I_{i1} \alpha_{i,\text{in}} \tag{61}$$

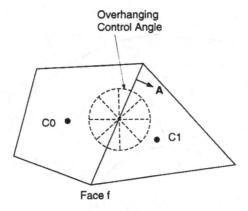

Figure 9 Control angle overhang in 2D

where I_{i0} and I_{i1} refer to the intensities in cells $C0$ and $C1$ in the direction i, and the terms "incoming" and "outgoing" are with respect to $C0$.

In 3-D, the overhang fractions $\alpha_{i,in}$ and $\alpha_{i,out}$ require integrations over control angle domains which are complex in shape. Murthy and Mathur [33] proposed a discrete pixelation method to compute these fractions. Here, the solid angle of interest is divided into $N_{\theta_p} \times N_{\phi_p}$ pixels. Each pixel is identified by a pixel direction s_{pi} written in terms of the pixel centroid angles (θ_p, ϕ_p). We may then write Eq. (60) as:

$$\alpha_{i,out} = \hat{e}_n \cdot \sum_{s_p \cdot \hat{e}_n > 0} S_p$$

$$\alpha_{i,in} = \hat{e}_n \cdot \sum_{s_p \cdot \hat{e}_n \leq 0} S_p \tag{62}$$

The summation is over all incoming or outgoing pixels. Also

$$S_p = \int_{\Delta\theta_p} \int_{\Delta\phi_p} s_p \sin\theta \, d\theta \, d\phi \tag{63}$$

Here the integration is over the pixel. Control angle overhang can thus be computed up to the pixel resolution. Computational effort may be minimized by only pixelating those control angles which exhibit overhang. A detailed treatment of various boundary conditions may be found in [33, 41, 42] and is not repeated here.

It was found in [33] that the approach of Chai et al. was adequate for domains with gray-diffuse boundaries, though pixelation offered small improvements with coarse angular discretizations when used at boundaries. However, it was found that accounting accurately for control angle overhang was critical in problems where ray rotation effects are important, for example at rotationally-periodic boundaries [41]. Another situation in which control angle overhang must be accounted for is reflection at arbitrarily-oriented reflecting boundaries and refraction at Fresnel interfaces [42]. Here again, ray rotation must be correctly captured to account for the

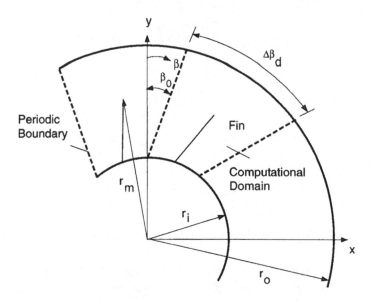

Figure 10 Rotationally periodic domain

redistribution of radiant energy with angle. Furthermore, at Fresnel interfaces, the reflectivity rises sharply at critical incidence. The use of pixelation allows the accurate integration of reflected radiant energy.

To illustrate this, we consider participating radiation in the cylindrical annulus formed by two cylinders of radius r_i and r_o, as shown in Fig. 10. The inner cylinder has vanes attached to it, spaced at an angle $\Delta\beta_d$, and angled at β_0 with respect to the radial direction. The vanes extend halfway across the annulus. The calculation domain consists of one rotationally periodic module of extent $\Delta\beta_d$ and is positioned as shown in Fig. 10 with respect to the global coordinate system. The annulus is filled with an absorbing and emitting medium at T_h, with an optical thickness κr_i. All walls are black and at $T = 0$. For the calculations shown here, β_0 is 20°, $\Delta\beta_d$ is 40°, $r_o/r_i = 2$ and $r_m/r_i = 1.5$. Comparisons are made with a ray-tracing solution; details are available in [41].

Figures 11(a) and (b) show a comparison of different angular discretizations and pixelations for $\kappa r_i = 1.0$, using a 40 × 40 quadrilateral mesh. Here, the dimensionless incoming heat flux, $q^*(= q/\sigma T_h^4)$, on the bottom wall, q^*, is plotted versus β. A measure of solution correctness is the continuity of q^* across the periodic boundary, which occurs at $\beta = 0$. For coarse angular discretizations using 1 × 1 pixelation (which is equivalent to the treatment in [32]) a jump in q^* is seen at $\beta = 0$, signaling errors in ray rotation at the periodic boundary. These errors persist even with an 8 × 8 angular discretization (Fig. 11(b)), but are substantially reduced for a 16 × 16 angular discretization. When 1 × 10 pixelation is used, the predicted q^* improves dramatically, with the maximum error under 3% for all the angular discretizations considered; for the 16 × 16 angular discretization, the maximum error is 0.76%.

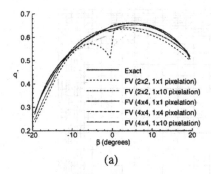
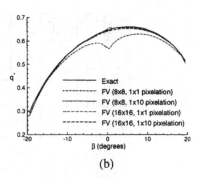

(a)   (b)

Figure 11 Rotationally periodic domain: dimensionless heat flux on bottom wall for different angular discretizations

5.2.4   Boundary Conditions. At a gray-diffuse boundary face b, the net incoming radiative flux is found using pixelation as described in Section 5.2.3

$$q_{b,\text{in}} = \sum_i I_{i0}\alpha_{i,\text{in}} \qquad (64)$$

where $\alpha_{i,in}$ is associated with the portion of the solid angle incoming to the boundary. The outgoing radiative flux is given by

$$q_{b,\text{out}} = (1 - \epsilon_b)\, q_{b,\text{in}} + \epsilon_b \sigma T_b^4 \qquad (65)$$

The net radiative heat flux from the boundary into the domain is the difference of the emission and incoming flux

$$q_{b,\text{rad}} = q_{b,\text{out}} - q_{b,\text{in}} \qquad (66)$$

The T_b^4 term in Eq. (65) is linearized to provide the appropriate coefficients in Eq. (33) for inclusion in the boundary heat balance. The boundary intensity I_b for all directions outgoing from the boundary face is obtained from

$$I_b = \frac{q_{b,\text{out}}}{\pi} \qquad (67)$$

5.2.5   Coupling with Energy Equation. Equation (52) is linearized as:

$$S_r = S_r^* + \left.\frac{\partial S_r}{\partial T}\right|_{T^*} (T - T^*) \qquad (68)$$

Here T^* is the current iterate of the temperature of the participating medium. Using $I_b = \sigma T^4/\pi$ in Eq. (52), the source term in the discrete energy equation for cell C0 is written as

$$S_r \Delta V_0 = \Delta V_0 \left(\kappa \sum_i \omega_i I_{i0} - 4\kappa \sigma T_0^4 \right) \qquad (69)$$

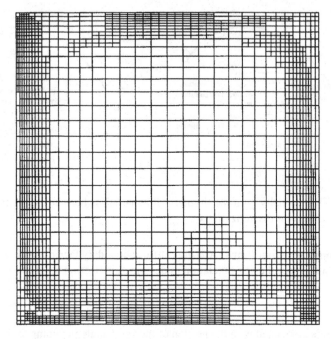

Figure 12 Coupled convection and radiation in square box: adapted mesh

Assuming I_{i0} independent of T and evaluating $\frac{\partial S_r}{\partial T}$ from the second term in Eq. (69), we may write

$$S_r \Delta V_0 = \Delta V_0 \left(\kappa \sum_i \omega_i I_{i0} + 12\kappa\sigma T_0^{*4} - 16\kappa\sigma T_0^{*3} T_0 \right) \tag{70}$$

The last term in Eq. (70) is included implicitly in the discrete energy equation; the other terms are included explicitly.

The method described here is applied to the problem of combined natural convection and participating radiation in a square box. Figure 12 shows the 2178-cell mesh used to discretize the domain. The left wall is cold and at T_c. The right wall is hot and at T_h. The top and bottom walls are insulated. Gravity points downward. The enclosure is filled with an absorbing and emitting gas. Computations are done for Rayleigh number, $Ra = 5 \times 10^6$, Prandtl number, $Pr = 0.72$, $T_0/(T_h - T_c) = 1.5$, $\kappa L = 1.0$, and Planck number, $Pl = k/(4\sigma T_0^3 L) = 0.02$. Here $T_0 = 0.5(T_h + T_c)$, and L is the length of the side. An initial 30×30 structured mesh is used to obtain a solution using a 2×4 angular discretization. The mesh is then adapted to the temperature field using hanging node adaption. Figure 13 shows plots of the normalized $u-$ and $v-$velocities along the vertical and horizontal centerlines, respectively. Also plotted are the results of Yucel et al. [43], obtained using the S_4 approximation (12 directions total). The comparison in all cases is good. Details may be found in [33].

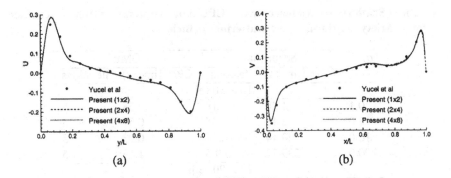

Figure 13 Coupled convection and radiation in square box: (a) U-velocity along vertical centerline and (b) V-velocity along horizontal centerline

5.2.6   Convergence Acceleration. The intensity equation in the direction i is linear in I_i and coupled to the intensities in the other directions through the scattering term in the volume of the domain, and due to reflection or refraction at boundaries and Fresnel interfaces. It is coupled to the energy equation through the emission term. Because the fluid flow and heat transfer solution procedure is sequential, it is customary to append a sequential procedure for solving the intensities in the directions i to the solution loop. Each intensity equation is solved using the multigrid procedure described in the next section, and the solution loop is iterated one governing equation at a time until convergence. This procedure has been found to work well for low to moderate optical thicknesses but slows down considerably as the optical thickness increases [44, 45]. In the thick limit, the energy and intensity equations become tightly coupled, and a sequential procedure is not optimal. When scattering dominates, different directions become tightly coupled to each other. Similarly, at Fresnel interfaces and specular boundaries, radiant energy is transferred from one direction to another due to reflection and refraction, again binding different directions tightly together. In these cases, other iteration methods are required to accelerate convergence. Chui and Raithby [44] suggested solving for multiplicative corrections about a mean intensity field in order to correct the directional intensities. A similar strategy was suggested by Fiveland [45] along with other strategies. The multiplicative correction strategy was found to fail when applied on meshes where the cell-based optical thickness was less than O(1); Fiveland was able to obtain substantial acceleration by restricting the correction to cell clusters with optical thickness of O(1).

Recently, Mathur and Murthy [46] developed the coupled ordinates method (COMET). Here the intensity and energy equations are solved simultaneously at a point using optimized direct methods. The point-coupled solution is used as a relaxation sweep in a multigrid procedure. Table 1 shows the improvement in CPU times and iterations using COMET. The problem considered is isotropic scattering in a square box of side L. All walls are at $T = 0$ except the bottom wall which is

Table 1 Scattering in a square cavity: CPU time and iterations for convergence for a variety of grid sizes and scattering coefficients

$\sigma_s L$	Sequential		COMET	
	CPU Time (s)	Iterations	CPU Time (s)	Iterations
		10×10 cells		
0.1	1.28	18	0.53	5
1.0	2.46	35	0.57	5
10.0	20.44	297	0.93	5
100.0	239.77	3385	1.15	5
		20×20 cells		
0.1	3.03	15	1.68	5
1.0	5.65	35	2.05	5
10.0	45.77	346	5.66	5
100.0	679.07	5080	10.04	5
		40×40 cells		
0.1	13.46	15	6.45	5
1.0	22.46	32	10.01	5
10.0	159.71	371	35.13	5
100.0	>2500	>6000	93.81	5
		80×80 cells		
0.1	58.28	14	24.31	5
1.0	100.90	29	41.44	5
10.0	659.78	371	157.13	5
100.0	>1e5	> 6000	593.63	5

at $T = 1000$. The medium does not emit or absorb radiation so that the radiation problem is decoupled from the thermal problem. All the intensity equations are, however, coupled to each other through the scattering term. A series of structured quadrilateral meshes is used, with $N_\theta \times N_\phi$ of 2×2. No pixelation is necessary because of the orthogonal geometry. The coupled approach results in significant improvement in convergence rates for the entire range of scattering coefficients studied. Speed-up factors increase dramatically with increased $\sigma_s L$ as well as with grid size. The number of iterations remain constant over the entire range of $\sigma_s L$ and grid size.

6   MULTIGRID SCHEME

The discretization procedures for fluid flow, heat transfer and radiation result in a set of nominally linear discrete algebraic equations. These equations are sparse but unstructured. They may be solved with a variety of techniques, including conjugate gradient and multigrid methods [16, 47]. Algebraic multigrid methods have proven useful for solving sparse, unstructured equation sets; when implemented in conjunction with agglomeration strategies which account for coefficient anisotropy, they provide a very efficient solution engine for a wide range of stiff physics.

6.1   Basic Scheme

The algebraic multigrid (AMG) method is well suited for unstructured meshes since it does not involve discretization of the governing equations on coarser grids. Instead, a hierarchy of coarse equation sets is constructed by grouping a number of fine level discrete equations. Residuals from a fine level relaxation sweep are "restricted" to form the source terms for the coarser level correction equations. The solution from the coarser equations in turn is "prolongated" to provide corrections at the finer level. This effective use of different grid sizes permits the reduction of errors at all wavelengths using relatively simple smoothing operators.

The procedure used in the present work is to visit each ungrouped fine level cell and group it with n of its neighboring ungrouped cells for which the coefficient M_{ij} is the largest [48]. Best performance is observed when the group size, n, is 2. The coefficients for the coarse level equations are obtained by summing up coefficients of the fine level equations:

$$M_{IJ}^{l+1} = \sum_{i \in G_I} \sum_{j \in G_J} M_{ij}^l \qquad (71)$$

where the superscripts denote the grid level and G_I is the set of fine level cells that belong to the coarse level group I. This results in a system of equations of the same form as the fine level (i.e, Eq. (23)), with $\frac{1}{n^{th}}$ the number of unknowns:

$$M_{IJ}^{l+1} \phi_J^{l+1} - \sum_{i \in G_I} R_i^l = 0 \qquad (72)$$

where R_i^l is the residual in the fine level equation at the current iteration

$$R_i^l = M_{ij}^l \phi_j^{*l} + S_i \qquad (73)$$

The process is repeated recursively until no further coarsening is possible. A variety of strategies, such as the V, W and Brandt cycles [49] may be used to cycle between the grid levels. The solution at any level is obtained by a Gauss-Seidel iterative scheme and is used to correct the current iterate at the next finer level. Thus, for all $i \in G_I$:

$$\phi_i^l = \phi_i^{*l} + \phi_I^{l+1} \qquad (74)$$

6.2   Agglomeration Strategy

The linear systems encountered in industrial heat transfer applications are frequently stiff. This stiffness is a result of a number of factors: large aspect-ratio geometries, disparate grid sizes typical of unstructured meshes, large conductivity ratios in conjugate heat transfer problems, and others. The agglomeration strategy employed to create the coarse level equations from the fine level equations is critical in determining the convergence rate of the solution scheme.

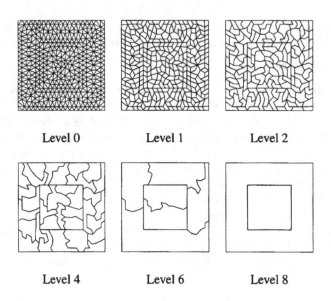

Level 0   Level 1   Level 2

Level 4   Level 6   Level 8

Figure 14 Conduction in composite domain: multigrid coarsening

Consider the situation depicted in Fig. 14. A composite domain consists of a low-conductivity outer region surrounding a highly conducting inner square domain. The ratio of conductivities is 1000; a ratio of this order would occur for a copper block in air. The temperature is specified on the four external walls of the domain. Convergence of typical linear solvers is inhibited by the large anisotropy of coefficients for cells bordering the interface of the two regions. Coefficients resulting from the diffusion term scale as $kA/\Delta x$, where A is a typical face area and Δx is a typical cell length scale. For interface cells in the highly conducting region, coefficients to interior cells are approximately three orders of magnitude bigger than coefficients to cells in the low-conducting region. However, Dirichlet boundary conditions, which set the level of the temperature field, are only available at the outer boundaries of the domain, adjacent to the low-conducting region. Information transfer from the outer boundary to the interior region is inhibited because the large-coefficient terms overwhelm the boundary information transferred through the small-coefficient terms. An agglomeration strategy which clusters cell neighbors with the largest coefficients results in the coarse levels shown in Fig. 14. At the coarsest level, the domain consists of a single cell in the high-conducting region, and another in the low-conducting region. The associated coefficient matrix has coefficients of the same order. The mean temperature of the inner region is set primarily by the multigrid corrections at this level and results in very fast convergence.

Another example is shown in Fig. 15. The problem involves orthotropic conduction in a triangular domain with temperature distributions given on all bound-

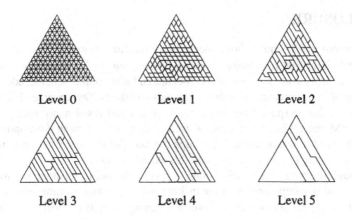

Level 0   Level 1   Level 2

Level 3   Level 4   Level 5

Figure 15 Orthotropic conduction: multigrid coarsening

aries [24]. The material has a conductivity $k_{\eta\eta} > 0$ in the η direction, aligned at $\pi/3$ radians from the horizontal; the conductivity $k_{\xi\xi}$ in the direction perpendicular to η is zero. Mesh agglomeration based on coefficient size results in coarse-level meshes aligned with η as shown. Since all faces with normals in the ξ direction have zero coefficients, the primary direction of information transfer is in the η direction. Thus, the coarse level mesh correctly captures the direction of information transfer.

We should note here that the ability to handle fully unstructured meshes in the multigrid procedure offers substantial convergence advantages over purely structured-grid approaches. In structured-grid methods, coarse levels are created by agglomerating complete grid lines in each grid direction. This has the advantage of preserving the grid structure at all coarse levels, permitting the use of the same line-by-line relaxation schemes as used on the finest level. However, such an agglomeration procedure does not necessarily result in optimal multigrid acceleration in general situations since coefficient anisotropies are not always aligned along lines.

Algebraic multigrid methods used with sequential solution procedures have the advantage that the agglomeration strategy can be equation-specific; the discrete coefficients for the specific governing equation can be used to create coarse mesh levels. Since the coarsening is based on the coefficients of the linearized equations, it also changes appropriately as the solution evolves. This is especially useful for non-linear and/or transient problems. In some heat-transfer applications, however, the mutual coupling between the governing equations is the main cause of convergence degradation. Geometric or FAS multigrid methods that solve the coupled problem on a sequence of coarser meshes may offer better performance in such cases.

7 CLOSURE

Unstructured mesh finite volume methods are reaching maturity, and their use in large-scale industrial calculations is rapidly increasing. In this chapter, we have attempted to provide an overview of a cell-centered finite volume scheme, addressing a variety of issues specific to unstructured mesh methods. Our coverage is far from exhaustive; due to space limitations, we have not addressed many related topics in the FEM and CVFEM literature, as well as unstructured mesh developments in the high-speed flow community. We hope the material presented here provides an adequate starting point for delving into this expanding field.

Though there has been substantial success in the use of unstructured meshes, ever-increasing computer speeds are making many new and exciting opportunities available to the CFD researcher. Foremost among these is the increasing fidelity with which turbulence is being modeled; advances in this area have a profound importance for industrial CFD. Large Eddy Simulation (LES) is an increasingly viable candidate. If LES is to be used in industrial problems, unstructured mesh methods capable of providing the requisite accuracy must be devised. The development of higher-order spatial and temporal schemes for use with hybrid unstructured meshes is therefore an important area for future investigation. As memory becomes cheap and plentiful, the increased use of coupled multigrid methods for large-scale industrial flows is inevitable. Improved geometric multigrid methods are a promising area for new work. It would also be useful to extend these methods to other physics, for example, to Eulerian multiphase flows. As more and more complex systems are included in large-scale CFD analyses, the desire to compute coupled fluid flow, heat transfer, structural motion and deformation will inevitably arise. Progress has already been made in this direction [6], but much remains to be done to devise robust coupling algorithms, especially for large structural deformations.

Ultimately, the end goal of industrial CFD analyses is to improve design and production. To use the results of CFD calculations to maximum advantage, substantial effort must be dedicated to the development of CFD-based design and optimization methods. The underlying CFD solvers must be made truly hands-off so they can actually be used as engines in optimization codes. Though individual CFD runs will be made sufficiently fast by the increases in processor speed, optimization will require tens, or even hundreds of runs to adequately populate the design space. Algorithms suitable for coarse to moderate grained parallel machines are required. Thus, despite significant success during the last decade, the quest for ever-faster solver methodologies, better physical models, and more accurate numerics is likely to be with us well into the foreseeable future.

8 ACKNOWLEDGMENTS

We wish to acknowledge the use of Fluent Inc.'s solver FLUENT/UNS, and its mesh generators PreBFC and TGrid, in the results presented here.

9 REFERNCES

1. S.R. Mathur and J.Y. Murthy, A Pressure Based Method for Unstructured Meshes, *Numer. Heat Transfer*, vol. 31, pp. 195–216, March 1997.

2. S.R. Mathur and J.Y. Murthy, Pressure Boundary Conditions for Incompressible Flows Using Unstructured Meshes, *Numer. Heat Transfer*, vol. 32, pp. 283–298, October 1997.

3. O. C. Zienkiewicz, *The Finite Element Method*, McGraw-Hill, New York, 1980.

4. Y. Jiang and A.J. Przekwas, Implicit, Pressure-Based Incompressible Navier-Stokes Equations Solver for Unstructured Meshes, AIAA-94-0305, 1994.

5. L. Davidson, A Pressure Correction Method for Unstructured Meshes with Arbitrary Control Volumes, *Int. J. Numer. Meth. Fluids*, vol. 22, pp. 265-281, 1996.

6. I. Demirdzic and S. Muzaferija, Numerical Method for Coupled Fluid Flow, Heat Transfer and Stress Analysis Using Unstructured Moving Meshes With Cells of Arbitrary Topology, *Comput. Meth. Appl. Mech. Eng.*, vol. 125, pp. 235-255, 1995.

7. R. Löhner, K. Morgan, J. Peraire, and M. Vahdati, Finite Element Flux Corrected Transport (FEM-FCT) for the Euler and Navier–Stokes Equations, *Int. J. Numer. Meth. Fluids*, vol. 7, pp. 1093–1109, 1987.

8. D. Mavriplis and A. Jameson, Solution of the Navier-Stokes Equations on Triangular Meshes, *AIAA J.*, vol. 28, pp. 1415–1425, 1990.

9. B. R. Baliga and S. V. Patankar, A Control-Volume Finite Element Method for Two-Dimensional Fluid Flow and Heat Transfer, *Numer. Heat Transfer*, vol. 6, pp. 245–261, 1983.

10. C. Prakash, An Improved Control Volume Finite Element Method for Heat and Mass Transfer, *Numer. Heat Transfer*, vol. 9, pp. 253-276, 1986.

11. G.E. Schneider and M. J. Raw, Control-Volume Finite-Element Method for Heat Transfer and Fluid Flow Using Co-Located Variables–1. Computational Procedures, *Numer. Heat Transfer*, vol. 11, pp. 363–390, 1987.

12. T. J. Barth, *Aspects of Unstructured Grids and Finite-Volume Solvers for the Euler and Navier-Stokes Equations*. Special Course on Unstructured Grid Methods for Advection Dominated Flows, AGARD Report 787, 1992.

13. A. Jameson, T. J. Baker, and N. P. Weatherhill, Calculation of Inviscid Flow Over a Complete Aircraft, AIAA-86-0103, 1986.

14. D. Kwak, J.L. Chang, S.P. Shanks, and S.R. Chakravarthy, A Three-Dimensional Incompressible Navier-Stokes Flow Solver Using Primitive Variables, *AIAA J.*, vol. 23, pp. 390–396, 1985.

15. J.M. Weiss and W.A. Smith, Preconditioning Applied to Variable and Constant Density Time-Accurate Flows on Unstructured Meshes, AIAA-94-2209, 1994.

16. V. Venkatakrishnan and D. Mavriplis, Implicit Solvers for Unstructured Meshes, In *AIAA 10th CFD Conference Proceedings*, pages 115–124, 1991.

17. M. Raw, Robustness of Coupled Algebraic Multigrid for the Navier-Stokes Equations, AIAA 96-0297, 1996.

18. R. Webster, An Algebraic Multigrid Solver for Navier-Stokes Problems, *Int. J. Numer. Meth. Fluids*, vol. 18, pp. 761–780, 1994.

19. R. Jyotsna and S.P. Vanka, Multigrid Calculation of Steady, Viscous Flow in a Triangular Cavity, *J. Comp. Phys.*, vol. 122, pp. 107-117, 1995.

20. I. Demirdzic, Z. Lilek, and M. Peric, Fluid Flow and Heat Transfer Test Problems For Non-Orthogonal Grids: Bench-Mark Solutions, *Int. J. Numer. Meth. Fluids*, vol. 15, pp. 329–354, 1992.

21. S. V. Patankar, *Numerical Heat Transfer and Fluid Flow*, McGraw-Hill New York, 1980.

22. K.C. Karki and S.V. Patankar, Pressure Based Calculation Procedure for Viscous Flows at All Speeds in Arbitrary Configurations, *AIAA J.*, vol. 27, pp. 1167-1174, 1989.

23. J.Y. Murthy and S.R. Mathur, A Conservative Numerical Scheme for the Energy Equation, *J. Heat Transfer*, vol. 120, pp. 1081–1085, November 1998.

24. J.Y. Murthy and S.R. Mathur, Computation of Anisotropic Conduction Using Unstructured Meshes, *J. Heat Transfer*, vol. 120, pp. 583–591, August 1998.

25. J. R. Howell, Thermal Radiation in Participating Media: The Past, the Present and Some Possible Futures, *J. Heat Transfer*, vol. 110, pp. 1220 – 1226, 1988.

26. M. F. Modest, *Radiative Heat Transfer*, Series in Mechanical Engineering, McGraw Hill, 1993.

27. M. P. Menguc and R. Viskanta, Radiative Heat Transfer in Combustion Systems, *J. Quant. Spectrosc. Radiat. Transfer*, vol. 33, pp. 533 – 549, 1985.

28. F. C. Lockwood and N. P. Shah, A New Radiation Solution Method for Incorporation in General Combustion Prediction Procedures, In *Proceedings of the 18th Symposium (Intl.) on Combustion*, pages 1405 – 1414. The Combustion Institute, 1981.

29. W. A. Fiveland and J. P. Jessee, Finite Element Formulation of the Discrete Ordinates Method for Multidimensional Geometries, *J. Thermophys. Heat Transfer*, vol. 8, pp. 426 – 433, 1994.

30. G. D. Raithby and E. H. Chui, A Finite-Volume Method for Predicting Radiant Heat Transfer in Enclosures with Participating Media, *J. Heat Transfer*, vol. 112, pp. 415 – 423, 1990.

31. E. H. Chui and G. D. Raithby, Computation of Radiant Heat Transfer on a Non-Orthogonal Mesh Using the Finite-Volume Method, *Numer. Heat Transfer*, vol. 23, pp. 269 – 288, 1993.

32. J.C. Chai, G. Parthasarathy, S.V. Patankar, and H.S. Lee, A Finite-Volume Radiation Heat Transfer Procedure for Irregular Geometries, AIAA 94-2095, June 1994.

33. J.Y. Murthy and S.R. Mathur, Finite Volume Method for Radiative Heat Transfer Using Unstructured Meshes, *J. Thermophys. Heat Transfer*, vol. 12, pp. 313–321, Jul-Sep 1998.

34. F.P. Incropera and D.P. DeWitt, *Fundamentals of Heat Transfer*, John Wiley and Sons, 1981.

35. M.W. Glass, CHAPARRAL: A Library for Solving Large Enclosure Radiation Heat Transfer Problems, Sandia National Laboratories, August 1995.

36. A.F. Emery, O. Johannson, M. Lobo, and A. Abrous, A Comparative Study of Methods for Computing the Diffuse Radiation Viewfactors for Complex Structures, AIAA 88-2223, 1988.

37. M.F. Cohen and D.P. Greenberg, The Hemi-Cube: A Radiosity Solution for Complex Environments, *Computer Graphics*, vol. 19, pp. 31–40, 1985.

38. B. G. Carlson and K. D. Lathrop, Transport Theory – The Method of Discrete Ordinates. In H. Greenspan, C. Kelber, and D. Okrent, editors, *Computing Methods in Reactor Physics*, pages 171–266. Gordon and Breach, New York, 1968.

39. J. C. Chai and J. P. Moder, Spatial-Multiblock Procedure for Radiation Heat Transfer, In A. Gopinath, P. D. Jones, J. Syed-Yagoobi, and K.A. Woodbury, editors, *HTD-Vol.332, Proceedings of the ASME Heat Transfer Division*, volume 1, pages 119 – 128. ASME, 1996.

40. J. Y. Murthy and S. R. Mathur, Radiative Heat Transfer in Axisymmetric Geometries Using an Unstructured Finite Volume Method, *Numer. Heat Transfer*, vol. 33, pp. 397–416, 1998.

41. S.R. Mathur and J.Y. Murthy, Radiative Heat Transfer in Periodic Geometries Using a Finite Volume Scheme, *J. Heat Transfer*, vol. 121, pp. 357–364, May 1998.

42. J.Y. Murthy and S.R. Mathur, A Finite Volume Scheme for Radiative Heat Transfer in Semi-Transparent Media, AJTE99:6293,presented at the ASME-JSME Thermal Engineering Joint Conference, San Diego, CA, March 1999.

43. A. Yucel, S. Acharya, and M. L. Williams, Natural Convection and Radiation in a Square Enclosure, *Numer. Heat Transfer (Part A)*, vol. 15, pp. 261–278, 1989.

44. E. H. Chui and G. D. Raithby, Implicit Solution Scheme to Improve Convergence Rate in Radiative Transfer Problems, *Numer. Heat Transfer*, vol. 22, pp. 251-272, 1992.

45. W.A. Fiveland and J.P. Jessee, Acceleration Schemes for the Discrete Ordinates Method, HTD-Vol. 315, 1995 National Heat Transfer Conference–Volume 13, pp. 11–19, 1995.

46. S.R. Mathur and J.Y. Murthy, A Point-Coupled Multigrid Acceleration Scheme for Radiative Heat Transfer, AIAA-99-3365, January 1999.

47. B. R. Hutchinson and G. D. Raithby, A Multigrid Method Based on the Additive Correction Strategy, *Numer. Heat Transfer*, vol. 9, pp. 511-537, 1986.

48. R. D. Lonsdale, An Algebraic Multigrid Scheme for Solving the Navier-Stokes Equations on Unstructured Meshes, In *Proceedings of the 7th International Conference on Numerical Methods in Laminar and Turbulent Flow*, pages 1432–1442, Stanford University, CA, 1991.

49. A. Brandt, Multi-Level Adaptive Solutions to Boundary Value Problems, *Math. Comput.*, vol. 31, pp. 333-390, 1977.

THREE

SPECTRAL ELEMENT METHODS FOR UNSTEADY FLUID FLOW AND HEAT TRANSFER IN COMPLEX GEOMETRIES: METHODOLOGY AND APPLICATIONS

C.H. Amon

1 INTRODUCTION

Computational Heat Transfer (CHT) has experienced exceptional advances due to the improved computer hardware combined with the development of advanced numerical techniques and algorithms over the last decade. Numerical simulation has emerged as an alternative and, sometimes, as the only approach to analyze in detail, complex thermal-fluid phenomena. However, CHT is still at the stage of intensive development, particularly in engineering applications, where most of the problems considered in the past involve significant simplifications regarding geometry, physics and parameter range. Many computational techniques for thermal-fluid problems have been proposed, tested, and refined, mainly for steady flow and time-averaged conservation laws, the latter for modeling transport phenomenon in turbulent flows. Recently, there has been an increasing trend toward simulation of more complex thermal-fluid phenomena with a level of complexity that is close to industrial applications. CHT is therefore becoming an emerging field, not only in fundamental research, but also as a design and analysis tool in engineering practice [1, 2].

The ability to simulate complex-geometry and complex-physics flows has grown rapidly in the last two decades because of the effort devoted to Computational Fluid Dynamics (CFD). In the zeroth-order approach to convective heat transfer, the temperature field is solved in the presence of a known velocity field, i.e., uncoupled momentum and energy equations. Since the velocity field is seldom known, the flowfield needs to be obtained first, making CFD an essential part of CHT. The numerical prediction of convective heat transfer usually requires the combined solution of the velocity and temperature fields that are governed by the equations of conservation of momentum, mass, and energy. Often, simplifying assumptions or models are necessary to make complicated problems tractable. However, if a fairly complete and accurate mathematical description of the main factors affecting the transport phenomena are retained, and the numerical algorithms are suitable to solve the mathematical equation, then the results can be considered an accurate computer simulation of the physical process. Therefore, there are two different issues to consider regarding the analysis of numerical uncertainties: first, the quantification of the mathematical modeling errors, and, second, the identification and estimation of the numerical errors of the computational scheme used to solve the governing equations modeling the physical phenomenon.

We can broadly classify the numerical schemes used for simulating thermal-fluid phenomena into Eulerian and Lagrangian schemes with regards to formulation and into spectral, finite difference, finite volume, boundary-element and finite element techniques with regards to discretization algorithms. These schemes vary in complexity, computing efficiency, numerical accuracy, and flexibility. Each numerical method offers different advantages and limitations for simulating a certain class of transport phenomena. However, no one technique appears to be superior in solving a broad range of problems. Although most of these widely used numerical schemes fall within one of these categories and their differences might have been initially quite clear, recent approaches have combined these traditional schemes, leading to hybrid algorithms with encouraging results for solving complex thermal-fluid phenomena as well as for effective utilization of modern computer architecture. Several hybrid methods have been proposed using high-order polynomial expansions local to finite element, called p or combined h-p finite elements [3], global elements [4, 5] and spectral elements [6-9].

In this chapter, we concentrate on a class of hybrid discretization, called spectral element method, which combines finite elements and spectral schemes to produce high-order spatial accuracy. Both approaches are based on the principle of weighted residuals, and the spectral element method utilizes variational projection operators in conjunction with local Chebyshev or Legendre polynomial expansions which exhibit exponential convergence to smooth solutions. Even though other polynomial series can be employed, Chebyshev and Legendre polynomials are most frequently used because they are complete orthogonal sets of eigenfunctions which come from the solution of Sturm-Liouville problems. Furthermore, their coefficients are easily evaluated with a recurrence formula. The spectral element method can also be classified as a domain decomposition technique that combines globally unstructured and locally structured spatial discretizations. The global decomposition in macroelements provides geometric flexibility, and the local structure permits an efficient high-order approximation by spectral expansions through Chebyshev collocation points.

2 MATHEMATICAL FORMULATION

We consider unsteady incompressible flows and forced convection in three-dimensional domains which are governed by the Navier-Stokes, mass conservation and energy equations. We denote by $D = D_f \cup D_s$ the three-dimensional computational domain, and by ∂D the computational boundary surfaces, composed of solid heat conductive walls, $\partial D_{s\text{-}f} = \partial D_f \cup \partial D_s$, and periodic surfaces, ∂D_p. The governing dimensionless equations for fluid flow, conjugate conduction/convection heat transfer with a volumetric heat source, and mass transfer is given by a system of partial differential equations for the solid and fluid domains written in the following form:

$$\frac{\partial \overline{V}}{\partial t} = \overline{V} \times \omega - \nabla \widetilde{\Pi} + \text{Re}^{-1} \nabla^2 \overline{V} \tag{1}$$

$$\nabla \cdot \overline{V} = 0 \tag{2}$$

$$\frac{\partial T}{\partial t} = (\text{Re} \cdot \text{Pr})^{-1} \nabla^2 T - \overline{V} \cdot (\nabla T) + S \tag{3a}$$

$$\frac{\partial c}{\partial t} = (\text{Re} \cdot Sc)^{-1} \nabla^2 c - \overline{V} \cdot \nabla c \quad \text{in D} \tag{3b}$$

where $\overline{V}(\overline{x},t) = u\hat{x} + v\hat{y} + w\hat{z}$ is the velocity; $\overline{x}$ and t represent space and time, respectively; $\widetilde{\Pi} = p + 1/2\, V^2$ is the dynamic pressure, $T(\overline{x},t)$ is the temperature field, and $\omega = \nabla \times \overline{V}$ is the vorticity, Re is the Reynolds number, Pr is the Prandtl number, and S corresponds to the nondimensional volumetric heat source; Sc is the Schmidt number and $c(\overline{x},t)$ is the concentration field. Buoyancy effects are considered negligible for the range of Reynolds numbers investigated, and viscous dissipation and radiation are likewise neglected in the energy equation. The boundary conditions for velocity in the Navier-Stokes equation are no-slip along the solid-fluid interfaces $\partial D_{s\text{-}f}$, and inflow/outflow or periodicity in the streamwise and spanwise directions. In the solid region D_s the velocities are zero in the energy equation and no-slip boundary conditions are satisfied. The appropriate boundary conditions will be discussed in the context of the individual applications presented in the later sections.

For periodic streamwise conditions and solid walls, the velocity boundary conditions are

$$\overline{V}(\overline{x},t) = 0 \quad \text{on } \partial D_s \tag{4}$$

$$\overline{V}(x + nL, y, z, t) = \overline{V}(x, y, z, t) \quad \text{on } \partial D_p \tag{5}$$

where L is the geometric periodicity and n an integer periodicity index. For the pressure we require

$$\tilde{\Pi}(\bar{x},t) = -f(t)x + \Pi(\bar{x},t) \tag{6}$$

$$\Pi(x + nL, y, z, t) = \Pi(x, y, z, t) \tag{7}$$

where $f(t)$ is the driving pressure gradient. Since in complex geometry flows or transitional flows the pressure gradient is unknown, it is preferable to impose the volume flow rate $Q(t)$ as

$$Q(t) = \int_{-W}^{W} \int_{\partial D_{Bottom}}^{\partial D_{Top}} u(x_0, y, z, t) dy dz \tag{8}$$

where 2W is the length in the spanwise direction.

For the forced convective heat transfer equation in the fluid domain D_f, the temperature boundary conditions can be either Dirichlet (temperature), Neumann (flux), mixed Robin (heat transfer coefficient), periodicity, or continuity of temperature and heat flux for conjugate conduction/convection problems.

Simplified thermal modeling procedures often reduce a forced convective problem to solving the heat equation within the solid domain D_s and imposing convective effects of the fluid through a heat transfer boundary condition at the solid-fluid interface s-f, Eq. (9).

$$h(T_{s-f} - T_{ref}) = -k_s \frac{\partial T_s}{\partial \hat{n}} \tag{9}$$

where h is the convective heat transfer coefficient and is a proportionality constant that models the ability of the fluid to remove heat at the solid-fluid interface. The proper reference temperature T_{ref} is problem-dependent and can be chosen as the ambient, bulk, adiabatic, or inlet temperatures. Reformulation of the problem necessitates that the heat transfer coefficient be obtained either empirically or numerically by solving the coupled system of Eqs. (1)-(3) only in the fluid domain D_f and imposing thermal boundary conditions along the solid-fluid boundary. This implies that the effect of heat conduction within the solid is replaced by idealized boundary conditions that assume the heat flow paths in the solid and decouple the resistance associated with conduction within the solid and convection within the fluid. However, in conjugate problems, neither the temperature nor the heat flux at the solid-fluid interface can be prescribed accurately a priori, especially in systems that involve intense heat transfer, multimaterial solid domains, and localized heat generation [10]. Therefore, the convective boundary condition, as previously described, may not provide accurate predictions. The appropriate thermal boundary conditions for conjugate conduction/convection [11] are continuity of heat flux and temperature at the solid-fluid interface and are termed boundary conditions of the fourth kind, Eqs. (10a)-(10b).

$$k_s \frac{\partial T_s}{\partial \hat{n}} = k_f \frac{\partial T_f}{\partial \hat{n}}$$ (10a)

$$T_{s_{\cdot-f}} = T_{f_{\cdot-f}}$$ (10b)

Different numerical techniques have been implemented to investigate conjugate conduction/convection and to demonstrate the effect of conjugation on thermal performance characteristics. However, most of the situations analyzed have been steady state with homogeneous solid domains and simple geometries. A two-dimensional steady conjugate study in a laminar boundary layer with a heat source at the solid-fluid interface was conducted using finite differences [12]. The time-dependent conjugate behavior of a semi-infinite slab exposed to uniform surface heating was studied using the unsteady surface element method [13]. The two-dimensional conjugate behavior of hydrodynamically, fully developed, laminar flow through a circular tube with thick walls and a finite heated length was investigated using a finite volume approach [14]. A semi-analytical approach that utilizes an integral formulation for the fluid domain and a finite volume formulation for the solid domain was successfully developed to study plates with discrete heat sources, which model surface-mounted electronic packages [15]. Mixed laminar convection from local heat-generating components was also studied using a simpler-based approach [16]. Time-dependent studies of multimaterial, local heat generating configurations using the spectral element method were conducted by Nigen and Amon [17, 18] for both laminar and transitional Reynolds numbers. This investigation contrasted thermal behavior characteristics for conjugate and convection-only representations of a simulated electronic package and demonstrated the significance of including time-dependency and conjugation.

3 DISCRETIZATION

3.1 Temporal Discretization

The time-discretization procedure consists of a fractional scheme for the semidiscrete formulation of the time-dependent term in the Navier-Stokes, energy and concentration equations. Intermediate velocities $\hat{V}$ and $\hat{\hat{V}}$, Eqs. (11)-(13), and temperature $\hat{T}$, Eqs. (16)-(17) are computed in a way that the left-hand side yields $\partial \overline{V} / \partial t$ whereas the right-hand side contains the contributions of the nonlinear, pressure, and viscous terms. The advantage of this time-splitting scheme is that it reduces the coupled system of Eqs. (1) and (2) into a system of separately solvable equations for the pressure and velocity enabling the application of different algorithms to different terms in the Navier-Stokes equations to obtain gains in efficiency. The error, due to the time-splitting scheme, scales as $[o(\Delta t^2) + o(\Delta t/Re)]$ and restricts the time-step size in applications seeking to simulate time-dependent transitional flows. The nonlinear convective term, Eq. (11), is treated explicitly to decrease the computer time required per step because of the need to solve a nonlinear problem at each time step. The viscous term, Eq. (13), is treated implicitly to avoid unreasonable time-step restrictions due to the stiffness of diffusion problems [5] and the high resolution of Chebyshev spectral approximations near the boundary of the elements [19]. The dynamic pressure Π is calculated in Eq. (14) so that the velocity satisfies the incompressible condition of Eq. (2) even though Π does not appear in this equation. The time-stepping procedure is given by the following steps:

Nonlinear step:

$$\frac{\hat{V}^{n+1} - \bar{V}^n}{\Delta t} = \sum_{i=0}^{2} \beta_i (\bar{V}^{n-i} \times \omega^{n-i}) + f \quad \text{in } D \quad (11)$$

Pressure step:

$$\frac{\hat{\hat{V}}^{n+1} - \hat{V}^{n+1}}{\Delta t} = -\nabla \Pi \quad \text{in } D \quad (12a)$$

$$\nabla \cdot \hat{\hat{V}}^{n+1} = 0 \quad \text{in } D \quad (12b)$$

$$\hat{\hat{V}} \cdot \hat{n} = 0 \quad \text{on } \partial D \quad (12c)$$

Viscous step:

$$\frac{\bar{V}^{n+1} - \hat{\hat{V}}^{n+1}}{\Delta t} = \frac{1}{Re} \nabla^2 \bar{V}^{n+1} \quad \text{in } D \quad (13)$$

where the superscript n refers to time step. The first step, Eq. (11) represents the explicit treatment of the nonlinear convective term by a third order Adams-Bashforth method, where the coefficients are $\beta_0 = 23/12$, $\beta_1 = -16/12$, and $\beta_2 = -5/12$. This scheme introduces low dispersion errors and contains a relatively large portion of the imaginary axis within the absolute stability region of the scheme.

In the second step, Eqs. (12a)-(12c), the effect of the pressure is included and incompressibility is satisfied. By taking the divergence of Eq. (12a) and imposing Eq. (12b), we obtain the following Poisson equation for the pressure:

$$\nabla^2 \Pi = \nabla \cdot \frac{\hat{V}^{n+1}}{\Delta t} \quad \text{in } D \tag{14a}$$

This elliptic equation is then solved implicitly subject to

$$\nabla \Pi \cdot \hat{n} = \frac{\hat{V}^{n+1} \cdot \hat{n}}{\Delta t} \quad \text{on } \partial D_s \tag{14b}$$

The imposition of an inviscid-type boundary condition, proposed first by Deville and Orszag [19], introduces errors of $o(\Delta t/\mathrm{Re})$ that are important only at very low Reynolds number flows.

Finally, at the third fractional step, for the Navier-Stokes equation, Eq. (13), the viscous corrections are handled implicitly using a Crank-Nicolson scheme and the no-slip boundary conditions are imposed, giving

$$\left(\nabla^2 - \frac{2}{\Delta t}\mathrm{Re}\right)\left(\overline{V}^{n+1} + \overline{V}^n\right) = -\frac{2}{\Delta t}\mathrm{Re}\left(\hat{V}^{n+1} + \overline{V}^n\right) \quad \text{in } D \tag{15}$$

The nonlinear convective term, treated explicitly, is the only source that imposes stability conditions for the scheme (Courant-Friedrech-Lewy condition number), since the pressure and viscous contributions are treated implicitly by Euler backward or Crank-Nicolson schemes, which are unconditionally stable, resulting in an efficient and robust inversion of the global system matrices. Consequently, the time-step size is constrained by both accuracy in the time-splitting formulation and stability of the explicit scheme. For high Reynolds number flows in the turbulent flow regime, the viscous term can be treated explicitly since stability conditions for the convective contributions, Eq. (11), are as severe as the ones for the diffusive contributions, Eq. (13).

For the energy and mass transfer equation, Eqs. (3a)-(3b), we use a similar semi-implicit time-stepping scheme with two steps. The first step is an explicit third-order Adams-Bashforth step for the convective and source volumetric heat generation terms, and the second step is an implicit Crank Nicolson step for the conductive terms. The semi-discrete equations for $T^n(x) = T(x, n\Delta t)$ are then

$$\hat{T}^{n+1} - T^n = -\Delta t \sum_{q=0}^{2} \beta_q \nabla \cdot (\overline{V}T)^{n-1} + S \tag{16}$$

$$T^{n+1} - \hat{T}^{n+1} = -\frac{\Delta t}{2\mathrm{Re} \cdot Pr}\left[\nabla^2 (T^{n+1} + T^n)\right] \tag{17}$$

and boundary conditions are imposed on Eq. (17).

The numerical approach consists of integrating the continuity and Navier-Stokes equations for the fluid portion of the domain and then integrating the energy and mass transfer equations for both the solid and fluid domains. This procedure is iterated in time until either an asymptotically-steady, time-periodic or converged state is reached.

3.2 Multi-dimensional Spatial Discretization

Once we obtain the semidiscrete temporal equations, we proceed with the spatial discretization using a spectral element-Fourier decomposition of the three-dimensional computational domain. For a homogeneous geometry in the z direction, periodic boundary conditions and symmetry are consistent with the governing equations (1) and (2). Therefore, the velocity and pressure can be represented as two-dimensional (x,y) components with Fourier expansions in the homogeneous z direction,

$$
\begin{bmatrix} u(x,t) \\ v(x,t) \\ w(x,t) \\ \Pi(x,t) \end{bmatrix} = \begin{bmatrix} u^{(2)}(x,y,t) \\ v^{(2)}(x,y,t) \\ w^{(2)}(x,y,t) \\ \Pi^{(2)}(x,y,t) \end{bmatrix} + \sum_{m=1}^{M} \begin{bmatrix} u_m(x,y,t)\cos(m\beta z) \\ v_m(x,y,t)\cos(m\beta z) \\ w_m(x,y,t)\sin(m\beta z) \\ \Pi_m(x,y,t)\cos(m\beta z) \end{bmatrix}
\tag{18}
$$

where β is the wave number in the spanwise direction. We use cosine and sine expansions instead of exponentials because of the symmetry in the z direction and the reality conditions of the velocity and pressure in the physical space. For the analysis of two-dimensional flows, the v_m are identically zero; for linear three-dimensional stability analysis, a single infinitesimally small spanwise mode is included; and for three-dimensional flows, M is chosen so as to include all excited spanwise scales.

To impose the flow-rate condition $Q(t)$ in Eq. (8), let us assume for simplicity the flow direction to be the x direction. Then, for incompressible flow, at station x_0,

$$
Q(t) = \iint u(x_0,y,t)\,dydz = \iint \left[u^{(2)}(x_0,y,t) + \sum_{m=1}^{M} u_m(x_0,y,t)\cos(m\beta z) \right] dydz
$$

$$
= \iint u^{(2)}(x_0,y,t)\,dydz + \sum_{m=1}^{M} \int u_m(x_0,y,t)\,dy \int \cos(m\beta z)\,dz
\tag{19}
$$

where

$$
\int_{-W}^{W} \cos(m\beta z)\,dz = 0, \qquad m \neq 0
$$

Therefore, only the mean streamwise velocity $u^{(2)}$ contributes to the net flow rate, obtaining

$$Q(t) = 2W \int_{\partial D_{Bottom}}^{\partial D_{Top}} \cdot u^{(2)}(x_0, y, t) \, dy = Q_{2D}(t) \cdot 2W \tag{20}$$

It suffices to impose at any x_0 station

$$Q_{2D}(t) = \int_{\partial D_{Bottom}}^{\partial D_{Top}} u^{(2)}(x_0, y, t) \, dy \tag{21}$$

which is done in a preprocessing stage, before the time-stepping procedure.

3.2.1 Fourier expansions. The Fourier expansions, Eq. (18), are then inserted into the semidiscrete equations, Eqs. (11), (14), and (15). To demonstrate this procedure, we consider the elliptic operator corresponding to the pressure step, Eq. (14a), and substituting Eq. (18), we obtain

$$\sum_{m=1}^{M} (\nabla^2 - m^2\beta^2) \Pi_m \cos(m\beta z) = \frac{1}{\Delta t} \left(\sum_{m=1}^{M} \frac{\partial \hat{u}_m}{\partial x} + \frac{\partial \hat{v}_m}{\partial y} + m\beta \hat{w}_m \right) \cos(m\beta z) \tag{22}$$

where $\nabla^2 = \partial^2/\partial x^2 + \partial^2/\partial y^2$ and Π_m are the Fourier coefficients for the spanwise direction. First, we follow a Galerkin approach in z, multiplying Eq. (22) by $\cos(k\beta z)$, integrating and applying orthogonality property to obtain the following equation for the Fourier coefficients Π_m [20]:

$$(\nabla^2 - m^2\beta^2) \Pi_m = \frac{1}{\Delta t} \left(\frac{\partial \hat{u}_m}{\partial x} + \frac{\partial \hat{v}_m}{\partial y} + m\beta \hat{w}_m \right) \tag{23}$$

$$m = 1, 2, ..., M$$

We can now proceed to discretize the two-dimensional components of the velocity and pressure in the x-y plane using a variational spectral element discretization.

3.2.2 Spectral element discretization. In a two-dimensional spectral element discretization, the computational domain is partitioned into K non-overlapping four-sided macroelements given by $\Re^K = [a^K, b^K]$. Each element is then isoparametrically mapped from the physical $\bar{x} = (x, y)$ space to the local (r,s) coordinate system. The geometry, pressure, and velocity are represented as a tensor product of high-order Lagrangian interpolants through Gauss-Lobatto Chebyhev collocation points, defined as

$$\left[\bar{x},\bar{V},\Pi\right](r,s) = \sum_{k=1}^{K} \sum_{i=0}^{N_1} \sum_{j=0}^{N_2} (\bar{x},\bar{V},\Pi)_{ij}^{k} h_i(r) h_j(s) \tag{24}$$

where $h_i(r)$ and $h_j(s)$ are local Lagrangian interpolants that satisfy $h_i(\zeta_j) = \delta_{ij}$ at the local (r,s) coordinates, and δ_{ij} is the Kronecker delta symbol. The discrete space is defined in terms of the spectral element discretization parameters (K, N_1, N_2), where K is the number of spectral elements, and N_1 and N_2 are the degrees of the piecewise high-order polynomials in the r and s directions, respectively. Isoparametric mappings are used to transform general curvilinear domains into standard domains as illustrated in Fig. 1. To insure rapid convergence of the resulting expansions, the local and physical collocation points are chosen to be the Gauss-Lobatto Chebyshev points, defined as $x_j^i = -\cos \pi j / N^i$. Also, the choice of these collocation points results in closed-form analytical expressions for the quadratures involved in the computations. The Langrangian interpolants in Eq. (24) are expanded as

$$h_m(\zeta) = \frac{2}{N} \sum_{n=0}^{N} \frac{1}{\bar{c}_m \bar{c}_n} T_n(\zeta_m) T_n(\zeta) \tag{25}$$

where $m = i,j$ and $\zeta = r,s$, and T_n are the Chebyshev polynomials defined as

$$T_n(\cos\theta) = \cos n\theta \tag{26}$$

and

$$\bar{c}_m = \begin{cases} 1 & m \neq 0, N \\ 2 & m = 0, N \end{cases} \tag{27}$$

In the numerical simulations presented in this paper, we choose the same resolution in both spatial directions, i.e., $N = N_1 = N_2$. However, in practice, this does not need to be the case. For general complex geometry computational domains, we can extend the two-dimensional spectral element discretizations to three-dimensions by employing hexahedral spectral elements. An alternative approach for flows with one homogeneous or periodic direction consists in discretizing this direction through equally-spaced planes, using the Fourier expansions introduced in Section 3.2.1.

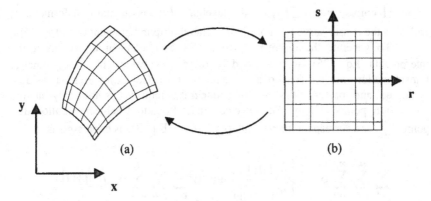

Figure 1 Four-sided spectral macroelement domain (a) curvilinear macroelement, and (b) standard spectral macroelement.

To illustrate the two-dimensional spectral element discretization, we consider Eq. (23) corresponding to the pressure step discretized by Fourier expansions in the spanwise direction. To simplify the notation, let us set $g = (\partial \hat{u}_m / \partial x + \partial \hat{v}_m / \partial y + m\beta \hat{w}_m)/\Delta t$ where all of the variables are known from the nonlinear step calculations. Then, it can be written as a modified Helmholtz equation of the form

$$\left(\nabla^2 - m^2 \beta^2\right)\Pi_m = g \quad \text{in D}$$
$$m = 1,...,M \tag{29}$$

subject to homogeneous Dirichlet boundary conditions, $\Pi = 0$ on ∂D. Equation (29) is discretized using variational spectral elements in the x-y plane. The variational formulation recognizes the equivalence between solving the differential Eq. (29) and maximizing the functional

$$I_m(\Pi) = \int_D \left[-\frac{1}{2} \nabla \Pi_m \cdot \nabla \Pi_m - \frac{m^2 \beta^2}{2}(\Pi_m)^2 - \Pi_m g \right] dA \tag{30}$$

The spectral element discretization corresponds to numerical quadrature of the variational form, Eq. (30), restricted to the discrete space defined in terms of the parameters (K, N_1, N_2).

In local $(r$-$s)$ coordinates, Eq. (30) is written as

$$I_m(\Pi) = \sum_{k=1}^{K} \int_{-1}^{1} \int_{-1}^{1} \left(-\frac{1}{2}\frac{\hat{\nabla}\Pi \cdot \hat{\nabla}\Pi}{|J|} - \frac{|J|m^2\beta^2}{2}\Pi^2 - |J|\Pi g \right) dr\,ds \tag{31}$$

where $\prod$ corresponds to $\prod_m$, J is the Jacobian of the elemental transformations, and $\hat{\nabla}$ is the conservative form of the Jacobian multiplied gradient operator [20].

To generate the discrete equations for each element k, we insert the interpolants Eq. (25) and the nodal collocation values of the geometric transformation into the functional. Then, we use a Galerkin-weighted residual technique and perform the resulting integrals, requiring stationarity at the collocation points. Using the selected Gauss-Lobatto Chebyshev collocation points ζ_{pq}^k and corresponding weights $\rho_{pq} = \rho_p \rho_q$, Eq. (31) is expressed as

$$\sum_{k=1}^{K} \sum_{p=0}^{N_1} \sum_{q=0}^{N_2} \rho_{pq} J_{pq}^k \left[\frac{\partial \prod}{\partial x_j} \frac{\partial \prod}{\partial x_j} \right]_{\zeta_{pq}^k} + m^2 \beta^2 \sum_{k=1}^{K} \sum_{p=0}^{N_1} \sum_{q=0}^{N_2} \rho_{pq} J_{pq}^k \left[\prod \prod \right]_{\zeta_{pq}^k}$$

$$= -\sum_{k=1}^{K} \sum_{p=0}^{N_1} \sum_{q=0}^{N_2} \rho_{pq} J_{pq}^k \left[\prod g \right]_{\zeta_{pq}^k} \tag{32}$$

The Jacobian J_{pq}^k of the transformation from physical to local coordinates is calculated from the partial derivatives of the geometric isoparametric transformation r_x, r_y, s_x, and s_y.

Once the local basis function is selected, the spectral element approximation for $\prod^k$ is

$$\prod^k = \sum_{n=0}^{N_1} \sum_{m=0}^{N_2} \prod_{mn}^k h_m(r) h_n(s), \qquad \forall_{m,n} \in (0,...,N_1), \ (0,...,N_2) \tag{33}$$

where $\prod_{mn}^k$ are the expansion coefficients and also the local nodal values of $\prod$. The geometry is also represented via similar tensorial products with the same order polynomial degree, i.e.,

$$(x,y)^k = \sum_{n=0}^{N_1} \sum_{m=0}^{N_2} \left(x_{mn}^k, y_{mn}^k \right) h_m(r) h_n(s), \qquad \forall_{m,n} \in (0,...,N_1), \ (0,...,N_2) \tag{34}$$

where x_{mn}^k and y_{mn}^k are the global physical coordinates at the node (mn) in the k element.

To construct the discrete matrix of the global system, we insert Eq. (33) into Eq. (32) and perform direct stiffness summation [21] adding at the boundary nodes the contributions from the neighboring elements, obtaining

$$\sum_{k=1}^{K}{'} \sum_{m=0}^{N_1} \sum_{n=0}^{N_2} (\phi_{ijmn}^k + m^2 \beta^2 J_{ij}^k B_{im}^k B_{jn}^k) \prod_{mn}^k = -\sum_{k=1}^{K}{'} \sum_{m=0}^{N_1} \sum_{n=0}^{N_2} J_{ij}^k B_{im}^k B_{jn}^k g_{mn}^k \tag{35}$$

where $\sum{}'$ denotes direct stiffness summation, and

$$\phi_{mn}^k = \rho_{pq} J_{pq}^k [(r_x)_{pq}^2 \mathcal{D}_{pi} \mathcal{D}_{pm} \delta_{nq} + (s_x)_{pq}^2 \mathcal{D}_{qj} \mathcal{D}_{qn} \delta_{mp} + (r_x s_x)_{pq}^2 \mathcal{D}_{pi} \mathcal{D}_{qn} \delta_{mp} +$$

$$(r_x s_x)_{pq} \mathcal{D}_{qj} \mathcal{D}_{qm} \delta_{nq}] \mathcal{D}_{ij} = \frac{\partial h_j}{\partial \zeta}(\zeta_i), \qquad B_{ij} = \int_{-1}^{1} h_i(\zeta) h_j(\zeta) d\zeta, \qquad \zeta = r,s \qquad (36)$$

The spectral solutions are C^0 across the boundaries of the elements with interfacial continuity constraints imposed only by the variational formulation without requiring any explicit patching at the elemental interfaces. Therefore, there is a weak coupling between dependent variables for neighboring elements. This results in banded, relatively sparse matrices, which are critical regarding computational efficiency of the method in terms of memory requirements and processing time.

Adequate mesh resolution is verified by comparing the temperature, concentration and flow characteristics using different order of local expansions and/or macroelement discretizations. The control of spatial resolution and the high degree of accuracy associated with this technique makes it well-suited for studying conjugate problems with localized heat generation and multimaterial solid domains, especially those with large variation in material properties.

4 APPLICATIONS

In this section, we present direct numerical simulations of several time-dependent fluid flows including mass and heat transport.

4.1 Flows in Abdominal Aortic Aneurysms

The first example corresponds to relatively complicated steady and pulsatile axisymmetric flows with curvilinear boundaries encountered in abdominal aortic aneurysms. Abdominal Aortic Aneurysms (AAAs) are localized balloon-shaped expansions commonly found in the infrarenal segment of the abdominal aorta, between the renal arteries and the iliac bifurcation. While the cause and nature of AAAs is still an important matter of debate, abdominal aortic aneurysm rupture is the 15th leading cause of death in the United States, affecting patients over 55 years of age, typically 2-4% of elderly males. As the overall mortality rate following aneurysm rupture may exceed 90% [22] determining the risk factors that may have an important role in aneurysm growth and rupture has become an integrated multidisciplinary task oriented towards obtaining an agreement on the pathogenesis and evolution of AAAs. It is well known now that arterial diseases, present in local irregular geometries, are the result of a combination of complex biochemical processes that take place in the vascular wall at the cellular level as well as hemodynamic factors resulting from the interaction of blood flow and the endothelium, the innermost cellular monolayer of the cardiovascular system. Recent investigations related to AAA phenomena are based on four major areas: (i) clinical studies focused on the etiology and screening of AAAs, as well as the determination of possible risk factors related to chemical alterations occurring in the cellular matrix of the endothelium; (ii) experimental and numerical studies focused on the simulation of physiological hemodynamics in aneurysm models; (iii) experimental and numerical studies based on stress-strain analysis and wall mechanics of the aneurysm wall, and (iv) in-vitro cell biology investigations and numerical simulations of cell models that attempt to correlate the hemodynamic patterns found in arterial models with the clinical evidence known for arterial diseases. Several numerical studies of steady and pulsatile flows through aneurysms are found in the literature. Guzmán and Amon [23] and Amon, et al. [24] have utilized spectral element discretizations to study the temporal flow evolution of laminar, transitional and chaotic flows in converging-diverging channels using a geometry similar to that represented in Fig. 2. Direct numerical simulations of non-Newtonian flow through double-aneurysm models have been conducted under pulsatile conditions, resulting in flow patterns and wall shear stresses that were underestimated for otherwise Newtonian flow behavior [25, 26]. Validation of pulsatile non-Newtonian flow simulations has been done by means of three asymptotic cases that take into account geometric irregularities in the model [27]. Currently, Finol and Amon [28, 29] are performing spectral element simulations of blood flow through AAA models. The objective of these simulations is to provide new insights to hemodynamic indicators that have not yet been evaluated when quantifying disturbed flow conditions, both steady and pulsatile, at the aneurysm wall.

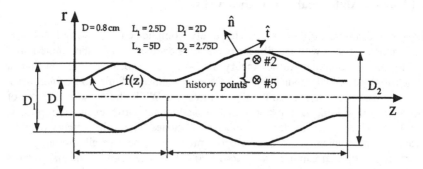

Figure 2 Representation of the axisymmetric model of the two-aneurysm abdominal aorta.

4.1.1 Mathematical formulation. The geometry of the abdominal aorta with two aneurysms is shown in Fig. 2. Two converging-diverging regions define this geometry, the physical model of which has been previously used by Guzmán, et al. [25, 27]. We consider both steady and pulsatile, incompressible, homogeneous, Newtonian flow in a two-aneurysm rigid-wavy-walled axisymmetric model. The deformed wall is represented by two sine functions as follows:

$$f(z) = \begin{cases} \left(\dfrac{D_1 - D}{4} \right) \left[1 + \sin\left(\dfrac{2\pi z}{L_1} - \dfrac{\pi}{2} \right) \right] + \dfrac{D}{2} & 0 \leq z \leq L_1 \\[4mm] \left(\dfrac{D_2 - D}{4} \right) \left[1 + \sin\left(\dfrac{2\pi(z - L_1)}{L_2} - \dfrac{\pi}{2} \right) \right] + \dfrac{D}{2} & L_1 \leq z \leq L_2 \end{cases} \tag{37}$$

The momentum and continuity equations in axisymmetric coordinates that govern these flows are given by

$$\rho\left(\frac{\partial u_z}{\partial t} + \bar{V} \cdot \nabla u_z \right) = -\frac{\partial p}{\partial z} + \left(\frac{\partial \tau_{zz}}{\partial z} + \frac{1}{r}\frac{\partial}{\partial r}(r\tau_{rz}) \right) \tag{38a}$$

$$\rho\left(\frac{\partial u_r}{\partial t} + \bar{V} \cdot \nabla u_r \right) = -\frac{\partial p}{\partial r} + \left(\frac{\partial \tau_{rz}}{\partial z} + \frac{1}{r}\frac{\partial}{\partial r}(r\tau_{rr}) \right) \tag{38b}$$

$$\frac{1}{r}\frac{\partial}{\partial r}(ru_r) + \frac{\partial u_z}{\partial z} = 0 \tag{38c}$$

where $\bar{V}(\bar{x},t) = u_r \hat{r} + u_z \hat{z}$ is the velocity vector, and τ_{zz}, τ_{rr} and τ_{rz} are the components of the two-dimensional stress tensor. The boundary conditions for the velocity $\bar{V}(\bar{x},t)$ are nonslip at the walls, symmetry at the centerline, fully developed parabolic profile at the inlet, and zero-traction outflow condition at the exit. Blood flow is simulated for average blood properties [30] with molecular viscosity $\mu = 0.00319$ Pa·s and density $\rho = 1,050$ kg/m^3. The governing equations are nondimensionalized by the factor D/2; hemodynamic parameters evaluated at the arterial wall are nondimensionalized using their corresponding magnitudes obtained for Poiseuille flow. The governing equations, subject to the appropriate boundary conditions, are solved numerically using a spectral element method for the spatial discretization with local Legendre polynomial expansions [6, 7].

The calculation of the local Wall Shear Stress Gradient (WSSG) is based on the predictor equation proposed by Lei and Kleinstreuer [31] at the cellular level, corrected from their previous work [32]:

$$WSSG = \sqrt{\left(\frac{\partial \tau_w}{\partial \hat{t}}\right)^2 + \left(\frac{\partial \tau_w}{\partial \hat{n}}\right)^2} \qquad (39)$$

where $\hat{t}$ and $\hat{n}$ are the local tangential and normal directions to the wall, as shown in Fig. 2. The nondimensional WSSG is obtained using the factor $2\tau_{wo}/D$, which is the gradient of shear stress at the wall for Poiseuille flow:

$$WSSG^* = \frac{WSSG}{2\tau_{wo}/D} \qquad (40)$$

4.1.2 Steady flow. Numerical results are obtained at Reynolds numbers over the range $10 \leq Re \leq 2265$. The Reynolds number is based on the undilated vessel diameter (diameter of the model at $z = 0$) and the average velocity at the entrance of the vessel. In the constant flow simulations, the inlet velocity profile is parabolic, corresponding to a fixed inlet flow rate. A time-dependent initial value code is used to find the solution for steady flow, starting from arbitrary initial conditions. The pressure at the exit of the two-aneurysm model is set to 70 mmHg, which is the time-average late diastolic pressure in the infrarenal segment of the human abdominal aorta [33]. Therefore, all the pressure results are relative to this value.

Typical laminar flow streamlines for the range of Reynolds number $10 \leq Re \leq 2265$ are shown in Fig. 3. At $Re = 10$, corresponding to a flow rate of 0.012 L/min, no flow separation occurs, and the main stream of fluid fills completely each of the vessel dilatations in a forward flow pattern. The converging-diverging shape of the model produces successive convective decelerations and accelerations of the flow that result in very small axial velocities near the wall at the center of each aneurysm. The onset of flow separation is found to occur within the range $24.74 < Re < 25.21$. For $Re = 25$, negative axial velocities in the order of magnitude of 10^{-5} cm/s are obtained close to the wall. Figures 3(b) through 3(f) show streamline plots that demonstrate a characteristic flow pattern: an inner core flow that passes through each dilation and two main surrounding regions of flow recirculation and separation. The symmetrically recirculating regions inside the arterial expansions are characterized by an upper subregion of reversed flow, which interacts with the AAA wall, and a lower subregion of forward flow which is sheared by the fluid core. For $Re \geq 500$, the fluid moving in reverse direction occupies most of the aneurysm sac volume, while the forward moving fluid in the vortex fills in a very small portion of the expansion. This is represented by a contraction of the streamlines in the distal half of each aneurysm, to the right and below the center of the vortex, and an expansion of the streamlines in the proximal half, especially close to the wall. Consequently, the upper subregion is essentially a zone of slowly moving particles of fluid, nearly stagnant, of high distal and low proximal shear due to blood flow-endothelium interaction. At each aneurysm enlargement, streamline separation occurs proximally. The dividing streamline represents particles of fluid that, after separation, continue flowing distally, along the undisturbed moving core. Particles traveling right above the dividing streamline enter the AAA sac, and upon reaching the distally located boundary layer reattachment point, reverse their direction, flowing backwards along the wall, towards the separation point. Therefore, fluid along the wall moves against a pressure gradient and after traveling the longitudinal perimeter of each aneurysm, the core flow, which is faster, separates from the wall and forms a laminar main stream with two or three annular regions of flow recirculation.

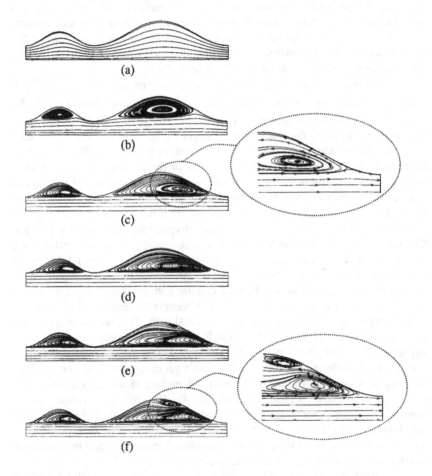

Figure 3 Streamlines for laminar steady flow at (a) Re = 10, (b) Re = 100, (c) Re = 500, (d) Re = 1000, (e) Re = 1750, and (f) Re = 2265. The direction of the flow is from left to right.

Figures 3(b) through 3(f) also show that as the Reynolds number is increased, the center of the recirculating flow regions moves downstream, distally, and also downward, closer to the main stream. This causes the displacement of the boundary layer separation points further upstream, and the reattachment points further downstream, increasing the volume occupied by the vortices within the aneurysm sacs. In the range of $500 < Re \leq 2265$, however, the movement of the vortex center occurs towards the distal edge of each aneurysm, but not closer to the main stream. This results in a fairly constant core flow volume through the model for the higher flow rates (0.57 L/min $< Q \leq 2.59$ L/min). For $Re \geq 1750$, an induced secondary vortex is formed in the large aneurysm. This small vortex, which increases in size and intensity for $Re = 2265$, is clockwise-rotating, indicating the presence of positive wall velocity gradients. The formation of a second recirculation region, trapped between the AAA wall and the main vortex, occurs for a range of Reynolds numbers for which transition to turbulence has been observed experimentally. Therefore, an intermittent transition regime characterizes the results obtained for $1750 \leq Re \leq 2265$.

The shear stress gradient distribution at the arterial wall is shown in Fig. 4. It is nondimensionalized by the $WSSG$ obtained in a Poiseuille flow, $2\tau_{wo}/D$, for which D is the undilated diameter of the aorta. The peak $WSSG^*$ in the small aneurysm is 6.40 times the shear stress gradient for an undilated aorta; similarly, the peak $WSSG^*$ in the large aneurysm is 7.55 times the initial $WSSG$. A healthy aorta would then be exposed to focal variations of shear stress direction 10 times larger than those found in the center of an aneurysm. However, it is the distal end of an arterial expansion that experiences the high levels of $WSSG$. If individual endothelial cells are exposed to these levels of $WSSG$s when an aneurysm grows in an arterial branch, then the zero-tension hypothesis proposed by Lei and Kleinstreuer [31] for the mechanisms of cell response can be interpreted for a segment of an injured arterial wall in the following way: A normal, healthy endothelium maintains a non-zero state of stress, which is the result of its interaction with blood flow and blood's cellular components, causing axial and circumferential tension, normal stresses and shear stresses; for steady flow, a uniform shear stress field exists along the arterial wall. This results in a uniform but non-zero $WSSG$ field ($WSSG^* = 10$ in Fig. 4), due to gradients of shear stress perpendicular to the wall. When this uniform $WSSG$ field is disrupted as a consequence of deformities in the endothelium's geometry, levels of low and high gradients of shear stresses coexist within focal regions of the arterial wall. For low levels of $WSSG$s ($WSSG^* < 10$ in Fig. 4), disturbed flow conditions are diminished and the endothelium tries to regain its integrity at these locations. For high levels of $WSSG$s ($WSSG^* > 10$ in Fig. 4), there is a stronger viscous interaction at the wall, which promotes thrombus formation inside the aneurysm [34], further increasing the risk of rupture of the injured site.

Normal endothelial cells experience a non-zero intercellular tension condition, which exists under a uniform wall shear stress field and a zero $WSSG$ [31]. However, while shear stress is constant at a healthy endothelium under steady flow conditions, there is a gradient of shear stress perpendicular to the wall. Since Eq. (39) defines the gradient of shear stress at the wall as a function of both $\partial \tau_w / \partial \hat{t}$ and $\partial \tau_w / \partial \hat{n}$, WSSG cannot be zero or negative, under normal conditions, even for quantification of cell responses.

4.1.3 Pulsatile flow.   For pulsatile flow conditions, the inlet mean cross flow velocity is time-dependent and the volumetric flow rate has an oscillatory nature, as shown in Fig. 5. We represent the cardiac waveform by a discrete Fourier series based on twenty-six experimental points presented by Nazemi, et al. [35]. The time dependency of the mean cross flow velocity is imposed by the following Fourier representation:

$$\overline{u_m}(t) = A_0 + \sum_{k=1}^{N}\left(A_k \cdot \cos 2\pi kt + B_k \cdot \sin 2\pi kt\right) \tag{41}$$

where N = 8, the number of harmonics used. The natural frequency of the pulsatile waveform is set to $\omega = 2\pi$ rad/s, with a period $T_p = 1$ s, as shown in Fig. 5. The Womersley number, $\alpha = (D/2)\sqrt{\omega \nu}$, which characterizes the flow frequency, the geometry of the model and the fluid viscous properties, is $\alpha = 5.75$. The amplitude coefficient ($\beta = Re_{max}/Re_m$) of the flow is $\beta = 2.64$, and the peak systolic flow occurs at $t = 0.247$ s. The mean Reynolds number is calculated as $Re_m = D\overline{u_m}/\nu$, where ν is the blood kinematic viscosity, $\overline{u_m}$ is the time-average inflow mean velocity, from which $\overline{Q} = \pi(D^2/4)\overline{u_m}$, the time-average volumetric flow rate.

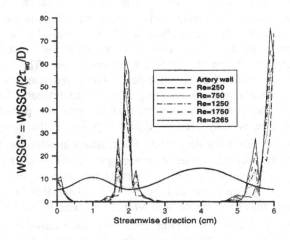

Figure 4 Distribution of nondimensional shear stress gradient along the wall of the two-aneurysm model.

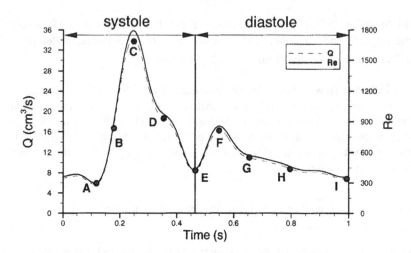

Figure 5 Pulsatile volumetric flow rate and Reynolds number for Re_m = 680. Flow stages A, B,..., I are of particular importance for the evaluation of hemodynamic parameters. Peak systolic flow occurs at t = 0.247 s and diastolic phase begins at t = 0.465 s.

Numerical simulations for pulsatile flow are performed at mean Reynolds numbers over the range $10 < Re_m < 680$. Re_m is defined as the time-average Reynolds number obtained by integrating the inlet Reynolds number over the pulsatile cycle. The instantaneous Reynolds number, $Re_i(t_j)$, is based on the spatially-averaged inlet velocity at time t_j. The pressure at the end of the two-aneurysm model is set to 70 mmHg, and all the pressure results presented are relative to this value. Results are obtained at 1/50 s intervals, but only selected flow stages considered to be the ones that represent the most important hemodynamic changes during the cycle are presented here. These stages are obtained at the last cycle of the asymptotically converged temporal solution, which is reached after a transient resulting from the application of the initial condition. Convergence is achieved when the flow becomes time periodic; this is verified by analyzing velocity plots at different points of the computational domain, as shown in Fig. 6 for Re_m = 100. For transient simulations, the initial value code solves the fully discrete set of governing equations at each time step by means of iterative solvers and tensor-product sum-factorization techniques.

The vortex dynamics induced by pulsatile blood flow in AAAs is characterized by means of a sequence of different flow stages in one period of the cardiac pulse. Figure 7 shows streamline plots for representative Re_m = 680 at nine flow stages in one cycle. The following five distinct flow phases depict the Reynolds-dependent and AAA-size-dependent vortex structures:

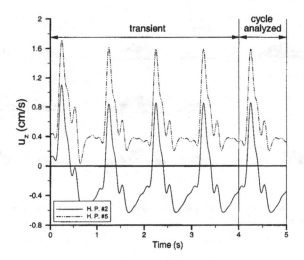

Figure 6 Temporal evolution of the axial velocity for $Re_m = 100$ at history points #2 and #5 (shown in Fig. 2) of the computational mesh.

(1) *Early systolic acceleration* which involves sweeping of vortices, left from the previous pulse, out of the deformed regions, resulting in an attached flow pattern. Vortices in the small AAA are swept earlier than in the large AAA.

(2) *Late systolic acceleration* is characterized by attached flow for low mean Reynolds numbers (Re_m). For high Re_m, the decrease in the temporal acceleration of the fluid upon reaching peak flow causes flow separation in the proximal ends of both aneurysms. The annular vortex forms earlier in the small AAA.

(3) *Systolic deceleration* is characterized by vortex growth and translation of vortex centers downstream. For high Re_m, secondary clockwise-rotating vortices are induced at the center of both aneurysms, once the main recirculation regions grow to full size. The end of systole itself is detailed as follows:

   • For low Re_m, flow reattachment only occurs in the distal end of the large AAA.

   • For moderate Re_m, the flow does not reattach to the wall.

   • For high Re_m, the flow reattaches at the center of both AAAs, while the secondary vortices are trapped against the wall.

(4) *Early diastole*, during which partial sweeping of the vortex structures left from systole is obtained, driven by a favorable pressure gradient. The recirculation regions are reduced considerably in size, as the flow attempts to reattach to the wall.

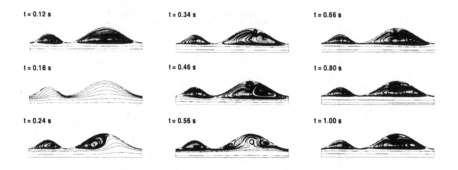

Figure 7 Streamlines for laminar pulsatile flow at $Re_m = 680$. The direction of the flow is from left to right.

(5) *Late diastole* involves fairly constant full-size vortices within both AAAs, similar to a steady flow pattern. For high Re_m, a secondary vortex coexists with the main recirculation region at the center of the large AAA.

In-vivo disturbed flow in localized regions of the cardiovascular system commonly subject to diagnosis of arterial diseases are typically associated with vortex structures, non-uniform fluid shear stress and high wall pressure. The focal occurrence of this hemodynamic disturbance has been correlated in-vitro with the alignment and migration of endothelial cells, as well as changes in their metabolic functions, which include cell division rates, protein-protein interactions, and cytosolic free calcium concentrations, among others. Non-uniform time-average shear stress measured experimentally at sites where irregular geometries take place in the arterial tree have led to the concept of shear stress gradient as a hemodynamic parameter of potential importance for explaining flow-induced arterial wall pathology and morphological changes in the endothelial lining. Therefore, we present the gradient of fluid shear stress evaluated in the tangential and normal directions to the arterial wall as a relevant hemodynamic force that influences the vortex dynamics of pulsatile blood flow in AAAs. The Wall Shear Stress Gradient (*WSSG*) distribution for $Re_m = 680$ is shown in Fig. 8. The low, almost constant shear stress at the center of the AAA walls produces a constant *WSSG* of near-zero magnitude during the pulsatile cycle at these locations. At the proximal and distal ends of each aneurysm, however, the oscillatory behavior of the *WSSG* distribution is characterized by spatial variations at the sites where large velocity gradients occur. The regions where high positive and low negative shear stresses coexist due to flow reattachment are subjected to high *WSSG* values which are maximum at $t = 0.247$ s, when peak flow is obtained. High levels of *WSSG* are obtained in the small aneurysm during the accelerating phases of the cycle (early systole and early diastole), while the large aneurysm is subject to high *WSSGs* during the decelerating phases. This is explained by the fact that during temporal acceleration (consider 0.16 s $\leq t \leq 0.22$ s, for example), vortex shedding and flow separation occurs earlier in the small aneurysm, producing higher shear stresses at the distal end. Alternatively, temporal deceleration induces secondary recirculation regions within the large aneurysm (e.g., $0.30 \text{ s} \leq t \leq 0.46 \text{ s}$), which result in an additional change of sign in the shear stress at the wall close to the point of flow reattachment. At peak values of flow rate where temporal acceleration is essentially zero ($t = 0.247 \text{ s}$ and $t = 0.555 \text{ s}$), AAA size determines the magnitude of the shear stress distribution, resulting in higher WSSGs in the large aneurysm.

While the role of wall shear stress gradients in specific arterial diseases is not thoroughly understood yet, it is known that hemodynamic disturbance influences the endothelial lining of the cardiovascular system, and that the endothelium does not seem to be affected at an early stage of lesion development [36, 37]. Our numerical results for pulsatile flow point to an intermediate stage of AAA growth for which disturbed flow conditions exist. The quantification of this disturbance results in three distinct regions of flow development: (i) very low wall shear stresses at the center of an aneurysm; (ii) high wall shear stress gradients in the distal end of each AAA; and (iii) oscillating wall shear stresses and wall shear stress gradients at any site of the expanded arterial segment. The oscillatory nature of the WSSG at high mean volumetric flow rates, once the aneurysm has begun to grow, produces peak values of a periodic, pulsating hemodynamic force that may be responsible for severe sites of injury to the endothelium over a relatively long period of time.

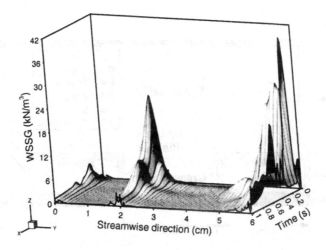

Figure 8 Wall Shear Stress Gradient (WSSG) distribution for $\text{Re}_m = 680$ as a function of time and axial location.

4.2 Heat Removal from Electronic Components: Convection-Only vs. Conjugate Conduction/Convection Predictions

The next application considers the heat removal by forced convection from surface-mounted electronic components on Printed Circuit Boards (PCBs). Figure 9 depicts a schematic of a periodic grooved-channel geometry with the configuration and material composition of the electronic components. Numerical simulations of the time-dependent convection-only and conjugate conduction/convection heat transfer are performed to ascertain the influence that conjugate effects have upon the convective heat transport. The conjugate results are contrasted with convection-only results obtained from simulating the fluid domain for the same geometry with uniform heat-flux boundary conditions along the grooved wall. The periodic grooved channels, formed by the electronic components and PCBs, are able to sustain Tollmien-Schlichting channel instabilities [38, 39]. The passive flow destabilization, induced by the spatially periodic disturbances introduced by the electronic components induces oscillatory flows, Tollmien-Schlichting traveling waves in the bypass region, and vortex ejections from the groove region which are flow mechanisms responsible for enhancing the overall mixing and, hence, improving the heat transfer performance [40]. Complex supercritical flow structure produces a time-repetitive sequence of convective exchange between the groove and the bypass regions. The periodic disruption of the shear layer coupled with the separation flow phenomenon at the downstream groove corner induces ejection of the hot fluid from the downstream component face into the bypass flow [18].

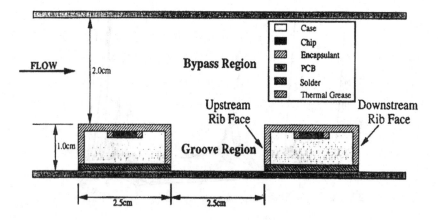

Figure 9 Schematic of the electronic component configuration and flow regions.

4.2.1 Convection simulations. We present first convection-only simulations where uniform heat flux at the solid-fluid interface is specified as the boundary condition. To satisfy this boundary condition, the wall temperature must adjust in accordance with the temperature of the fluid in the near-wall region of the boundary layer.   Therefore, the temperature distribution along the wall is governed by a combination of the local fluid temperature and the wall shear stress, both being strongly related to upstream effects. The effect of the flow regime on the time-averaged difference between the surface and bulk-mean temperatures (ΔT) is presented in Fig 10. The distributions of ΔT along the top surface of the electronic component (rib) are quite similar for all the Reynolds numbers. However, the largest ΔT is exhibited by the near critical Reynolds number case (Re = 550), followed by the subcritical (Re = 261) and then the supercritical cases (Re = 600, 693). Within the subcritical flow regime the wake from the preceding rib is not fully homogenized. Therefore, the thermal boundary layer initiates with a finite thickness.   As the boundary layer grows along the rib surface, its temperature increases to compensate for the energy diffused away from the wall. In the supercritical case, the wake is more completely homogenized, requiring a lower surface temperature at the leading edge of the rib to satisfy the constant flux boundary condition.   Additionally, the supercritical bypass flow induces higher diffusion rates along the top surface of the rib, resulting in a more rapid increase in wall temperature than in the subcritical cases.

As the flow becomes supercritical, the ΔT is substantially reduced due to the transverse momentum and time-dependent nature of the supercritical bypass flow. In fact, all of the surfaces along the grooved wall experience a reduction in temperature. The waviness of the supercritical flow structure results in transverse convective transport, inducing large-scale mixing and a reduction of the effects associated with the wake.

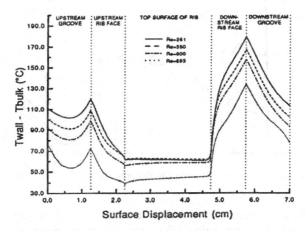

Figure 10 T_{wall} - T_{bulk} versus surface displacement for convection-only simulations.

The variations in convective behavior along the grooved surface can be visualized by examining the local distribution of the time-averaged Nusselt number, shown in Fig. 11, for Reynolds numbers Re in the subcritical (Re = 261) near critical (Re = 550) and supercritical (Re = 600, 693) flow regimes. Along the top surface of the rib, the Nusselt numbers decrease slightly with Reynolds number throughout the subcritical regime. However, this behavior reverses in the supercritical flow regime, as the Nusselt number increases significantly with Reynolds number. The Nusselt number within the groove region continuously increases with Reynolds number for both subcritical and supercritical flows, suggesting that the recirculating flow patterns within the groove significantly minimize any resistance associated with upstream effects.

In summary, the supercritical flow structure increases convective heat transport through three mechanisms. First, the traveling waves in the bypass region induce fluctuations in the velocity components. These fluctuations periodically increase the shear stress substantially above subscritical values, resulting in increased diffusion at the solid-fluid interface. Second, the supercritical bypass flow contains fluctuations in normal velocity, which partially homogenize the temperature distribution within the bypass flow and wake. Last, the supercritical flow structure disrupts the shear layer separating the groove and bypass flows. This disruption results in direct convective exchange between the groove and bypass flows, increasing surface transport rates. Therefore, the supercritical flow structure increase heat transport by both increasing diffusion at the solid-fluid interface and inducing large-scale mixing within and between the groove and bypass regions.

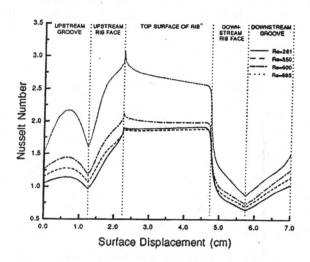

Figure 11 Time-averaged Nusselt number distribution along the grooved-channel wall, for subcritical and supercritical flows for convection simulations.

4.2.2 Conjugate conduction/convection simulations. In the conjugate conduction/convection simulations, heat is generated within the chip (Fig. 9), conducted through the electronic component and convected into the cooling fluid. The local flow characteristics in the grooved channel are the same as in the convection-only simulations with uniform heat-flux boundary conditions. However, the local thermal characteristics along with the composition and distribution of heat generation within the solid region dictate the local convective performance. Consequently, because of the additional internal heat resistances, in the electronic component, the surface temperature distribution (Fig. 12) exhibits a different pattern than in the convection-only simulations (Fig. 10). The concentrated heat generation in the chip produces a maximum ΔT directly above the location of the chip. This differs from the convection-only simulations in which the top surface of the rib displays the lowest ΔT and the groove surface the highest. In fact, for the conjugate case, the groove surface displays the lowest ΔT.

Contrary to the convection-only case, in conjugate conduction/convection, the ΔT increases from the subcritical to the supercritical flow regime, for the range of Reynolds numbers explored. In the convection-only case, the uniform flux boundary condition applied at the solid-fluid interface mandates that the product of the heat transfer coefficient and the temperature difference be constant. At supercritical Reynolds numbers, higher heat transfer coefficients result because of the combination of better flow homogenization and higher diffusive rates. Therefore, the ΔT must decrease to satisfy the uniform heat-flux boundary condition. However, the conjugate problem must satisfy a different boundary condition, namely continuity of flux and temperature at the solid-fluid interface.

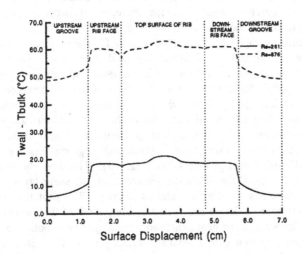

Figure 12 T_{wall} -T_{bulk} versus surface displacement for conjugate simulations.

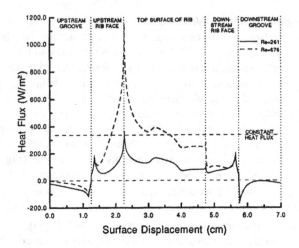

Figure 13 Time-averaged heat flux distribution along the solid-fluid interface for subcritical (Re = 261) and supercritical (Re = 676) flows for conjugate conduction/convection simulations, and constant heat.

As the Reynolds number increases within the range of this study, more heat is convected from the package surface (Fig. 13), causing larger internal temperature gradients. To satisfy the continuity of flux boundary condition, the temperature gradient within the fluid must similarly increase. Physically, as heat is removed from the package by the cooling fluid, the bulk temperature increases. Therefore, the surface temperature must likewise increase in order to maintain the temperature gradient at the solid-fluid interface.

The time-averaged heat flux distribution for the conjugate simulation is shown in Fig. 13 for subcritical (Re = 261) and supercritical (Re = 676) flow regimes. In the subcritical regime, the shear layer separating the groove and bypass flows precludes convective exchange of fluid. Therefore, the fluid recirculating within the groove, which is heated along the upstream and downstream component faces, becomes hotter than the local surface temperature at the groove bottom. This temperature gradient causes heat to be transferred from the fluid to the groove surface, producing negative heat fluxes in the upstream and downstream groove faces (Fig. 13). However, the bypass flow convectively cools the fluid within the groove when the shear layer is disrupted and the flow becomes supercritical and oscillatory. The heat flux distribution along the top surface of the rib is affected by the growth of the thermal boundary layer and the location of the heat generation, leading to an exponential decline in the downstream direction with a local rise directly over the location of the chip [18, 10]. Therefore, time-dependent forced convective transport phenomena, on complex, multimaterial solid domains with conjugate conduction/convection, exhibit different thermal performance than the one obtained by assuming uniform heat-flux boundary conditions. These results illustrate the importance of accounting for conjugate conduction/convection effects in the design and analysis of heat exchanger configurations such as those found in the cooling of electronic components.

4.3 Mass Transfer in Pulsating Flow Blood Oxygenators

For the final application, we consider three-dimensional simulations of pulsatile flows with oxygen transport. These simulations have been performed in the context of developing intravenous membrane oxygenators to complement the function of the natural lung for patients with the Acute Respiratory Distress Syndrome (ARDS). This is an illness that affects people's lungs characterized by a progressive interstitial edema, a diminished preliminary compliance and a decreased diffusion capability. Treatments based on mechanical ventilation, extracorporeal oxygenation and intravenous oxygenation have been developed with the goal of providing the patients with necessary respiratory support, until their lungs recover the normal gas exchange function [41-43]. Intravenous oxygenation is one of the treatments that have been proposed as a potentially attractive therapy in-patients with ARDS [44, 45]. The success of this treatment is based on its ability to reach levels of oxygen and carbon dioxide exchange appropriate for the metabolic requirements and make the intravenous oxygenation more clinically effective, without restricting cardiac return and seriously affecting the venous hemodynamics [44-46].

The Intravenous Membrane Oxygenator (IMO) device, shown in Fig. 14(a), is being currently tested at the University of Pittsburgh Artificial Lung program with the goal of reaching oxygen and carbon dioxide exchange of 50% of the metabolic requirements [45-47]. This device, which is inserted within the vena cava, has a centrally positioned elongated balloon within a bundle of hundreds of hollow fiber membranes. The balloon, made of elastic and non-permeable polyurethane walls, is inflated and deflated to a given amplitude and frequency by externally-pumped helium. The oxygen and carbon dioxide are transported longitudinally within the hollow fibers. The number of fibers within the IMO device varies between 700 and 1100 depending on the fiber size. Early in-vitro and in-vivo experiments [45-47] have indicated that balloon pulsation enhances IMO gas exchange by increasing fluid convection around the fiber bundle and that the additional convective flow depends on the amplitude and frequency of the balloon pulsations. In-vitro studies with a free fiber bundle, performed by Federspiel et al. [46], have shown that the critical frequency to maintain a sufficient oxygen flow rate and a good condition for volume amplitude of balloon pulsations is about 160 beats/min (bpm). The same experiments, performed in water, have shown that the gas exchange performance increases to an asymptotic stage as the pulsation frequency increases to 160 bpm. These results have allowed them to conclude that the pulsating balloon mechanism for enhancing convective mixing into the vena cava system most likely will be a key component to increase the efficiency of gas transfer at the fiber level. Recent in-vitro experiments by Federspiel et al. [48] with a constrained fiber bundle have indicated that this fiber configuration represents a better alternative in terms of pressure drop and gas exchange than the free fiber bundle. Guzmán and Amon [49, 50] have performed spectral element simulations and parametric investigations of the IMO device, to determine the flow mixing characteristics and mass transfer enhancement mechanisms induced by cross-pulsating blood flows, for both stationary and pulsating balloon conditions.

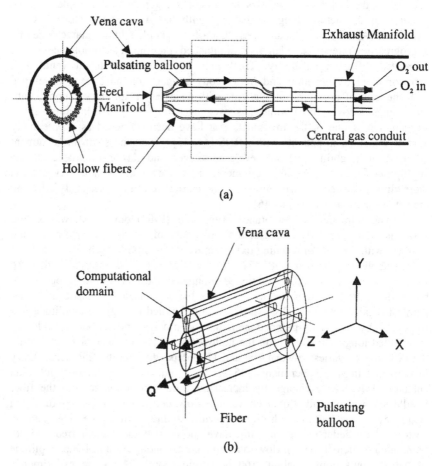

Figure 14 Intravenous Membrane Oxygenator (IMO) device: (a) schematic representation within the vena cava; and, (b) computational domain.

A schematic representation of the physical model of the IMO device is shown in Fig. 14(a), where for clarity, only a few fibers are shown. In the current simulations, the flow characteristics correspond to a hydrodynamically fully-developed flow regime. The three-dimensional (3D) computational model shown in Fig. 14(b) ignores inlet and exit geometric effects of the IMO device and assumes that the oxygen transport process occurs far from the feed and exhaust manifolds. This model also considers that the balloon inflates and deflates uniformly in the radial direction (i.e., normal to the surface balloon) and that the fibers remain stationary as in the in-vitro experiments, with their longitudinal axes parallel to the streamwise z-direction. The computational model assumes that there are 128 fibers uniformly distributed between the vena cava and balloon. Therefore, the physical space is divided into 128 regions of equal dimensions, where the flow and the oxygen transport characteristics repeat in each region. The computational domain is composed by one of these regions and is bounded by the physical boundaries of the vena cava, balloon and fiber, and by the fluid boundaries in the spanwise x- and streamwise z-directions. Notice that the streamwise length of this computational domain of $L = 4$ cm is shorter than the streamwise length (25 cm approximately) of the physical model.

The computational study considers a time-dependent, incompressible and Newtonian blood flow within the vena cava. For the operating conditions of a stationary balloon, the velocity field corresponds to a steady-state regime; whereas, for the pulsating balloon, the velocity field is time dependent because of the balloon motion. Two fundamental processes govern the oxygen transport mechanism: bulk convection and molecular diffusion. Oxygen is transported by the bulk convection in the blood that flows through the vessel and by molecular diffusion from the blood across vessel walls to the bulk flow. Almost all the oxygen is carried in reversible combination within the hemoglobin contained in the blood cells. In this study, the oxygen transport within the blood is described by a model of non-reacting species transport given by the conservation of species equation. Thus, the governing equations 1, 2 and 3(b), describe the blood flow and the oxygen transport within the Intravenous Membrane Oxygenator for the two operating conditions of this study: a stationary balloon and a pulsating balloon. Figure 15(a) shows the spectral element discretization of the computational domain; whereas, Figs. 15(b) and (c) illustrate mesh details and nodal points near the fiber surface where high velocity and oxygen concentration gradients are expected. The IMO model length is 4 cm, the fiber diameter is 250 μm and the vena cava diameter is 2.5 cm. Simulations performed with a stationary balloon consider the action of a streamwise pressure gradient only; whereas, for a pulsating balloon, a time-sinusoidal motion of the balloon wall along with the streamwise pressure gradient originates the flow. The balloon wall motion boundary condition defines a time-dependent computational domain problem [51].

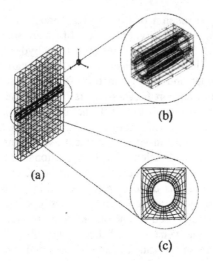

Figure 15 Spectral element discretization: (a) computational domain; (b) 3D macroelements close to the fiber; and, (c) collocation points.

Figures 16(a) and (b) show streamwise u_z-velocity profiles and velocity vector representations, respectively, for a stationary balloon condition in the Reynolds number range of Re = 5.7 - 455.2. The Reynolds number is defined as Re = $D_{vc} \bar{u} / v$, where $\bar{u}$ is the mean velocity in the vena cava, D_{vc} is the vena cava diameter and v is the kinematic viscosity. For successive increases of the Reynolds number the streamwise u_z-velocity profiles remain parabolic in shape between the vena cava and fiber and balloon wall. The only effect of the higher Re is to increase the mean velocity of the velocity profiles, but not its parabolic shape [49, 52].

For a stationary balloon, the vector representations show that the flow is parallel to the streamwise direction with absence of secondary flows and recirculation regions in the Reynolds number range of these simulations. Therefore, for a stationary balloon, the flow remains parallel, bi-directional, and parabolic in shape and without secondary flows. Figure 17 shows oxygen concentration profiles at the inlet of the computational domain, for the Reynolds numbers range of 5.7 - 455.2. Simulations have been performed with an oxygen concentration value of $c_w = 1.0 \times 10^{-5}$ ml-O_2/ml. The oxygen concentration profiles and their gradients at the fiber wall remain similar in shape and size as the Reynolds number increases. The oxygen concentration transport occurs very close to the fiber wall by a diffusion mechanism. The streamwise convection does not affect significantly the transport of oxygen to the bulk blood flow. Therefore, the oxygen diffusive mechanism to the blood flow saturates at relatively low Reynolds numbers. Thus, higher volumetric flow rates in the streamwise direction do not significantly increase the oxygen transport process efficiency [50, 51].

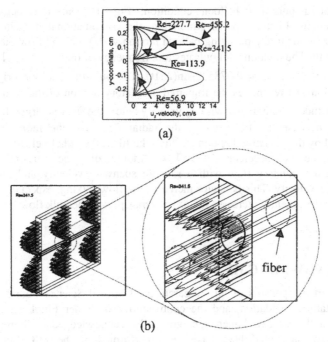

(a)

(b)

Figure 16 Velocity representation for a stationary balloon for the Reynolds number range of Re = 5.7 - 455.2: (a) streamwise u_z-velocity profiles; and, (b) velocity vectors and close up for Re = 341.5.

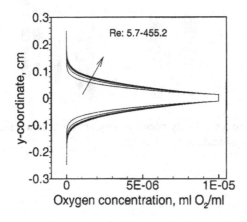

Figure 17 Oxygen concentration profiles for a stationary balloon for the Reynolds number range of Re = 5.7 - 455.2.

For a pulsating balloon condition, Fig. 18 shows velocity vector representations. In this type of simulation, the computational domain deforms continuously due to the pulsating balloon motion with no blood flow through the balloon wall. The balloon wall pulsates to a frequency of 60 beats/min (1 Hz) and a radial velocity of $u_y = 0.078539 \cdot \sin(2\pi 1 t)$. The pulsating balloon originates a fluid motion that resembles the impaired oscillatory motion of the balloon wall. The magnitude of radial crosswise u_y-velocity below the fiber is larger than above the fiber because the fiber damps the radial flow and the momentum flux originated by the pulsating balloon. Around the fiber, the radial velocity increases in magnitude and accelerates the flow. Additionally, the pulsating balloon originates a secondary flow with a random spanwise velocity and a pulsatile streamwise velocity. This behavior increases the flow mixing, especially near the fiber and consequently increases the oxygen transport to the bulk flow [51, 52].

5 ACKNOWLEDGMENTS

The support of sponsors of this research, the National Science Foundation and The Whitaker Foundation, and the contributions of Ender Finol, Dr. Amador Guzmán and Dr. Jay Nigen are gratefully acknowledged. Some of the computations presented here have been performed at the PSC and NCSA Supercomputing Centers.

Figure 18 Instantaneous velocity vector representations for a pulsating balloon at two times of the sinusoidal balloon motion.

6 REFERENCES

1. C.H. Amon and B.B. Mikic, Computational Methodologies as a Research Tool in Complex Forced Convection Heat Transfer Systems: Present Status and Future Expectations, *Computers and Computing in Heat Transfer Science and Engineering*, CRC Press, Begell House, Boca Raton, FL, pp. 61-86, 1993.

2. W. Najayama and K.T. Yang, *Computers and Computing in Heat Transfer Science and Engineering*, CRC Press, Begell House, Boca Raton, FL, 1993.

3. I. Babuska and M.R. Dorr, *Numerical Mathematics*, Vol. 37, No. 2, pp. 257-277, 1981.

4. C. Canuto, M.Y. Hussaini, A. Quarteroni, and T.A. Zang, *Spectral Methods in Fluid Dynamics*, Springer-Verlag, New York, 1987.

5. J.P. Boyd, Chebyshev and Fourier Spectral Methods, *Lecture Notes in Engineering*, Springer-Verlag, Berlin, 1989.

6. A.T. Patera, A Spectral Element Method for Fluid Dynamics: Laminar Flow in a Channel Expansion, *Journal of Computational Physics*, Vol. 54, pp. 468-488, 1984.

7. C.H. Amon, Spectral Element-Fourier Method for Transitional Flow in Complex Geometries, *AIAA Journal*, Vol. 31, No. 1, pp. 42-48, 1993.

8. G.E. Karniadakis and R.D. Henderson, Spectral Element Methods for Incompressible Flows, *The Handbook of Fluid Dynamics*, CRC Press, pp. 29.1-29.41, 1998.

9. Y. Maday and A.T. Patera, Spectral Element Methods for the Navier-Stokes Equations, *ASME, State of the Art Surveys in Computational Mechanics*, 1987.

10. J.S. Nigen and C.H. Amon, Effect of Material Composition and Localized Heat Generation on Time-Dependent Conjugate Heat Transport, *International Journal of Heat and Mass Transfer*, Vol. 38, No. 9, pp. 1565-1576, 1995.

11. A.V. Luikov, Conjugate Convective Heat Transfer Problems," *International Journal of Heat and Mass Transfer*, Vol. 17, pp. 257-265, 1974.

12. A. Brosh, D. Degani and S. Zalmanovich, Conjugated Heat Transfer in a Laminar Boundary Layer with Heat Source at the Wall, *Journal of Heat Transfer*, Vol. 104, pp. 90-95, 1982.

13. K.D. Cole and J.V. Beck, Conjugated Heat Transfer from a Strip Heater with the Unsteady Surface Element Method, *Journal of Thermophysics and Heat Transfer*, Vol. 1, No. 4, pp. 348-354, 1987.

14. A. Campo and C. Schuler, Heat Transfer in Laminar Flow Through Circular Tubes Accounting for Two-Dimensional Wall Conduction, *International Journal of Heat and Mass Transfer*, Vol. 31, No. 11, pp. 2251-2259, 1988.

15.   J.R. Culham, T.F. Lemczyk, S. Lee and M.M. Yovanovich, META — A Conjugate Heat Transfer Model for Air Cooling of Circuit Boards with Arbitrarily Located Heat Sources, *Heat Transfer in Electronic Equipment*, American Society of Mechanical Engineers-HTD, Vol. 171, pp. 117-126, 1991.

16.   S. Ray and J. Srinivasan, Analysis of Conjugate Laminar Mixed Convection Cooling in a Shrouded Array of Electronic Components, *International Journal of Heat and Mass Transfer*, Vol. 35, No. 4, pp. 815-822, 1992.

17.   J.S. Nigen and C.H. Amon, Conjugate Forced Convective Effects of Time Dependent Recirculating Flows in Grooved Channels, *Fundamentals of Forced Convection Heat Transfer*, edited by M.A. Ebadian and P.H. Oosthuizen, American Society of Mechanical Engineers-HTD, Vol. 210, pp. 91-98, 1992.

18.   J.S. Nigen and C.H. Amon, Time-Dependent Conjugate Heat Transfer Characteristics of Self-Sustained Oscillatory Flows in a Grooved Channel, *Journal of Fluid Engineering*, Vol. 116, pp. 499-507, 1994.

19.   M. Deville and S.A. Orszag, *Lecture Notes in Mathematics*, Vol. 771, Springer-Verlag, Berlin, 1980.

20.   C.H. Amon, Heat Transfer Enhancement and Three-Dimensional Transition by a Spectral Element-Fourier Method, Sc.D. Thesis, Massachusetts Institute of Technology, Cambridge, MA, 1988.

21.   G. Strang and G. Fix, An Analysis of the Finite Element Method, Prentice-Hall, Englewood Cliffs, NJ, 1973.

22.   C. Ernst, Abdominal Aortic Aneurysm, *The New England Journal of Medicine*, Vol. 328, No. 16, pp. 1167-1172, 1993.

23.   A.M. Guzmán and C.H. Amon, Dynamical Flow Characterization of Transitional and Chaotic Regimes in Converging-Diverging Channels, *Journal of Fluid Mechanics*, Vol. 321, pp. 25-57, 1996.

24.   C.H. Amon, A.M. Guzmán, and B. Morel, Lagrangian Chaos, Eulerian Chaos, and Mixing Enhancement in Converging-Diverging Channel Flows, *Physics of Fluids*, Vol. 8, No. 5, pp. 1192-1206, 1996.

25.   A.M. Guzmán, N.O. Moraga and C.H. Amon, Pulsatile Non-Newtonian Flow in a Double Aneurysm, *Proceedings of the 1997 ASME International Mechanical Engineering Congress and Exposition*, pp. 87-88, 1997a.

26.   N.O. Moraga, A.M. Guzmán and C.E. Rosas, Mecánica de Fluidos No Newtonianos de Flujo Transiente en Tubería con Sección Transversal Variable en el Espacio, *VI Congreso La Ingeniería en la Industria del Cobre*, Universidad de Antofagasta, Chile, pp. 167-175, 1997.

27.   A.M. Guzmán, N.O. Moraga, G. Muñoz and C.H. Amon, Pulsatile Non-Newtonian Flow in a Converging-Diverging Tube, *AIChE Symposium Series*, Vol. 93, No. 314, pp. 288-294, 1997b.

28.   E. Finol and C. Amon, Blood Flow in Abdominal Aortic Aneurysms: The Two-Aneurysm Model, Part I — Steady Flow, *ASME Journal of Biomechanical Engineering*, 1999a, (to be published).

29.   E. Finol and C. Amon, Blood Flow in Abdominal Aortic Aneurysms: The Two-Aneurysm Model, Part II – Pulsatile Flow, *ASME Journal of Biomechanical Engineering*, 1999b (to be published).
30.   E. Albritton, Standard Values in Blood, *United States Air Force*. Wright Air Development Center, pp. 5-7, 1951.
31.   M. Lei and C. Kleinstreuer, The Zero-Tension Hypothesis for the Mechanism of Atherogenesis and the Wall Shear Stress Gradient (*WSSG*) Predictor Equation, *Proceedings of the 1996 ASME International Mechanical Engineering Congress and Exposition*, pp. 211-212, 1996.
32.   M. Lei, C. Kleinstreuer, and G. Truskey, Numerical Investigation and Prediction of Atherogenic Sites in Branching Arteries, *ASME Journal of Biomechanical Engineering*, Vol. 117, pp. 350-357, 1995.
33.   C. Mills, I. Gabe, J. Gault, D. Mason, J. Ross Jr., E. Braunwald, and J. Shillingford, Pressure-Flow Relationships and Vascular Impedance in Man", *Cardiovascular Research*, Vol. 4, pp. 405-417, 1970.
34.   D. Bluestein, L. Niu, R. Schoephoerster, and M. Dewanjee, Steady Flow in an Aneurysm Model: Correlation between Fluid Dynamics and Blood Platelet Deposition, *ASME Journal of Biomechanical Engineering*, Vol. 118, pp. 280-286, 1996.
35.   M. Nazemi, C. Kleinstreuer and J. Archie Jr., Pulsatile Two-dimensional Flow and Plaque Formation in a Carotid Artery Bifurcation, *Journal of Biomechanics*, Vol. 23, No. 10, pp. 1031-1037, 1991.
36.   R. Nerem, D. Harrison, R. Taylor, and R. Alexander, Hemodynamics and Vascular Endothelial Biology, *Journal of Cardiovascular Pharmacology*, Vol. 21, Suppl. 1, pp. S6-S10, 1993a.
37.   R. Nerem, Hemodynamics and the Vascular Endothelium, *ASME Journal of Biomechanical Engineering*, Vol. 115, pp. 510-514, 1993b.
38.   C.H. Amon, D. Majumdar, C.V. Herman, F. Mayinger, B.B. Mikic and D.P. Sekulic, Numerical and Experimental Studies of Self-Sustained Oscillatory Flows in Communicating Channels, *International Journal of Heat and Mass Transfer*, Vol. 35, No. 11, pp. 3115-3129, 1992.
39.   C.H. Amon and A.T. Patera, Numerical Calculation of Stable Three-Dimensional Tertiary States in Grooved-Channel Flow, *Physics Fluids*, Vol. 1, No. 12, pp. 2005-2009, 1989.
40.   C.H. Amon and B.B. Mikic, Numerical Prediction of Convective Heat Transfer in Self-Sustained Oscillatory Flows, *Journal of Thermophysics and Heat Transfer*, Vol. 4, No. 2, pp. 239-246, 1990.
41.   K.M., High, M.T. Snider, R. Richard, G.B. Russell, J.K. Stene, D.B. Campbell, T.X. Aufiero, and G.A. Thieme, Clinical Trials of an Intravenous Oxygenator in Patients With Adult Respiratory Distress Syndrome," *Anesthesiology*, Vol. 77, pp. 856-863, 1992.
42.   F.L. Fazzalari, R.H. Bartlett, M.R. Bonnell, and J.P. Montoya, An Intrapleural Lung Prosthesis: Rationale, Design, and Testing, *Artificial Organs*, Vol. 18, pp. 801-805, 1994.

43.   S. Ichiba, and R.H. Bartlett, Current Status of Extracorporeal Membrane Oxygenation for Severe Respiratory Failure, *Artificial Organs*, Vol. 20, pp. 120-123, 1996.

44.   J.D. Mortensen, Intravascular Oxygenator: A New Alternative Method for Augmenting Blood Gas Transfer in Patients With Acute Respiratory Failure, *Artificial Organs*, Vol. 16, pp. 75-82, 1992.

45.   B.G. Hattler, P.C. Johnson, P.J. Sawzik, F.D. Shaffer, M. Klain, L.W. Lund, G.D. Reeder, F.R. Walters, J.S. Goode, and H.S. Borovetz, Respiratory Dialysis: A New Concept in Pulmonary Support, *American Society for Artificial Internal Organs Journal*, Vol. 38, pp. M322-M325, 1992.

46.   W.J. Federspiel, T. Hewitt, M.S. Hout, F.R. Walters, L.W. Lund, P.J. Sawzik, G.D. Reeder, H.S. Boravetz, and B.G. Hattler, Recent Progress in Engineering The Pittsburgh Intravenous Membrane Oxygenator, *ASAIO Journal*, Vol. 42, 1996, pp. M435-M442.

47.   B.G. Hattler, G.D. Reeder, P.J. Sawzik, L.W. Lund, F.R. Walters, A.S. Shah, J. Rawleigh, J.S. Goode, M. Klain, and H.S. Borovetz, Development of an Intravenous Membrane Oxygenator (IMO): Enhanced Intravenous Gas Exchange Through Convective Mixing of Blood Around Hollow Fiber Membranes, *Artificial Organs Journal*, Vol. 18, pp. 806-812, 1994.

48.   W.J. Federspiel, M.S. Hout, T.J. Hewitt, L.W. Lund, S.A. Heinrich, P. Litwak, F.R. Walters, G.D. Reeder, H.S. Boravetz, and B.G. Hattler, Development of a Low Flow Resistance Intravenous Oxygenator, *ASAIO Journal*, Vol. 43, pp. M725-M730, 1997.

49.   A.M. Guzmán, and C.H. Amon, Mass Transfer Performance Evaluations of an Intravenous Membrane Oxygenator, *Proc. of the 1998 ASME International Mechanical Engineering Conference and Exposition*, HTD-Vol. 362; BED-Vol. 40, *Advances in Heat and Mass Transfer in Biotechnology*, S. Clegg (Ed.) pp. 149-154, 1998.

50.   A.M. Guzmán, and C.H. Amon, Flow and Mass Transfer Characteristics of an Intravenous Membrane Oxygenator: A Computational Study, *Journal of Computer Methods in Biomechanics and Biomedical Engineering*, 1999, in press.

51.   A.M. Guzmán, and C.H. Amon, Mass Transfer Enhancement in an Intravenous Membrane Oxygenator Induced by a Pulsating Balloon, *1999 ASME International Mechanical Engineering Conference and Exposition*, Nashville, TN, November 1999.

52.   A.M. Guzmán, C.H. Amon, W.J. Federspiel, and B.G. Hattler, Spectral-Element Simulations of the Flow Kinematics in an Intravenous Membrane Oxygenator, *Proc. of the Fourth World Congress on Computational Mechanics*, Vol. 2, p. 1011. Also published in *Computational Mechanics, New Trends and Applications*, pp. 1-14, Idelsohn, Onate & Dvorkin (Eds.), CIMNE, 1998.

FOUR

FINITE-VOLUME METHOD
FOR RADIATION HEAT TRANSFER

J.C. Chai
S.V. Patankar

1 INTRODUCTION

1.1 Purpose of the Chapter

This chapter presents a finite-volume (FV) method for computing radiation heat transfer processes. The main ingredients of the calculation procedure were presented by Chai et al. [1]. The resulting method has been tested, refined and extended to account for various geometrical and physical complexities.

1.2 Scope and Limitations

The number of publications on radiation heat transfer procedures has increased rapidly over the past few years. As a result, a comprehensive review of all literature dealing with the subject will not be attempted here. In order to provide a *focus* to the paper, only the FV method is described in detail. The discrete-ordinates (DO) method and its variants will be discussed when their inclusion is needed. The paper focuses on the *method* rather than on its applications.

It should be pointed out that it is possible that many significant papers might be omitted from this chapter. However, any such omissions are unintentional and do not imply any judgement as to the quality and usefulness of the papers.

As mentioned earlier, this chapter focuses on the method of solution rather than on the different physical models. As a result, nongray models, which are very important for a variety of applications, are not discussed here. Combined mode heat transfer (combined diffusion, convection and radiation) is also not considered in this article.

It is *not* the objective of this article to criticize any method. However, the advantages and disadvantages of various approaches will be discussed.

1.3   Outline of the Chapter

This chapter is divided into twelve sections. The equations governing radiative heat transfer, boundary conditions and other related relations are presented in Section 2. A discussion on the similarity between the radiative transfer equation (RTE) and the general convection-diffusion equation is presented in Section 3. This discussion shows that the RTE is a special case of the general convection-diffusion equation. Before presenting the FV method for radiation heat transfer, a discussion of the flux, DO and FV methods is given in Section 4. Section 5 shows the discretization of the spatial and angular domains. The RTE is converted into a set of algebraic equations in Section 6. Other details related to the discretization of the RTE are also discussed in this section.

Section 7 highlights a few approaches for the treatment of irregular geometries. A slight inconvenience that can arise in the modeling of irregular geometries, called the control-angle overlap, is discussed in Section 8. Advanced spatial differencing schemes are discussed in Section 9. This is followed by a discussion on the various shortcomings of the FV method. Section 11 gives a brief discussion of two recent developments in the FV method. A list of work performed by various researchers is included in Section 12 to conclude this chapter.

2   GOVERNING EQUATIONS AND RELATED QUANTITIES

The equations governing the "transport" of radiant energy, boundary conditions, scattering phase function and other related quantities are presented in this section.

2.1   Radiative Transfer Equation

The RTE for a gray medium can be written as

$$\frac{dI(\vec{r},\hat{s})}{ds} = -\beta(\vec{r})I(\vec{r},\hat{s}) + \kappa(\vec{r})I_b(\vec{r}) + \frac{\sigma(\vec{r})}{4\pi}\int_{4\pi} I(\vec{r},\hat{s}')\Phi(\hat{s}',\hat{s})d\Omega' \tag{1}$$

From Eq. (1), it is clear that radiant intensity I depends on spatial position $\vec{r}$ and angular direction $\hat{s}$. The solution of Eq. (1) requires the specification of the boundary conditions. Although the method described in this chapter can handle non-diffuse and semitransparent surfaces, only opaque diffuse surfaces are described for simplicity.

2.2 Boundary Condition for an Opaque Diffuse Surface

The radiant intensity leaving an opaque diffuse surface contains emitted and reflected energy. This can be written as

$$I(\vec{r},\hat{s}) = \varepsilon(\vec{r})I_b(\vec{r}) + \frac{\rho(\vec{r})}{\pi} \int_{\hat{s}'\bullet\hat{n}<0} I(\vec{r},\hat{s}')|\hat{s}'\bullet\hat{n}|d\Omega' \tag{2}$$

Equation (2) provides the boundary intensity for the RTE.

2.3 Scattering Phase Function

The scattering phase function Φ in the RTE describes how radiant energy is scattered by a participating medium. Scattering can be classified into two categories. These are isotropic and anisotropic scattering. Isotropic scattering scatters energy equally into all directions. Anisotropic scattering can be further divided into backward and forward scattering. Backward scattering scatters more energy into the backward directions, while forward scattering scatters more energy into the forward directions. Scattering phase functions satisfy

$$\int_{4\pi} \Phi(\hat{s}',\hat{s})d\Omega' = 4\pi \tag{3}$$

2.4 Radiation Heat Transfer Relations

A few useful quantities are defined in this subsection for ease of reference. The incident radiation is defined as

$$G(\vec{r}) = \int_{4\pi} I(\vec{r},\hat{s})d\Omega \tag{4}$$

The radiative heat flux in direction i is defined as

$$q_i(\vec{r}) = \int_{2\pi} I(\vec{r},\hat{s})(\hat{s}\bullet\hat{i})d\Omega \tag{5}$$

where $\hat{i}$ is the unit vector pointing in the i direction. For example, the radiative heat flux in the x direction is

$$q_x(\vec{r}) = \int_{2\pi} I(\vec{r},\hat{s})(\hat{s}\bullet\hat{e}_x)d\Omega \tag{6}$$

The divergence of the radiative heat flux is

$$\nabla\bullet q = \kappa[4\pi I_b(\vec{r}) - G(\vec{r})] \tag{7}$$

Equation (7) defines an important quantity in combined mode heat transfer as well as in radiation-dominated processes. In the absence of a heat source/sink, a system is in radiative equilibrium if other modes of heat transfer are absent. Under such condition, $\nabla \bullet q = 0$, and the temperature of the medium can be obtained from Eq. (7). In combined mode heat transfer processes with a participating medium, $\nabla \bullet q$ is the radiation source term in the energy equation.

Before proceeding with the formulation of a numerical procedure for radiation heat transfer, the similarity between the general transport equation encountered in fluid flow and heat-transfer-related processes and the RTE is described.

3   THE CONVECTION-DIFFUSION EQUATION AND THE RADIATIVE TRANSFER EQUATION

The FV method has been formulated for various fluid flow and heat-transfer-related processes. The transport of momentum and energy is described by partial differential equations which express the conservation of momentum and energy. These transport equations can be considered as special cases of a general transport equation. The steady-state form of the general transport equation for two-dimensional Cartesian coordinates can be written as

$$\rho u \frac{\partial \phi}{\partial x} + \rho v \frac{\partial \phi}{\partial y} = \frac{\partial}{\partial x}\left(\Gamma \frac{\partial \phi}{\partial x}\right) + \frac{\partial}{\partial y}\left(\Gamma \frac{\partial \phi}{\partial y}\right) + S \tag{8}$$

This steady-state transport equation consists of three terms: the *convection* term, the *diffusion* term and the *source* term. In Eq. (8), ϕ is the general dependent variable, Γ is the generalized diffusion coefficient, and S is the source term. In the absence of diffusion ($\Gamma = 0$), Eq. (8) can be written as

$$\rho u \frac{\partial \phi}{\partial x} + \rho v \frac{\partial \phi}{\partial y} = S \tag{9}$$

When the ρu and ρv are constants, Eq. (9) can be further reduced to

$$\alpha \frac{\partial \phi}{\partial x} + \gamma \frac{\partial \phi}{\partial y} = S \tag{10}$$

where $\alpha = \rho u$ and $\gamma = \rho v$. If the source term S is written as $S = S_1 + S_2 \phi$, the general transport equation then becomes

$$\alpha \frac{\partial \phi}{\partial x} + \gamma \frac{\partial \phi}{\partial y} = S_1 + S_2 \phi \tag{11}$$

The RTE (Eq. 1) can be written as

$$\frac{dI}{ds} = -\beta I + S \tag{12}$$

where the last two terms of Eq. (1) are combined to form S. For two-dimensional Cartesian coordinates, Eq. (12) can be written as

$$\mu \frac{\partial I}{\partial x} + \xi \frac{\partial I}{\partial y} = -\beta I + S \tag{13}$$

where μ and ξ are the direction cosines in the x- and y-directions, respectively.

Comparing Eq. (11) and Eq. (13), the similarity between the general transport equation and the RTE is obvious. The RTE is a special case of the general transport equation with $\Gamma = 0$ and constant ρu and ρv.

This similarity implies that (1) anyone capable of modeling the general transport equation can model the RTE (Eq. 1), and (2) various differencing schemes for the general transport equation can be used in the modeling of the RTE.

4   THE FLUX, DISCRETE-ORDINATES AND FINITE-VOLUME METHODS

A brief overview of the flux, DO and FV methods is given in this section. Since radiant intensities have to be resolved in both the angular and spatial domains, a complete discretization procedure should show these two discretizations. However, the main difference in the flux, DO and FV methods lies in the treatment of the *angular* space. Therefore, this section focuses on the discretization of the *angular* space. Spatial discretization practices will be discussed in a later section. A detailed evaluation of the relations among the three methods is given at the end of this section.

4.1   The Flux Method

The flux method, also known as the Schuster-Schwarzschild approximation, was proposed by Schuster [2] and Schwarzschild [3] for one-dimensional radiative transfer. The actual intensities (Fig. 1a) are approximated by dividing the 4π solid angle into two solid angles (one in each coordinate direction). The magnitude of the radiative intensity over the positive coordinate direction is assumed uniform. The magnitude of the radiant intensity in the negative coordinate direction is also assumed to be uniform but is allowed to be different from the magnitude of the intensity in the positive coordinate direction. Radiant energy is allowed to travel in *all* directions within the positive and negative coordinate directions. This approximation is depicted in Fig. 1b and can be written as

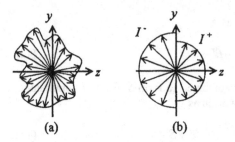

Figure 1 Radiant intensity distribution: (a) actual intensity, (b) two-flux method.

$$I = \begin{cases} I^+ & \hat{n}_z \bullet \hat{s} > 0 \quad \text{positive coordinate direction} \\ I^- & \hat{n}_z \bullet \hat{s} < 0 \quad \text{negative coordinate direction} \end{cases} \qquad (14)$$

For isotropic scattering ($\Phi = 1$), the RTE for the positive coordinate direction can be written as

$$\frac{dI^+}{dz} = -\beta I^+ + \kappa I_b + \frac{\sigma}{4\pi}\left(I^+ + I^-\right) \qquad (15)$$

Integrating Eq. (15) over the positive coordinate hemisphere gives

$$\frac{dI^+}{dz}\int\limits_{\hat{n}_z \bullet \hat{s} > 0}(\hat{n}_z \bullet \hat{s})d\Omega = \left[-\beta I^+ + \kappa I_b + \frac{\sigma}{2}\left(I^+ + I^-\right)\right]\int\limits_{\hat{n}_z \bullet \hat{s} > 0}d\Omega \qquad (16a)$$

or

$$\frac{1}{2}\frac{dI^+}{dz} = -\beta I^+ + \kappa I_b + \frac{\sigma}{2}\left(I^+ + I^-\right) \qquad (16b)$$

A similar equation for the negative coordinate direction can be written as

$$\frac{1}{2}\frac{dI^-}{dz} = -\beta I^- + \kappa I_b + \frac{\sigma}{2}\left(I^+ + I^-\right) \qquad (16c)$$

In summary, the two-flux method divides the *angular* space into two solid angles (one in each coordinate direction) in which the *magnitudes* of the radiant intensities are assumed constant. Radiation is allowed to travel in *all* directions within each solid angle (see Fig. 1b). This approach reduces the RTE to two ordinary differential equations, which can be solved using any convenient method. If the spatial domain is discretized into a finite number of control volumes, a discretization equation can be formulated for each control volume, and appropriate solution procedures can be employed to solve the resulting set of algebraic equations.

4.2   The Discrete-Ordinates Method

Most solution procedures for radiation heat transfer, including the DO method, were developed for astrophysics and neutron transport applications. Khalil and Truelove [4] and Fiveland [5] adopted the DO method to model radiative heat transfer processes and reported DO solutions for radiation heat transfer problems in 1977 and 1982 respectively. Over the past few years, there has been a tremendous increase in the number of papers on the modeling of radiation heat transfer using the DO method. Due to its popularity, it is appropriate to examine the DO method and compare it with the flux and FV methods.

Chandrasekhar [6] proposed the DO method in 1960. It was realized that the two-flux method could not accurately model anisotropic scattering with the two-solid-angle discretization practice. In the DO approximation, the actual radiation field (Fig. 2a) is divided into a finite number of discrete directions (Fig. 2b). The RTE at a *discrete* direction for one-dimensional problems can be written as

$$\mu^l \frac{dI^l}{dz} = -\beta I^l + \kappa I_b + \frac{\sigma}{4\pi} \sum_{l'=1}^{L} I^{l'} \overline{\Phi}^{l'l} w^{l'} \qquad (17)$$

There is no solid-angle integration. For nonscattering media in a black enclosure, radiant intensity along any direction can be calculated directly from Eq. (17) *without* solid-angle reference. These angular directions can, in principle, be chosen arbitrarily. However, several procedures have been developed to generate quadrature sets (ordinate directions and angular weights) that integrate the radiant energy and inscattering terms accurately. Therefore, exact locations of these angular directions are chosen such that the products of the angular directions and their weights satisfy certain moment constraints. These can be obtained by considering

$$\int_{4\pi} I d\Omega = \sum_{l=1}^{L} I^l w^l \qquad (18)$$

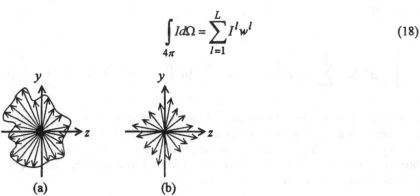

(a)   (b)

Figure 2 Radiant intensity distribution: (a) actual intensity, (b) DO method.

For ease of presentation, consider a one-dimensional problem where the intensity is invariant with respect to the azimuthal angle. Equation (18) can then be written as

$$\int_{-1}^{1} I d\mu = \sum_{l=1}^{L} I^l w^l \tag{19}$$

If I is a Kth-order polynomial in μ, Eq. (19) can be written as

$$\int_{-1}^{1} \left(a_0 + a_1 \mu + a_2 \mu^2 + \cdots + a_k \mu^k \right) d\mu =$$

$$\sum_{l=1}^{L} \left[a_0 + a_1 \mu^l + a_2 \left(\mu^l \right)^2 + \cdots + a_k \left(\mu^l \right)^k \right] w^l \tag{20}$$

One possible way to ensure that Eq. (20) is satisfied is to impose the following equalities:

$$\int_{-1}^{1} a_0 d\mu = 2a_0 = \sum_{l=1}^{L} a_0 w^l \quad \Rightarrow \quad \sum_{l=1}^{L} w^l = 2 \tag{21a}$$

$$\int_{-1}^{1} a_1 \mu d\mu = 0 = \sum_{l=1}^{L} a_1 \mu^l w^l \quad \Rightarrow \quad \sum_{l=1}^{L} \mu^l w^l = 0 \tag{21b}$$

$$\int_{-1}^{1} a_2 \mu^2 d\mu = \frac{2}{3} a_2 = \sum_{l=1}^{L} a_2 \left(\mu^l \right)^2 w^l \quad \Rightarrow \quad \sum_{l=1}^{L} \left(\mu^l \right)^2 w^l = \frac{2}{3} \tag{21c}$$

$$\vdots$$

$$\int_{-1}^{1} a_k \mu^k d\mu = \sum_{l=1}^{L} a_k \left(\mu^l \right)^k w^l \quad \Rightarrow \quad \sum_{l=1}^{L} \left(\mu^l \right)^k w^l = \frac{1}{k+1} \left[1 - (-1)^{k+1} \right] \tag{21d}$$

Equation (21d) is the general full-range k^{th} moment equation. These moments are also important if the scattering phase function in the RTE is to be evaluated correctly. Most quadrature sets do not integrate the scattering phase function correctly. As a result, phase functions are normalized to ensure that Eq. (3) is satisfied. This normalization can be written as

$$\frac{1}{4\pi} \sum_{l'=1}^{L} \Phi^{l'l} w^{l'} = 1 \tag{22}$$

Fiveland [7] and Truelove [8] showed that the half-range first moment is dominant in the calculation of wall heat fluxes. This can be written as

$$\int_0^1 a_1 \mu d\mu = \frac{1}{2} a_1 = \sum_{l=1}^{L/2} a_1 \mu^l w^l \quad \Rightarrow \quad \sum_{l=1}^{L/2} \mu^l w^l = \frac{1}{2} \quad (23)$$

It is common to use level-symmetric, equal-weight quadrature sets (called the S_n quadrature), which are $90°$ rotationally invariant [7, 8]. When lower-order quadrature sets are used, only limited numbers of moments are satisfied. Since the weights are chosen by dividing the 4π solid angle into discrete weights of equal size, the zeroth moment given by Eq. (21a) is always satisfied when the S_n quadrature sets are used. However, it is difficult to find a general quadrature set that integrates a general phase function accurately. This problem is avoided in the FV method.

4.3 The Finite-Volume Method

The FV method for radiation heat transfer presented in the literature has formulated the discretization equation by integration over both spatial control volume and angular control (solid) angle. For the purpose of this discussion, spatial discretization is deferred and only angular discretization is considered. A typical control (solid) angle is shown in Fig. 3.

Integrating the RTE over a control angle (Fig. 3) gives

$$\int_{\Delta\Omega^l} \frac{dI}{ds} d\Omega = \int_{\Delta\Omega^l} (-\beta I + \kappa I_b) d\Omega + \frac{\sigma}{4\pi} \int_{\Delta\Omega^l} \int_{4\pi} (I'\Phi) d\Omega' d\Omega \quad (24)$$

For a one-dimensional problem, the left side of Eq. (24) can be written as

$$\int_{\Delta\Omega^l} \frac{dI}{dz} (\hat{n}_z \bullet \hat{s}) d\Omega = \int_{\Delta\Omega^l} (-\beta I + \kappa I_b) d\Omega + \frac{\sigma}{4\pi} \int_{\Delta\Omega^l} \int_{4\pi} (I'\Phi) d\Omega' d\Omega \quad (25)$$

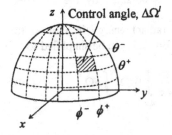

Figure 3 Control (solid) angle for the FV method.

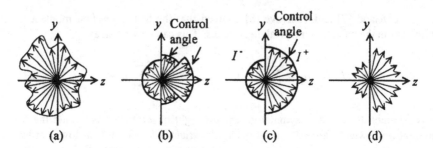

Figure 4 Radiant intensity distribution: (a) actual intensity, (b) FV method, (c) two-control-angle FV discretization, (d) discrete direction discretization.

In the control volume approach, the intensity is assumed constant within a control angle (Fig. 4b). Equation (25) can be simplified to

$$\frac{dI}{dz}D_{cz}^{l} = \left(-\beta I + \kappa I_{b}\right)\Delta\Omega^{l} + \frac{\sigma}{4\pi}\Delta\Omega^{l}\int_{4\pi}\left(I'\Phi\right)d\Omega' \qquad (26)$$

where

$$D_{cz}^{l} = \int_{\Delta\Omega^{l}}\left(\hat{n}_{z}\bullet\hat{s}\right)d\Omega \qquad\qquad \Delta\Omega^{l} = \int_{\Delta\Omega^{l}}d\Omega \qquad (27)$$

When two control angles (one in each coordinate direction), as shown in Fig. 4c, are used to discretize the angular space in an isotropically scattering medium, Eq. (26) becomes

$$\frac{1}{2}\frac{dI^{+}}{dz} = -\beta I^{+} + \kappa I_{b} + \frac{\sigma}{2}\left(I^{+} + I^{-}\right) \qquad (28a)$$

$$\frac{1}{2}\frac{dI^{-}}{dz} = -\beta I^{-} + \kappa I_{b} + \frac{\sigma}{2}\left(I^{+} + I^{-}\right) \qquad (28b)$$

These two equations are *identical* to the equations obtained by the two-flux method.

For an isotropically scattering medium, if radiant energy is allowed to travel along *discrete* directions as shown in Fig. 4d, Eq. (25) can be simplified to

$$\frac{dI^{l}}{dz}\mu^{l}\int_{\Delta\Omega^{l}}d\Omega = \left(-\beta I^{l} + \kappa I_{b}\right)\int_{\Delta\Omega^{l}}d\Omega + \frac{\sigma}{4\pi}\int_{4\pi}I'd\Omega'\int_{\Delta\Omega^{l}}d\Omega \qquad (29)$$

or

$$\mu^l \frac{dI^l}{dz} = -\beta I^l + \kappa I_b + \frac{\sigma}{4\pi} \sum_{l=1}^{L} I^{l'} w^{l'} \tag{30}$$

which is the RTE for the DO method (Eq. 17). When a quadrature set that satisfies the half-range first moment is used, Eq. (30) reduces *mathematically* to the two-flux equation.

4.4   Closure

From the above discussion, for one-dimensional problems, the FV method is a higher-order flux method. The discretization equation is formulated by integrating the RTE over a discrete solid angle. In both the two-flux and FV methods, the magnitude of the radiant intensity is assumed uniform over a control angle. Radiation is allowed to travel in *all* directions within a solid angle. When two control angles (one in each coordinate direction) are used, the FV method *always* reduces to the two-flux method.

In multi-dimensional problems, the philosophy of the FV method is slightly different from the philosophy of the four-flux or six-flux methods. The FV method *usually* does not divide the angular space along the four or six coordinate directions as in the flux method. However, the angular discretization practice employed by the four-flux or six-flux methods can of course be accommodated by the FV method.

In the DO method, radiant energy is allowed to travel along *discrete* directions [6]. There is no solid angle integration. For non-scattering media in a black environment, radiant intensity along any direction can be calculated directly without solid angle reference. These angular directions can, in principle, be chosen arbitrarily. However, several procedures have been developed to generate quadrature sets (ordinate directions and angular weights) that integrate the radiant energy and inscattering terms accurately. When a S_2 quadrature set which satisfied the half-range first moment is used in one-dimensional problems, the resulting RTE is *mathematically* similar to that of the two-flux and the FV (with two control angles) methods. However, these equations are obtained using very different principles. The RTE for the DO method can be obtained from the FV method if the radiant energy is restricted to travel along *discrete* directions.

5   DOMAIN DISCRETIZATION

The numerical method described here is based on the control volume approach. Discretization equations are formulated by integrating the RTE over control volumes and control (solid) angles. Control volumes and control angles are subdivisions of the spatial and angular spaces respectively. The next subsections describe these subdivisions.

5.1   Control Volumes and Grid Points

For ease of discussion, a two-dimensional spatial domain is used in this section. Figure 5a shows a structured grid for a rectangular domain. Figure 5b shows an unstructured grid for the same geometry. The term unstructured is used here to refer to computational grids without any structured order in the x- and y-directions. It is understood that with proper indexing, an unstructured grid procedure can be used on the grid shown in Fig. 5a.

The control volume boundaries (dashed lines) are drawn first. Grid points are then placed at the geometric centers of the control volumes. Control volume faces should be designed to capture "discontinuities" (see shaded regions of Figs. 5a and 5b) in physical properties, boundary conditions and sources.

5.2   Control Angles

Similar to the spatial discretization, the angular space is discretized by placing control (solid) angle boundaries throughout the 4π solid angle. Although unstructured control angles can be used with the FV method, only structured control angles are discussed in this chapter.

Figure 6a shows a possible angular discretization. The simplest angular discretization is to divide the angular space into $N_\theta \times N_\phi$ control angles with equally spaced $\Delta\theta$ and $\Delta\phi$. The size of these control angles can be adjusted to capture the physics of the problem at hand. For example, collimated incidence can be captured by designing a control angle with small $\Delta\theta$ and $\Delta\phi$. Figure 6b shows how collimated incidence can be captured using the present method.

6   DERIVATION OF THE DISCRETIZATION EQUATION

The discretization equation is the counterpart of the general differential equation (Eq. 1). It is obtained by integrating the RTE over a typical control volume (Fig. 7a) and control angle (Fig. 7b). Before proceeding with the formulation of the discretization equation, it is important to examine the various possibilities in the definition of the angular direction $\hat{s}$.

6.1   Angular Direction

A radiation direction is defined using a set of base vectors. There are at least three common alternatives. These are (1) Cartesian base vectors, (2) cylindrical base vectors, and (3) spherical base vectors. When Cartesian spatial grids are used, the choice of base vectors is obvious. However, when non-Cartesian spatial grids are encountered, cylindrical base vectors are the natural choice for cylindrical spatial grids. Similarly, spherical base vectors are the choice for spherical spatial grids. It should be pointed out that the orientation of the Cartesian base vectors does not change with spatial location; thus, a set of fixed values for (θ, ϕ) defines the *same* direction at any spatial location. The orientation of the cylindrical and spherical base vectors changes with the spatial location. Since the RTE describes the change in radiant intensity along a *straight-line* path, an additional term called the angular redistribution term appears when non-Cartesian base vectors are used. Moder et al. [9] presented a detailed discussion of the angular redistribution term.

Since the purpose of this chapter is to present a procedure which is applicable to *all* spatial grid systems (including non-orthogonal and unstructured grid systems), the radiation direction is defined using the *Cartesian* base vectors. Therefore, the angular direction (Fig. 8) can be described by the unit vector

$$\hat{s} = (\sin \theta \cos \phi)\hat{e}_x + (\sin \theta \sin \phi)\hat{e}_y + (\cos \theta)\hat{e}_z \tag{31}$$

Although this choice eliminates the angular redistribution term in the RTE, control angle overlap (also called control angle overhang [10]) can appear when non-Cartesian grids are used. This will be described in Section 8.

6.2   Mathematical Formulation

The RTE (Eq. 1) can be written as

$$\frac{dI(\vec{r}, \hat{s})}{ds} = -\beta(\vec{r})I(\vec{r}, \hat{s}) + S(\vec{r}, \hat{s}) \tag{32}$$

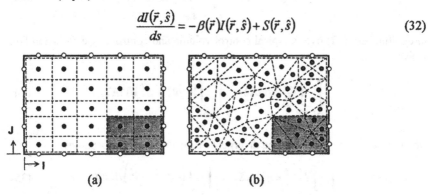

Figure 5 Spatial grids: (a) structured grid, (b) unstructured grid.

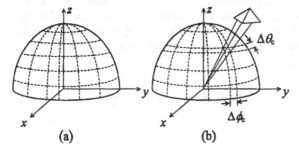

Figure 6 Angular grids: (a) typical, (b) collimated beam.

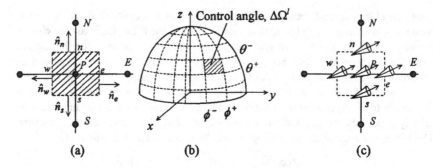

Figure 7 Typical (a) control volume, (b) control angle, (c) control angle orientation.

where the extinction coefficient and the source function are

$$\beta(\vec{r}) = \kappa(\vec{r}) + \sigma(\vec{r}) \tag{33a}$$

$$S(\vec{r}, \hat{s}) = \kappa(\vec{r}) I_b(\vec{r}, \hat{s}) + \frac{\sigma(\vec{r})}{4\pi} \int_{4\pi} I(\vec{r}, \hat{s}') \Phi(\hat{s}', \hat{s}) d\Omega' \tag{33b}$$

Integrating Eq. (32) over a typical control volume and control angle shown in Fig. 7 gives

$$\int_{\Delta\Omega^l} \int_{\Delta V} \frac{dI^l}{ds} dV d\Omega = \int_{\Delta\Omega^l} \int_{\Delta V} \left(-\beta I^l + S^l\right) dV d\Omega \tag{34}$$

where $I^l \equiv I(\vec{r}, \hat{s})$. Applying the divergence theorem, Eq. (34) becomes

$$\int_{\Delta\Omega^l} \int_{\Delta A} I^l \left(\hat{s}^l \cdot \hat{n}\right) dA d\Omega = \int_{\Delta\Omega^l} \int_{\Delta V} \left(-\beta I^l + S^l\right) dV d\Omega \tag{35}$$

In Eq. (35), $\hat{n}$ is the unit *outward* normal vector shown in Fig. 7a. The left side of Eq. (35) denotes the "inflow" and "outflow" of radiant energy across the control volume faces. The right side represents the attenuation and augmentation of energy within a control volume and control angle. In the FV method, the magnitude of the intensity is assumed constant over the control angle and control volume. Equation (35) can then be simplified to

$$\sum_{i=nb} I_i^l A_i \int_{\Delta\Omega^l} \left(\hat{s}^l \cdot \hat{n}_i\right) d\Omega^l = \left(-\beta I^l + S^l\right) \Delta V \Delta\Omega^l \tag{36}$$

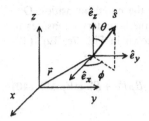

Figure 8 A typical angular direction.

where

$$S^l = \kappa I_b + \frac{\sigma}{4\pi} \sum_{l'=1}^{L} I^{l'} \overline{\Phi}^{l'l} \Delta \Omega^{l'} \qquad (37)$$

In Eq. (37), $\overline{\Phi}^{l'l}$ is the *average* scattering phase function from control angle l' to control angle l to be discussed in the next subsection. For the control volume and control angle orientation shown in Fig. 7c, Eq. (36) can be written as

$$A_e D_{ce}^l I_e^l + A_w D_{cw}^l I_w^l + A_n D_{cn}^l I_n^l + A_s D_{cs}^l I_s^l = \left(-\beta_P I_P^l + S_P^l\right) \Delta V_P \Delta \Omega^l \qquad (38)$$

where

$$D_{ce}^l = \int_{\Delta\Omega^l} \left(\hat{s}^l \bullet \hat{e}_x\right) d\Omega \qquad (39a)$$

$$D_{cn}^l = \int_{\Delta\Omega^l} \left(\hat{s}^l \bullet \hat{e}_y\right) d\Omega \qquad (39b)$$

$$D_{cw}^l = -D_{ce}^l, \qquad D_{cs}^l = -D_{cn}^l \qquad (39c)$$

$$\Delta\Omega^l = \int_{\Delta\Omega^l} d\Omega \qquad (39d)$$

$$S_P^l = \kappa_P I_{b,P} + \frac{\sigma_P}{4\pi} \sum_{l'=1}^{L} I_P^{l'} \overline{\Phi}^{l'l} \Delta\Omega^{l'} \qquad (39e)$$

$$A_e = A_w = \Delta y, \qquad A_n = A_s = \Delta x \qquad (39f)$$

$$\Delta V_P = \Delta x \Delta y \qquad (39g)$$

As seen in the above equation, the areas A_e and A_w are equal for Cartesian coordinates. Separate symbols are used for generality. A spatial differencing scheme is needed to relate the boundary intensities to the nodal intensities. One such scheme is the step or "upwind" scheme that sets the boundary intensities to the "upstream" nodal intensities. For the situation shown in Fig. 7c, Eq. (38) becomes

$$A_e D_{ce}^l I_P^l + A_w D_{cw}^l I_W^l + A_n D_{cn}^l I_P^l + A_s D_{cs}^l I_S^l = \left(-\beta_P I_P^l + S_P^l\right) \Delta V_P \Delta \Omega^l \quad (40)$$

Equation (40) can be written as

$$a_P^l I_P^l = a_W^l I_W^l + a_S^l I_S^l + b^l \quad\quad (41)$$

where

$$a_W^l = A_w \left| D_{cw}^l \right|, \quad\quad a_S^l = A_s \left| D_{cs}^l \right| \quad\quad (42a)$$

$$a_P^l = A_e D_{ce}^l + A_n D_{cn}^l + \beta_P \Delta V_P \Delta \Omega^l \quad\quad (42b)$$

$$b^l = S_P^l \Delta V_P \Delta \Omega^l \quad\quad (42c)$$

A more compact form of the discretization equation suitable for control angles pointing in all directions can be written as

$$a_P^l I_P^l = a_W^l I_W^l + a_E^l I_E^l + a_S^l I_S^l + a_N^l I_N^l + b^l \quad\quad (43)$$

where

$$a_E^l = \max\left(-A_e D_{ce}^l, 0\right), \quad\quad a_W^l = \max\left(-A_w D_{cw}^l, 0\right) \quad\quad (44a)$$

$$a_N^l = \max\left(-A_n D_{cn}^l, 0\right), \quad\quad a_S^l = \max\left(-A_s D_{cs}^l, 0\right) \quad\quad (44b)$$

$$a_P^l = \max\left(A_e D_{ce}^l, 0\right) + \max\left(A_w D_{cw}^l, 0\right) +$$

$$\max\left(A_n D_{cn}^l, 0\right) + \max\left(A_s D_{cs}^l, 0\right) + \beta_P \Delta V_P \Delta \Omega^l \quad\quad (44c)$$

$$b^l = S_P^l \Delta V_P \Delta \Omega^l \quad\quad (44d)$$

Other quantities are already defined in Eq. (39) and are not repeated here.

6.3   Scattering Phase Function

As mentioned in Section 2.3, the scattering phase function Φ must satisfy

$$\int_{4\pi} \Phi(\hat{s}', \hat{s}) d\Omega' = 4\pi \quad\quad (45)$$

When a phase function is known analytically, Eq. (45) can be evaluated analytically and satisfied exactly. The average phase function can then be calculated using

$$\overline{\Phi}^{l'l} = \frac{\int_{\Delta\Omega^{l'}} \Phi(\hat{s}', \hat{s}) d\Omega'}{\Delta\Omega^{l'}} \tag{46}$$

For complicated scattering phase functions, analytic evaluation of Eq. (45) can be computationally intensive or impossible. Although it is possible that the exact evaluation of $\int \Phi(\hat{s}', \hat{s}) d\Omega'$ might not be possible for certain scattering phase functions, the value of $\Phi^{l'l}$, which is the scattering from a discrete radiant direction $\hat{s}'$ into $\hat{s}$, must be known before a solution to the problem can be obtained. As a result, the approach presented here assumes that $\Phi^{l'l}$ is known.

When $\Phi^{l'l}$ is known, it is possible to approximate Eq. (45) using

$$\int_{4\pi} \Phi(\hat{s}', \hat{s}) d\Omega' \approx \sum_{l'=1}^{L} \Phi^{l'l} \Delta\Omega^{l'} \tag{47}$$

However, since $\Phi^{l'l}$ is the scattering from a discrete radiant direction $\hat{s}'$ into $\hat{s}$, the approximation will not satisfy Eq. (45) unless scattering is isotropic. As a result, phase function renormalization similar to the approach used in the DO method (Eq. 22) is required. An improved approach [1], which ensures the satisfaction of Eq. (45), is described next.

In this approach, the control angles $\Delta\Omega^{l'}$ and $\Delta\Omega^{l}$ are subdivided into smaller sub-control angles as shown in Fig. 9. The total energy scattered from $\Delta\Omega^{l'}$ into $\Delta\Omega^{l}$ is

$$\int_{\Delta\Omega^{l}} \int_{\Delta\Omega^{l'}} \Phi(\hat{s}', \hat{s}) d\Omega' d\Omega = \sum_{l_s=1}^{L_s} \sum_{l'_s=1}^{L'_s} \Phi^{l'_s l_s} \Delta\Omega^{l'_s} \Delta\Omega^{l_s} \tag{48}$$

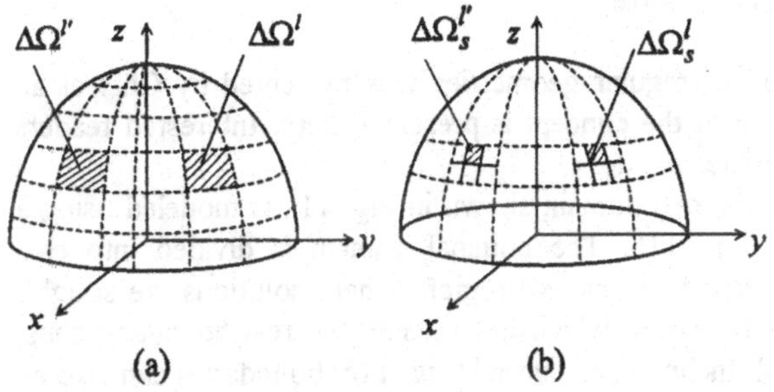

Figure 9 (a) Typical control angles, (b) possible sub-control angles.

where L'_s and L_s are the numbers of sub-control angles in $\Delta\Omega^{l'}$ and $\Delta\Omega^l$ respectively. For the example shown in Fig. 9b, $L'_s = L_s = 6$. The scattering phase functions $\Phi^{l'_s l_s}$ are evaluated along discrete radiant directions l'_s and l_s. The average scattering phase function is then calculated using

$$\overline{\Phi}^{l'l} = \frac{\displaystyle\int_{\Delta\Omega^l}\int_{\Delta\Omega^{l'}} \Phi(\hat{s}',\hat{s})d\Omega'd\Omega}{\Delta\Omega^l\Delta\Omega^{l'}} = \frac{\displaystyle\sum_{l_s=1}^{L_s}\sum_{l'_s=1}^{L'_s} \Phi^{l'_s l_s}\Delta\Omega^{l'_s}\Delta\Omega^{l_s}}{\Delta\Omega^l\Delta\Omega^{l'}} \tag{49}$$

Using this approach, Eq. (45) is satisfied accurately.

6.4   Solution Procedure

The FV discretization results in a set of algebraic equations with the radiant intensities as the unknowns. An iterative method is used to solve the resulting set of equations. Within each iteration, a marching order can be employed to efficiently solve the set of equations. For Cartesian grid problems, an efficient marching order for a control angle pointing in the first quadrant is shown in Fig. 10. Extensions to the other three quadrants are straightforward and are left to the exploration of interested readers.

7   TREATMENT OF IRREGULAR GEOMETRIES

The procedure outlined above can be used to solve radiative transfer problems that can be described using Cartesian grids. A procedure for three-dimensional cylindrical geometries was reported by Moder et al. [9]. Irregular geometries can be handled using the blocked-off region procedure, spatial-multiblock procedure, body-fitted grids, and unstructured grids. These treatments are discussed in this section.

7.1   Blocked-Off Region Procedure

A simple procedure to model irregular geometries was presented by Chai et al. [11]. Only a brief description of the concept is presented here. Interested readers should refer to the above article.

With this approach, the real domain shown in Fig. 11a is modeled using a nominal domain shown in Fig. 11b. The nominal domain is divided into two regions. These are (1) the *active* (unshaded) region, where solutions are sought, and (2) the *inactive* (shaded) region, which lies outside the real boundary; thus, solutions are not meaningful. Inclined and curved (Fig. 11c) boundaries can also be modeled using the proposed procedure.

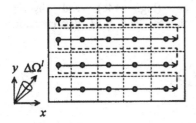

Figure 10 A possible marching order.

7.2 Body-Fitted Grid

Figure 12a shows a body-fitted grid for an irregular geometry. Chui and Raithby [12] and Chai et al. [13] reported radiation heat transfer solutions using body-fitted grids. Obstructions (Fig. 12b) can be modeled using the blocked-off region procedure described above. Although many irregular geometries can be handled using these procedures, body-fitted grids are difficult and sometimes impossible to generate for some engineering problems. The spatial-multiblock procedure and the unstructured grid procedure can be used to eliminate part of this problem.

7.3 Spatial-Multiblock

The spatial-multiblock procedure divides the solution domain into a finite number of spatial blocks where spatial grids can be generated easily. Figures 13a and 13b show a situation where spatial-multiblock is beneficial. When body-fitted grids are used (Fig. 13a), significant changes in the aspect ratio of the grids are encountered in regions with large area variation. Figure 13b shows a spatial-multiblock grid. Fine grids can be embedded into selected region(s) as shown. Another possible benefit of spatial-multiblock lies in the fact that simpler spatial grids can be used to model complex geometries. The T-shaped enclosure shown in Fig. 13b is modeled using *Cartesian* grids. Figure 13c shows a situation where fine spatial grids are needed at one corner of the solution domain. When a single-block procedure is used, unnecessarily fine grids are also used in part of the remaining domain. A spatial-multiblock procedure can be used to eliminate this problem. Figure 13d shows a sample spatial-multiblock procedure for this problem. Fine grids are employed at the appropriate region of the domain. A method that ensures the conservation of radiant energy between blocks was presented by Chai and Moder [14].

7.4 Unstructured Grid

Most spatial-multiblock procedures are an extension of structured grid procedures. There are many engineering problems where spatial-multiblock grids are difficult to generate. Unstructured grid procedures eliminate this deficiency. Figure 14a shows a structured body-fitted grid. The same spatial domain can be divided into a finite number of unstructured triangular (Fig. 14b) and hybrid quadrilateral/triangular (Fig. 14c) control volumes. The final discretization equation for the FV method can be written as

$$a_P^l I_P^l = \sum_{nb} a_{nb}^l I_{nb}^l + b^l \tag{50}$$

where

$$D_{c,nb}^l = \int_{\Delta\Omega^l} \left(\hat{s}^l \bullet \hat{n}_{nb}\right) d\Omega \tag{51a}$$

$$a_{nb}^l = \max\left(-A_{nb} D_{c,nb}^l, 0\right) \tag{51b}$$

$$a_P^l = \sum_{nb} \max\left(A_{nb} D_{c,nb}^l, 0\right) + \beta_P \Delta V_P \Delta\Omega^l \tag{51c}$$

$$b^l = S_P^l \Delta V_P \Delta\Omega^l \tag{51d}$$

The subscript nb represents the number of neighbors associated with the node point P. The total number of neighbors depends on both the dimensionality of the problem and the type of spatial computational grid. For the quadrilateral spatial grids shown in Fig. 14a, there are a total of four (4) neighbors for every node point. For the triangular grids shown in Fig. 14b, there are three (3) neighbors per node point. For the hybrid quadrilateral and triangular grids shown in Fig. 14c, there are either three (3) or four (4) neighbors for every nodal point. Finite-volume methods for unstructured grids were presented by Murthy and Mathur [10], Vaidya [15], and Liu et al. [16].

As a result of the choice made of the definition of the angular direction, control-angle overlap is encountered when the physical boundaries of the problem or the boundaries of the control volumes are not aligned with the Cartesian coordinates. Control-angle overlap is discussed in the next section.

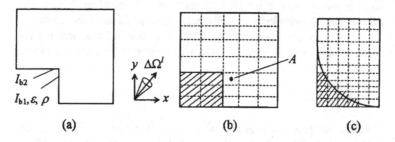

(a)   (b)   (c)

Figure 11 Irregular geometries: (a) real boundary condition, (b) simulated vertical and horizontal walls, and (c) simulated curved wall.

8   CONTROL-ANGLE OVERLAP

As discussed in Section 6.1, the angular direction is defined using the Cartesian base vectors (Eq. 31). This eliminates the undesirable angular redistribution term. However, control-angle overlap is encountered in most problems when the physical boundaries of the problem or the boundaries of the control volumes are *not* aligned with the Cartesian coordinates. Figure 15a shows possible control-angle overlaps at an internal control volume boundary and a physical wall boundary. Figure 15b depicts an enlarged view of a control-angle overlap at a wall. Conceptually, control-angle overlap is due to the presence of "complex geometries" in the *angular* space. Therefore, it can be handled using an extension of the procedure described in Section 7.1.

Chui and Raithby [12] reported an approach to handle control-angle overlap for two-dimensional geometries. The procedure subdivides a control angle into sub-control angles and associates radiant energy with the appropriate sub-control angles. In Fig. 15b, the control angle $\Delta\Omega^l$ is divided into two sub-control angles. The incoming energy is associated with $\Delta\Omega^{l+}$ and the outgoing energy is accounted for in $\Delta\Omega^{l-}$. For a two-dimensional problem, control-angle overlap can be handled by subdividing a control angle in the ϕ direction only as shown in Fig. 15b. The coefficients of the discretization equations are evaluated analytically, and appropriate adjustments are made to the discretization equations of control angle l and the control angle adjacent to the wall. For three-dimensional problems, a wall can bisect a control angle along an incline as shown in Fig. 15c. Therefore, it is not sufficient to subdivide the control angle in the ϕ direction only. Murthy and Mathur [10] extended the approach of Chui and Raithby [12] by subdividing the control angle into smaller sub-control angles. Figure 15d shows a possible discretization. For the situation shown, radiant energy leaves and enters the wall in eight (unshaded) and four (shaded) sub-control angles respectively. It should be pointed out that the approach taken here is conceptually similar to the one used in the blocked-off region procedure discussed in Section 7.1. This concept of subdividing a control angle was used to accurately calculate the scattering phase function as described in Section 6.3. Vaidya [15] presented another approach to handle control-angle overlap.

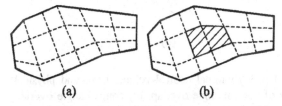

(a)   (b)

Figure 12 Body-fitted grids: (a) sample grid, (b) blocked-off region grid.

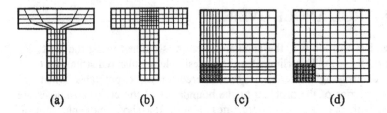

Figure 13 Spatial grids: (a) single-block, (b) multiblock, (c) single-block, (d) multiblock.

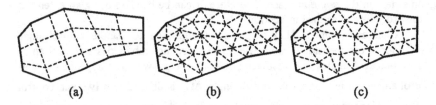

Figure 14 Possible control volumes: (a) quadrilateral, (b) triangular and (c) hybrid quadrilateral and triangular control volumes.

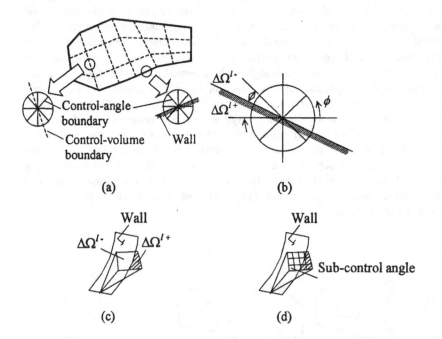

Figure 15 Control-angle overlap: (a) internal control-volume faces and physical boundaries, (b) enlarged view of control-angle overlap, (c) control-angle overlap for three-dimensional problems, (d) possible treatment of control-angle overlap.

The simplest approach is to do nothing. Moder et al. [9] and Chai et al. [13] obtained reasonable results for the problems tested using the "do-nothing" approach. Murthy and Mathur [10] also agreed that for most practical problems, the "do-nothing" approach produces reasonable results. Raithby [17] presented an excellent study of control-angle overlap errors.

9   SPATIAL DIFFERENCING SCHEMES

The step scheme is used in the discretization equations presented in Sections 6.2 and 7.4. Chai et al. [18] presented a study on spatial differencing schemes for radiation heat transfer. The study indicates that control volume boundary intensity should be calculated by tracing a beam to an appropriate "upstream" location where the intensity is known or can be approximated. A few "more accurate" spatial differencing schemes are discussed in this section. Most "more accurate" spatial differencing schemes are formulated as the step scheme with additional source or sink terms. The additional source or sink terms can be negative. This can lead to negative intensities.

Physically, radiant intensity is an always-positive variable. Therefore, negative intensity is physically unrealistic. The discretization equation should not allow the possibility of negative intensity as a solution. The always-positive variable treatment of Patankar [19] can be used to eliminate this possibility. The skewed "upwind" differencing scheme (SUDS) of Raithby [20] has been applied to radiation heat transfer problems and is discussed in the next section.

9.1   Skewed "Upwind" Differencing Schemes

Raithby and Chui [12] and Chai et al. [1] employed skewed "upwind" schemes to approximate the boundary intensities. This is an extension of the SUDS of Raithby [20]. In the SUDS, the value of the boundary intensity shown in Fig. 16 is calculated by tracing the intensity to an "upstream" location where the "upstream" intensity can be approximated. An appropriate profile, usually obtained from the exact solution of a simplified RTE, is assumed between the "upstream" point and the control volume boundary. The boundary intensity can then be obtained using the assumed profile.

Chai et al. [1] used the modified-exponential scheme [21] to evaluate the boundary intensities. For the situation shown in Fig. 16, the east boundary intensity is

$$I_e^l = I_P^l e^{-\left(\beta_m^l\right)_P d_e^l} + \left(\frac{S_m^l}{\beta_m^l}\right)_P \left(1 - e^{-\left(\beta_m^l\right)_P d_e^l}\right) \tag{52}$$

The distance d_e^l is shown in Fig. 16. The modified extinction coefficient β_m^l and the modified source function S_m^l are

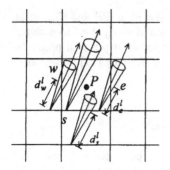

Figure 16 A possible control angle.

$$\beta_m^l = \beta - \frac{\sigma}{4\pi}\overline{\Phi}^{ll}\Delta\Omega^l \tag{53a}$$

$$S_m^l = \kappa I_b + \frac{\sigma}{4\pi}\sum_{l'=1,l'\neq l}^{L}I^{l'}\overline{\Phi}^{l'l}\Delta\Omega^{l'} \tag{53b}$$

The final discretization equation can be written as

$$a_P^l I_P^l = a_W^l I_W^l + a_E^l I_E^l + a_S^l I_S^l + a_N^l I_N^l + b^l \tag{54}$$

where

$$a_E^l = \max\left(-A_e D_{ce}^l e^{-\left(\beta_m^l\right)_E d_e^l},0\right) \tag{55a}$$

$$a_W^l = \max\left(-A_w D_{cw}^l e^{-\left(\beta_m^l\right)_W d_w^l},0\right) \tag{55b}$$

$$a_N^l = \max\left(-A_n D_{cn}^l e^{-\left(\beta_m^l\right)_N d_n^l},0\right) \tag{55c}$$

$$a_S^l = \max\left(-A_s D_{cs}^l e^{-\left(\beta_m^l\right)_S d_s^l},0\right) \tag{55d}$$

$$a_P^l = \max\left(A_e D_{ce}^l e^{-\left(\beta_m^l\right)_P d_e^l},0\right) + \max\left(A_w D_{cw}^l e^{-\left(\beta_m^l\right)_P d_w^l},0\right) +$$

$$\max\left(A_n D_{cn}^l e^{-\left(\beta_m^l\right)_P d_n^l},0\right) + \max\left(A_s D_{cs}^l e^{-\left(\beta_m^l\right)_P d_s^l},0\right) +$$

$$\left(\beta_P + \frac{S_2}{I_P^{l*}} \right) \Delta V_P \Delta \Omega^l \tag{55e}$$

$$b^l = \left(S_P^l + S_1 \right) \Delta V_P \Delta \Omega^l \tag{55f}$$

$$S_1 = \frac{1}{\Delta V_P \Delta \Omega^l} \sum_{nb} \max\left(-A_i D_{ci}^l, 0 \right) \left(\frac{S_m^l}{\beta_m^l} \right)_I \left(1 - e^{-\left(\beta_m^l \right)_I d_i^l} \right) \quad \begin{matrix} i = e, w, n, s \\ I = E, W, N, S \end{matrix} \tag{55g}$$

$$S_2 = \frac{1}{\Delta V_P \Delta \Omega^l} \left(\frac{S_m^l}{\beta_m^l} \right)_P \sum_{nb} \max\left(A_i D_{ci}^l, 0 \right) \left(1 - e^{-\left(\beta_m^l \right)_P d_i^l} \right) \quad i = e, w, n, s \tag{55h}$$

The always-positive variable treatment of Patankar [19] is used in Eq. (54). Other possible spatial differencing schemes are described next.

9.2 Other Spatial Differencing Schemes

Due to the similarity between the convection-diffusion equation and the RTE presented in Section 3, some of the differencing schemes developed for computational fluid dynamics (CFD) can be adapted to model the RTE.

The step scheme, although bounded, produces false scattering,[1] which decreases the accuracy of the solution. The diamond scheme can produce physically unrealistic overshoots and undershoots [18], which do not disappear with spatial grid refinement in multidimensional problems. A few bounded schemes used in CFD are the CLAM [22], MINMOD [23], MUSCL [24], and SMART [25] schemes. Darwish [26] and Darwish and Moukalled [27] presented systematic evaluations of these schemes. Recently, Jessee and Fiveland [28] applied some of the schemes tested by Darwish [26] to the DO method. Uniformly spaced Cartesian grids were used in the study.

There is still a need for a simple yet accurate spatial differencing scheme for the modeling of RTE.

10 RAY CONCENTRATION ERROR, RAY EFFECT AND FALSE SCATTERING

The FV method presented in this chapter, when used properly, is a flexible, efficient, economical and accurate procedure for radiation heat transfer. However, similar to other numerical methods, the FV method for radiation heat transfer is not without shortcomings. If the FV method is to be used as the procedure of choice, it is important that these shortcomings are well understood. A brief discussion of three types of error, namely, ray concentration error [17], ray effect

---

[1] False scattering is discussed in Section 10.

and false scattering [29], are discussed in this section. Readers interested in in-depth discussion of these errors are referred to the above references.

10.1 Ray Effect and Ray Concentration Errors

Ray effect arises from approximating a continuously varying angular nature of radiation with a discrete set of angular directions and is independent of spatial differencing practices. Once such an *angular* approximation is made, ray effect is encountered, even if a perfect spatial discretization practice is available or the exact solutions for the discrete directions are used. Raithby [17] presented an excellent study on the *ray concentration error*.

10.2 False Scattering

Figure 17 shows a simple problem where false scattering can be understood without unnecessary complications. It consists of a black enclosure with three cold walls and a hot left wall. For the purpose of this discussion, consider radiant energy travelling at 45° from the horizontal wall as shown in Fig. 17. For this simple enclosure, representative intensity distributions for nonscattering and scattering media are shown in Figs. 17a and 17b respectively.

It can be shown that selected spatial differencing schemes produce an intensity profile similar to that depicted in Fig. 17b even when the medium does not scatter energy. Therefore, the result gives an impression that the medium *scatters* energy. Since this effect is nonphysical and is due to the chosen spatial differencing scheme, it is called *numerical* or *false scattering*. Chai et al. [29] presented a detailed study on ray effect, false scattering and their interactions.

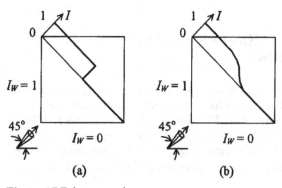

Figure 17 False scattering.

11 ADVANCED TOPICS

Two advanced topics, which will improve the capabilities of the FV method, are discussed briefly in this section.

11.1 Angular-Multiblock Procedure

For many optically thick problems, reasonably accurate solutions can be obtained with relatively coarse angular grids. For optically thin problems, however, fine angular grids are needed to reduce "ray effects." In many engineering problems, rapid changes in optical properties are encountered. These include, but are not limited to, radiation heat transfer in furnaces and in material processing. Presently, the same fine angular grids are also employed in the optically thick region. This represents a waste of computational efforts. An angular-multiblock procedure can be used to eliminate this problem. Figure 18a shows an interface between two angular blocks where radiant energy from two control angles is being transferred into three control angles. Figure 18b shows an enlarged view of the angular multiblock grids. Chai and Moder [30] presented a procedure which ensures the conservation of radiant energy between the two angular grids.

11.2 Procedure for Materials with Different Indices of Refraction

Radiation heat transfer in materials with different indices of refraction is encountered in many engineering applications. These include, but are not limited to, radiation heat transfer (1) in thermal barrier coatings, (2) in the processing of fiber optics, (3) in crystal growth applications, and (4) in rapid thermal processing.

The refractive indices can have a significant effect on the internal temperature distribution. Within the materials, internal emission of radiant energy depends on n^2, where n is the refractive index. Since the radiant energy leaving a surface cannot exceed that of its surroundings, there is significant *internal* reflection (Fig. 19a) at the interface when the refractive indices of the material and its surroundings differ significantly. There is also refraction of radiant energy (Fig. 19b) as radiation travels from one medium to another. Note the change in both the radiation direction and the size of the solid angle.

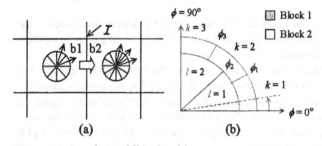

(a)   (b)

Figure 18 Angular multiblock grids.

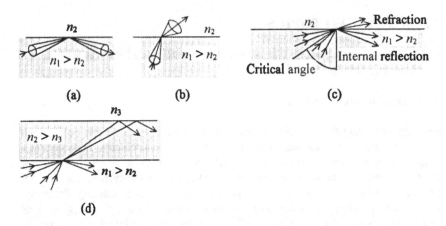

Figure 19 Effect of refractive indices: (a) total internal reflection, (b) refraction, (c) reduction in radiation beams, (d) total loss of radiation beams.

Due to the presence of internal reflection and changes in the size of the solid angles, there is a reduction in the number of radiation solid angles as radiation travels from one material to another. This is illustrated in Fig. 19c. Solid angles are not shown to avoid cluttering. For the situation depicted in Fig. 19d, there is no radiative heat transfer from material one to material three. This is clearly physically unrealistic. This is the result of approximating the continuous angular space with a finite number of solid angles. In addition to the above concerns, the reflectivity of a material can be a strong function of the incident angle. For example, when a Fresnel boundary is encountered, the reflectivity increases drastically near the critical angle.

As a result, a procedure must be able to account for (1) the rapid changes in the reflectivity near the critical angle, (2) the changes in the size of the solid angles, and (3) the loss of solid angles. Chai and Moder [31] presented a procedure to account for the above-mentioned attributes.

12   CONCLUDING REMARKS

The number of publications using the FV methods for radiation heat transfer has increased over the last few years. This section attempts to highlight some of these publications for completeness.

Raithby [32] presented an excellent study on the DO and FV methods. The similarities and differences between the two methods are discussed systematically.

Baek and co-workers presented radiative heat transfer solutions using the FV method for cylindrical [33] and three-dimensional irregular [34] geometries. Radiant heat transfer of a rocket plume [35] and combined mode heat transfer [36] were analyzed using the FV method.

Liu and co-workers published FV solutions for radiative transfer using body-fitted [37] and unstructured [16] grids. Comparisons between the FV and DO methods [37] were performed, and it was concluded that the FV method is as accurate and efficient as the DO method. With the added flexibilities over the DO method, the FV method was recommended.

Murthy and co-workers presented an unstructured grid FV method for radiation heat transfer [10]. They also extended their procedure to account for periodic boundaries [38]. Vaidya [15] also presented an unstructured grid FV method for radiation heat transfer.

Coelho et al. [39] extended the blocked-off region procedure [11] to model thin baffles. Solutions for nongray media using the FV method were reported by Parthasarathy et al. [40], Lee et al. [41], Kumar et al. [42, 43] and Vaidya [44]. Sokolov et al. [45] used the FV method to model radiative transfer of a thermal energy storage canister. Now that the FV method has been incorporated into commercial codes [10, 15, 16], it is likely that the FV method will be used to model a wide range of industrial problems in the future.

13   NOMENCLATURE

a	coefficient of the discretization equation
A	area of control volume faces
b	source term in the discretization equation
D_c^l	direction cosine integrated over $\Delta\Omega^l$
$\hat{e}_x, \hat{e}_y, \hat{e}_z$	unit vectors in x, y, and z directions
I	intensity
L	number of control angles
$\hat{n}$	unit outward normal vector
s	distance traveled by a beam
S	source function
u, v	velocity components in x, y directions
x, y, z	coordinate directions
β	extinction coefficient
ΔV	volume of control volume
$\Delta\Omega^l$	control angle
ε	wall emissivity
κ	absorption coefficient
μ, ξ	direction cosines in x, y directions
α, γ	mass flow rate per unit area (Eq. 10)
ρ	density or wall reflectivity
θ	polar angle measured from $\hat{e}_z$
σ	Stefan-Boltzmann constant or scattering coefficient
ϕ	azimuthal angle measured from $\hat{e}_x$ or general dependent variable

w	angular weight
Γ	generalized diffusion coefficient
Φ	scattering phase function

Subscripts

b	blackbody
E, W, N, S	east, west, north, and south neighbors of P
e, w, n, s	east, west, north, and south control-volume faces
P	control volume P
x, y, z	coordinate directions

Superscripts

| l, l' | angular directions |

14 REFERENCES

1. J.C. Chai, H.S. Lee, and S.V. Patankar, Finite Volume Method for Radiation Heat Transfer, *Journal of Thermophysics and Heat Transfer*, Vol. 8, No. 3, pp. 419-425, 1994.
2. A. Schuster, Radiation Through a Foggy Atmosphere, *Astrophysical Journal*, Vol. 21, pp. 1-22, 1905.
3. K. Schwarzschild, Uber das Gleichgewicht der Sonnenatmospharen, *Akademie der Wissenschaften in Goettingen, Nachrichten. Mathematisch-Physikalische Klasse*, Vol. 1, pp. 41-53, 1906.
4. E.E. Khalil and J.S. Truelove, Calculation of Radiative Heat Transfer in a Large Gas Fired Furnace, *Letters in Heat and Mass Transfer*, Vol. 4, pp. 353-365, 1977.
5. W.A. Fiveland, A Discrete Ordinates Method for Predicting Radiative Heat Transfer in Axisymmetric Enclosures, ASME Paper No. 82-HT-20, 1982.
6. S. Chandrasekhar, *Radiative Transfer*, Dover Publications, Inc., New York, 1960.
7. W.A. Fiveland, Discrete Ordinates Methods for Radiative Heat Transfer in Isotropically and Anisotropically Scattering Media, *Journal of Heat Transfer*, Vol. 109, pp. 809-812, 1987.
8. J.S. Truelove, Discrete-Ordinate Solutions of the Radiation Transport Equation, *Journal of Heat Transfer*, Vol. 109, pp. 1048-1051, 1987.
9. J.P. Moder, J.C. Chai, G. Parthasarathy, H.S. Lee, and S.V. Patankar, Nonaxisymmetric Radiative Transfer in Cylindrical Enclosures, *Numerical Heat Transfer*, Part B, Vol. 30, pp. 437-452, 1996.
10. J.Y. Murthy and S.R. Mathur, Finite Volume Method for Radiation Heat Transfer Using Unstructured Meshes, *Journal of Thermophysics and Heat Transfer*, Vol. 12, No. 3, pp. 313-321, 1998.

11.   J.C. Chai, H.S. Lee, and S.V. Patankar, Treatment of Irregular Geometries using a Cartesian Coordinates Finite-Volume Radiation Heat Transfer Procedure, *Numerical Heat Transfer*, Part B, Vol. 26, pp. 225-235, 1994.

12.   E.H. Chui and G.D. Raithby, Computation of Radiant Heat Transfer on a Non-Orthogonal Mesh Using the Finite-Volume Method, *Numerical Heat Transfer*, Part B, Vol. 23, pp. 269-288, 1993.

13.   J.C. Chai, G. Parthasarathy, H.S. Lee, and S.V. Patankar, Finite Volume Radiation Heat Transfer Procedure for Irregular Geometries, *Journal of Thermophysics and Heat Transfer*, Vol. 9, No. 3, pp. 410-415, 1995.

14.   J.C. Chai and J.P. Moder, Spatial-Multiblock Procedure for Radiation Heat Transfer, *Numerical Heat Transfer*, Part B, Vol. 31, pp. 277-293, 1997.

15.   N. Vaidya, Multi-Dimensional Simulation of Radiation Using an Unstructured Finite Volume Method, AIAA paper 98-0856, 1998.

16.   J. Liu, H.M. Shang, and Y.S. Chen, Development of an Unstructured Radiation Model Applicable for Two-Dimensional Planar, Axisymmetric, and Three-Dimensional Geometries, Presented at the 37th AIAA Aerospace Sciences Meeting & Exhibit, January 10 – 14, 1999, Reno, AIAA-99-0974.

17.   G.D. Raithby, Evaluation of Discretization Errors in Finite-Volume Radiant Heat Transfer Predictions, *Numerical Heat Transfer*, Part B, In press.

18.   J.C. Chai, H.S. Lee, and S.V. Patankar, Evaluation of Spatial Differencing Practices for the Discrete Ordinates Method, *Journal of Thermophysics and Heat Transfer*, Vol. 8, No. 1, pp. 140-144, 1994.

19.   S.V. Patankar, *Numerical Heat Transfer and Fluid Flow*, Hemisphere Publishing Corp., Washington, D.C., 1980.

20.   G.D. Raithby, Skew Upstream Differencing Schemes for Problems Involving Fluid Flow, *Computer Methods in Applied Mechanics and Engineering*, Vol. 9, pp. 153-164, 1976.

21.   J.C. Chai, H.S. Lee, and S.V. Patankar, Improved Treatment of Scattering Using the Discrete Ordinates Method, *Journal of Heat Transfer*, Vol. 116, No. 1, pp. 260-263, 1994.

22.   B. Van Leer, Towards the Ultimate Conservative Difference Scheme. II. Monotonicity and Conservation Combined in a Second Order Scheme, *Journal of Computational Physics*, Vol. 14, pp. 361-370, 1974.

23.   A. Harten, High Resolution Schemes for Hyperbolic Conservation Laws, *Journal of Computational Physics*, Vol. 49, pp. 357-393, 1983.

24.   B. Van Leer, Towards the Ultimate Conservative Difference Scheme. V. A Second-Order Sequel to Godunov's Method, *Journal of Computational Physics*, Vol. 23, pp. 101-136, 1977.

25.   P.H. Gaskell, and A.K.C. Lau, Curvature Compensated Convective Transport: SMART, a New Boundedness Preserving Transport Algorithm, *International Journal for Numerical Methods for Fluids*, Vol. 8, pp. 617-641, 1988.

26.   M.S. Darwish, A New High-Resolution Scheme Based on the Normalized Variable Formulation, *Numerical Heat Transfer*, Part B, Vol. 24, pp. 353-371, 1993.

27. M.S. Darwish and F.H. Moukalled, Normalized Variable and Space Formulation Methodology for High-Resolution Schemes, *Numerical Heat Transfer*, Part B, Vol. 26, pp. 79-96, 1994.

28. J.P. Jessee and W.A. Fiveland, Bounded, High-Resolution Differencing Schemes Applied to the Discrete Ordinates Method, *Journal of Thermophysics and Heat Transfer*, Vol. 11, No. 4, pp. 540-548, 1997.

29. J.C. Chai, H.S. Lee, and S.V. Patankar, Ray Effect and False Scattering in the Discrete Ordinates Method, *Numerical Heat Transfer*, Part B, Vol. 24, pp. 373-389, 1993.

30. J.C. Chai and J.P. Moder, Angular-Multiblock Procedure for Radiation Heat Transfer, Presented at the International Conference in Computational Heat and Mass Transfer, Gazimagusa, North Cyprus, 1999.

31. J.C. Chai and J.P. Moder, Radiation Heat Transfer Procedure for Materials with Different Indices of Refraction, In preparation.

32. G.D. Raithby, Discussion of the Finite-Volume Method for Radiation, and its Application using 3D Unstructured Meshes, *Numerical Heat Transfer*, Part B, Vol. 35, pp. 389-405, 1999.

33. M.Y. Kim and S.W. Baek, Analysis of Radiative Transfer in Cylindrical Enclosures using the Finite Volume Method, *Journal of Thermophysics and Heat Transfer*, Vol. 11, No. 2, pp. 246-252, 1997.

34. S.W. Baek, M.Y. Kim and J.S. Kim, Nonorthogonal Finite-Volume Solutions of Radiative Heat Transfer in a Three-Dimensional Enclosure, *Numerical Heat Transfer*, Part B, Vol. 34, pp. 419-437, 1998.

35. S.W. Baek, and M.Y. Kim, Analysis of Radiative Heating of a Rocket Plume Base with the Finite-Volume Method, *International Journal of Heat and Mass Transfer*, Vol. 40, No. 7, pp. 1501-1508, 1997.

36. M.Y. Kim, and S.W. Baek, Numerical Analysis of Conduction, Convection, and Radiation in a Gradually Expanding Channel, *Numerical Heat Transfer*, Part A, Vol. 29, pp. 725-740, 1996.

37. J. Liu, H.M. Shang, Y.S. Chen, and T.S. Wang, Prediction of Radiative Transfer in Genereal Body-Fitted Coordinates, *Numerical Heat Transfer*, Part b, Vol. 31, No. 4, pp. 423-439, 1997.

38. S.R. Mathur, and J.Y. Murthy, Radiative Heat Transfer in Periodic Geometries using a Finite Volume Scheme, *Journal of Heat Transfer*, Vol. 121, pp. 357-364, 1999.

39. P.J. Coelho, J.M. Goncalves, and M.G. Carvalho, Modelling of Radiative Heat Transfer in Enclosures with Obstacles, *International Journal of Heat and Mass Transfer*, Vol. 41, No. 4-5, pp. 745-756,1998.

40. G. Parthasarathy, J.C. Chai, and S.V. Patankar, A Simple Approach to Non-Gray Gas Modeling, *Numerical Heat Transfer*, Part B., Vol. 29, No. 1, pp. 113-124, 1995.

41. P.Y.C. Lee, K.G.T. Hollands, G.D. Raithby, Reordering the Absorption Coefficient within the Wide Band for Predicting Gaseous Radiant Exchange, *Journal of Heat Transfer*, Vol. 118, pp. 394-400, 1996.

42. G.N. Kumar, J.P. Moder, H.C. Mongia and C. Prakash, Development of a Three Dimensional Radiative Heat Transfer Computational Methodology for Aircraft Engine Combustors, Presented at the 36th Aerospace Sciences Meeting & Exhibit, Jan 12-15, AIAA-98-0855, 1998.
43. G.N. Kumar, H.C. Mongia and J.P. Moder, Validation of Radiative Heat Transfer Computations Module for National Combustion Code, Presented at the 34th Joint Propulsion Conference, July 12-15, AIAA-98-3985, 1998.
44. N. Vaidya, An Unstructured Finite Volume Method for Nongray Radiation with Conjugate Heat Transfer and Chemistry, Presented at the 37th AIAA Aerospace Sciences Meeting & Exhibit, January 10-14, 1999, Reno, AIAA-99-0973.
45. P. Sokolov, M. Ibrahim, and T. Kerslake, Computational Fluid Dynamics and Heat Transfer Modeling of Thermal Energy Storage Canisters for Space Applications, Presented at the 36th Aerospace Sciences Meeting & Exhibit, Jan 12-15, AIAA-98-1018, 1998.

FIVE

BOUNDARY ELEMENT METHODS FOR HEAT CONDUCTION

A.J. Kassab
L.C. Wrobel

1 INTRODUCTION

This chapter presents the basic principles of heat transfer modeling using boundary integral equations and numerical solutions using the boundary element method. The work described is restricted to heat conduction; the range of applications considered encompasses steady-state and transient problems, nonlinear problems such as those arising from temperature-dependent thermophysical properties and phase change, non-Fourier (hyperbolic) heat conduction and inverse analysis.

The boundary element method (BEM) is a well-established numerical technique for the solution of many engineering problems. Its theoretical basis, mathematical formulation and numerical implementation are discussed in a number of textbooks on the subject [1],[2],[3].

The technique basically consists of obtaining an integral equation equivalent to the original partial differential equation describing the problem and solving it numerically through a discretization procedure. Since the integral equation usually relates to boundary values only, the discretization is confined to the boundaries of the region, and the dimensionality of the problem is thus reduced by one.

The chapter starts by introducing the BEM as applied to steady-state heat conduction in homogeneous, isotropic media, with no internal heat generation. In this case, the mathematical model is simply Laplace's equation, and the presentation follows classical concepts of potential theory. Next, transient heat conduction is dealt with by using time-dependent fundamental solutions and a time-marching scheme. Then, several types of nonlinearities which are common in heat transfer problems are discussed and their BEM formulation described in some detail. The next section describes BEM formulations for the non-Fourier mode of heat conduction which is modeled by a hyperbolic equation. The final section discusses two inverse problems in heat conduction, namely the inverse geometric problem and the identification of unknown convective heat transfer coefficients.

2 HEAT CONDUCTION

The boundary integral representation of steady heat conduction problems defined over homogeneous, isotropic media, can be deduced from the classical theorems of Green; in fact, the boundary integral equation (BIE) is simply given by Green's third identity:

$$T(y) = \int_S \left(T^*(x,y)\frac{\partial T(y)}{\partial n_y} - T(y)\frac{\partial T^*(x,y)}{\partial n_y} \right) dS_y \tag{1}$$

where S is the domain boundary, n_y the outward-drawn normal, T temperature, and T^* the fundamental solution of the Laplace equation, given by [4]

$$T^*(x,y) = -\frac{1}{2\pi}ln(r) \tag{2}$$

for two-dimensional problems, with $r = |x - y|$. The above equation can be used to calculate the temperature at any point within the region once the boundary temperatures and heat fluxes are all known.

In order to obtain a boundary integral equation relating only to boundary values, the limit is taken when the point x tends to a point ξ on the boundary S. Because of the singularity of the function T^*, the following equation is obtained for a point ξ on the boundary:

$$C(\xi)T(\xi) = \int_S \left(T^*(\xi,y)\frac{\partial T(y)}{\partial n_y} - T(y)\frac{\partial T^*(\xi,y)}{\partial n_y} \right) dS_y \tag{3}$$

with the free coefficient $C(\xi)$ given by:

$$C(\xi) = \frac{\alpha}{2\pi} \qquad 1 \geq C(\xi) \geq 0 \tag{4}$$

and α is the internal angle subtended at point ξ. The boundary integral Eq. (3) can be solved numerically using the boundary element method. Application of the method requires two types of approximation: the first is geometrical, involving a subdivision of the boundary S into N_e elements S_j, such that

$$C(\xi)T(\xi) = \sum_{j=1}^{N_e} \int_{S_j} \left(T^*(\xi,y)q(y) - T(y)q^*(\xi,y) \right) dS_y \tag{5}$$

where, for simplicity of notation, we have called $q = \partial T/\partial n$ and $q^* = \partial T^*/\partial n$.

The second approximation required by the BEM is functional, necessary because although the integration along the entire boundary has been reduced to a summation of integrals over each element, we do not know how the temperature and heat flux vary within each element (one or the other function or their ratio is usually known from the boundary conditions of the problem, but not both). Thus, we approximate the variation of T and q within each element by writing them in terms of their values at some nodal points using suitable interpolation functions.

Similarly to the geometry, different interpolation functions (and corresponding number of nodes) can be used to represent the variation of the temperature and its normal derivative within each element. The simplest possible approximation is piecewise constant, which simply assumes that T and q are constant within each element and equal to their respective values at the midpoint. By introducing this approximation into Eq. (5), we obtain:

$$C(\xi)T(\xi) = \sum_{j=1}^{N_e} q_j \int_{S_j} T^*(\xi,y)dS_y - \sum_{j=1}^{N_e} T_j \int_{S_j} q^*(\xi,y)dS_y \qquad (6)$$

with T_j and q_j the values of T and q at node j (the midpoint of element j). Note that, in the case of constant elements, the number of nodes is equal to the number of elements. The above equation is still valid for any boundary point ξ. But because the problem has been reduced to evaluating a finite number N_e of unknowns, it is only necessary to generate the same number of equations. These equations will be generated by applying Eq. (6) to the same nodal points along the boundary, *i.e.*

$$C_i T_i = \sum_{j=1}^{N_e} q_j \int_{S_j} T^* dS_j - \sum_{j=1}^{N_e} T_j \int_{S_j} q^* dS_j \qquad (7)$$

for a nodal point i (the dependence of T^* and q^* on the position of source and field points is implicit in the above equation).

Calling

$$G_{ij} = \int_{S_j} T^* dS_j \qquad (8)$$

and

$$\hat{H}_{ij} = \int_{S_j} q^* dS_j \qquad H_{ii} = \hat{H}_{ij} + C_i \delta_{ij} \qquad (9)$$

with δ_{ij} the Kronecker delta, Eq. (7) can be written in the form

$$\sum_{j=1}^{N_e} H_{ij} T_j = \sum_{j=1}^{N_e} G_{ij} q_j \qquad (10)$$

for any nodal point i. If the above equation is now applied, using a collocation technique, to all nodal points along the boundary, then a system of equations is generated which can be written in matrix form as

$$\mathbf{HT} = \mathbf{GQ} \qquad (11)$$

where $\mathbf{H}$ and $\mathbf{G}$ are square $N_e \times N_e$ matrices of influence coefficients, and $\mathbf{T}$ and $\mathbf{Q}$ are vectors containing the nodal values of the temperature and its normal derivative. Once the boundary conditions of the problem are applied to the system in Eq. (11), the matrices can be reordered in the form

$$AX = F \tag{12}$$

in which all unknowns have been collected into the vector X, and vector F is the "load" vector. This system can be solved by standard direct schemes such as Gauss' elimination, noting that the system matrix A is full and non-symmetric, or by iterative methods for non-symmetric equations such as the general minimization of residuals method (GMRES) when large problems are encountered. Higher order models for $T(y)$ and $q(y)$ in Eq. (5) are routinely used in BEM, with quadratic interpolation the most common higher order approximation in use. Corresponding expressions for H and G can be found in Brebbia, Telles, and Wrobel[1].

The previous formulation can also be applied to heat conduction problems defined over orthotropic or anisotropic regions, using the appropriate fundamental solutions given by Chang et al.[5]. Recently, general boundary integral formulations valid for heterogeneous regions have been developed by Kassab and Divo[6], and Divo and Kassab[7], [8], [9].

3 TRANSIENT PROBLEMS

The present section discusses boundary element formulations for transient heat conduction using time-dependent fundamental solutions. These formulations can be viewed as a direct extension of potential theory since the proper fundamental solution of the diffusion equation is used to obtain an equivalent boundary integral equation. Numerical techniques are then employed to solve the integral equation in discrete form through a time-marching procedure.

The fundamental solution adopted here is a free-space Green's function which has been used by Morse and Feshbach[10] and by Carslaw and Jaeger[11], among others, to obtain analytical solutions to some simple problems. Chang et al. [5] and Shaw[12] were the first to apply this fundamental solution in the context of the direct BEM, but their emphasis was on the analytical rather than numerical aspects of the method. The formulation was later extended by Wrobel and Brebbia[13] to allow higher-order space and time interpolation functions to be included, thus making possible the analysis of practical engineering problems.

3.1 Boundary Integral Equation

This section discusses boundary element solutions to the transient heat conduction equation

$$\nabla^2 T + b = \frac{1}{\alpha} \frac{\partial T}{\partial t} \tag{13}$$

where t is time, T is temperature, α is the thermal diffusivity ($\alpha = k / \rho c$), k is the thermal conductivity, ρ density, c specific heat, and the term b accounts for internal heat generation. The problem definition is completed with the specification of boundary and initial conditions. The above equation can be recast as an integral equation over space and time, with the help of the corresponding fundamental solution; between the initial time t_o and the final time t_F, the integral equation for a source point ξ on the boundary S of the domain Ω is written as [1]

$$C(\xi)T(\xi, t_F) + \int_{t_o}^{t_F} \int_S \alpha T(y, t)\, q^*(\xi, y; t_F, t) dS_y dt =$$

$$\int_{t_o}^{t_F} \int_S \alpha q(y, t)\, T^*(\xi, y; t_F, t) dS_y dt + \int_{t_o}^{t_F} \int_\Omega \alpha b(y, t)\, q^*(\xi, y; t_F, t) d\Omega_y dt +$$

$$\int_\Omega \alpha T_o(y)\, T^*(\xi, y; t_F, t_o) d\Omega_y \qquad (14)$$

The fundamental solution T^* is a free-space Green's function, describing the temperature field generated by a unit heat source applied at point y at time t_o [10],[11], *i.e.*

$$T^* = \frac{1}{(4\pi\alpha\tau)^{s/2}} exp\left(-\frac{r^2}{4\alpha\tau} \right) \qquad (15)$$

$$q^* = \frac{2d}{(4\alpha\tau)^{(s+2)/2}\pi^{s/2}} exp\left(-\frac{r^2}{4\alpha\tau} \right) \qquad (16)$$

with $d = -\vec{n} \cdot \vec{r}$, $\vec{r}$ being the distance vector connecting source and field points, $\tau = t_F - t$, and s is the number of spatial dimensions of the problem.

The numerical solution of the boundary integral Eq. (14) requires space and time discretization. The time-marching scheme generally used re-starts the solution for each time step from the initial time t_o; in this way, the domain integral associated with the initial conditions can be avoided for the majority of practical situations. The scheme is very stable and allows the use of large time steps. However, the computational effort rapidly increases as time progresses due to the time-history dependence of the solution. Truncation algorithms have been developed to improve the computer efficiency of such an approach by calculating only approximately the influence of initial steps after some elapsed time. The interested reader is referred to [14],[15],[16].

3.2 Space and Time Discretization

Equation (14) presents two domain integrals due to initial conditions and internal sources. For simplicity, it will be assumed that there is no internal heat generation and that the initial temperature is constant. In this case, it is possible to rewrite the problem into an equivalent one with zero initial condition for the temperature difference $T - T_o$.

With the above simplifications, Eq. (14) can be rewritten for a point i in a smooth portion of S as

$$\frac{1}{2}T_{i,F} + \int_{t_o}^{t_F} \int_S \alpha T q^* dS dt = \int_{t_o}^{t_F} \int_S \alpha q T^* dS dt \qquad (17)$$

with $T_{i,F}$ being the temperature at node i at time t_F. Dividing the boundary S into N boundary elements and the time span $t_F - t_o$ into F time steps, the following discretized equation is obtained

$$\frac{1}{2}T_{i,F} + \sum_{f=1}^{F}\sum_{j=1}^{N} \int_{t_{f-1}}^{t_f} \int_{S_j} \alpha T q^* dS dt = \sum_{f=1}^{F}\sum_{j=1}^{N} \int_{t_{f-1}}^{t_f} \int_{S_j} \alpha q T^* dS dt \qquad (18)$$

Assuming that the boundary elements are constant in space and linear in time, the temperature variation on element j between time levels $f - 1$ and f is given by

$$T = \frac{T_{j,f-1}(t_f - t) + T_{j,f}(t - t_{f-1})}{\Delta t_f} \qquad (19)$$

with $\Delta t_f = t_f - t_{f-1}$. A similar expression can be written for q. Calling

$$H1_{ij,Ff} = \frac{1}{\Delta t_f} \int_{t_{f-1}}^{t_f} \int_{S_j} \alpha (t_f - t) q^* dS dt \qquad (20)$$

$$\hat{H}2_{ij,Ff} = \frac{1}{\Delta t_f} \int_{t_{f-1}}^{t_f} \int_{S_j} \alpha (t - t_{f-1}) q^* dS dt \qquad (21)$$

$$G1_{ij,Ff} = \frac{1}{\Delta t_f} \int_{t_{f-1}}^{t_f} \int_{S_j} \alpha (t_f - t) T^* dS dt \qquad (22)$$

$$G2_{ij,Ff} = \frac{1}{\Delta t_f} \int_{t_{f-1}}^{t_f} \int_{S_j} \alpha (t - t_{f-1}) T^* dS dt \qquad (23)$$

$$H2_{ij,Ff} = \hat{H}2_{ij,Ff} + \frac{1}{2}\delta_{ij}\delta_{Ff} \qquad (24)$$

Eq. (18) becomes

$$\sum_{f=1}^{F}\sum_{j=1}^{N}(H1_{ij,Ff}T_{j,f-1} + H2_{ij,Ff}T_{j,f}) =$$
$$\sum_{f=1}^{F}\sum_{j=1}^{N}(G1_{ij,Ff}q_{j,f-1} + G2_{ij,Ff}q_{j,f}) \qquad (25)$$

Writing the above equation at all boundary nodes $i(1 \leq i \leq N)$ using a collocation technique, the following system of equations is obtained

$$\sum_{f=1}^{F}(\mathbf{H}1_{F,f}\mathbf{T}_{f-1} + \mathbf{H}2_{F,f}\mathbf{T}_f) = \sum_{f=1}^{F}(\mathbf{G}1_{F,f}\mathbf{Q}_{f-1} + \mathbf{G}2_{F,f}\mathbf{Q}_f) \qquad (26)$$

The temperature and flux at each node are known for all time levels previous to t_F, so that the above equation can be rewritten in the form

$$\mathbf{H}2_{F,F}\mathbf{T}_F = \mathbf{G}2_{F,F}\mathbf{Q}_F + \mathbf{S}_F \qquad (27)$$

with

$$\mathbf{S}_F = -\sum_{f=1}^{F-1}(\mathbf{H}1_{F,f}\mathbf{T}_{f-1} + \mathbf{H}2_{F,f}\mathbf{T}_f) - \mathbf{H}1_{F,F}\mathbf{T}_{F-1} +$$

$$\sum_{f=1}^{F-1}(\mathbf{G}1_{F,f}\mathbf{Q}_{f-1} + \mathbf{G}2_{F,f}\mathbf{Q}_f) + \mathbf{G}1_{F,F}\mathbf{Q}_{F-1} \qquad (28)$$

Once boundary conditions at time t_F are applied, the above system can be reordered and solved for any time level.

4 NONLINEAR PROBLEMS

This section deals with BEM applications to nonlinear heat conduction problems. The nonlinearities can be classified in many different ways. Based on the mathematical formulation of the corresponding boundary-value problems, the present classification encompasses four groups as follows:

1. Nonlinear materials, as in the case of temperature-dependent thermal conductivity and/or heat capacity. This feature gives rise to a nonlinear partial differential equation;
2. Nonlinear boundary conditions, caused by heat radiation or temperature-dependent heat transfer coefficient;
3. Nonlinear sources, characteristic of some kind of chemical reaction taking place within the solid medium;
4. Moving boundary problems, as in the case of phase change.

In what follows, a brief review of the most important BEM formulations dealing with all the above features is attempted. It is shown that the majority of nonlinear phenomena occurring in engineering practice can be efficiently treated by the method.

4.1 Material Nonlinearities

Initially, we consider problems in which the thermal conductivity and heat capacity of the medium are temperature-dependent. The mathematical formulation of the corresponding boundary-value problem produces a nonlinear equation which can be written, for the case of steady state with no internal heat generation, in the form

$$\frac{\partial}{\partial x}\left(k\frac{\partial T}{\partial x}\right) + \frac{\partial}{\partial y}\left(k\frac{\partial T}{\partial y}\right) + \frac{\partial}{\partial z}\left(k\frac{\partial T}{\partial z}\right) = 0 \qquad (29)$$

An approach which removes the nonlinearity associated with the temperature dependence of the conductivity is through the application of Kirchhoff's transform, which replaces the original nonlinear differential equation by a linear one in the transform space [11]. The basic idea is to construct a new variable $U = U(T)$ such that

$$\frac{dU}{dT} = k(T) \qquad (30)$$

or, in integral form,

$$U = \int_{T_o}^{T} k(T)dT \qquad (31)$$

where T_o is an arbitrary reference value. It can be seen by inspection that the direct substitution of the expression in (30) into Eq. (29) produces a Laplace equation for the transform variable U. The boundary conditions of the problem also need to be transformed. Boundary conditions of the Dirichlet and Neumann types are linear in the transform space, but Robin boundary conditions (*i.e.* convection) become nonlinear. Thus, an iterative scheme is still needed in this formulation, but all nonlinearities are now transferred to the boundary conditions.

The Kirchhoff transform formulation has been successfully used by many authors [17]-[19]. An efficient algorithm for the inverse Kirchhoff transformation which approximates the $k(T)$ dependence by a piecewise linear spline has been proposed independently by Azevedo and Wrobel[18] and Bialecki and Nhalik[19]. The important case of multizoned media with nonlinear material behavior was also examined by the above authors [18],[19]. The added difficulty in this case is that discontinuities arise in the values of the transformed temperature across the interfaces between different regions when continuity of temperature is imposed. Thus, an iterative algorithm is necessary to adjust the transformed temperatures along the internal boundaries of the various subregions so that the physical requirement of temperature continuity is satisfied at the final stage. The resulting set of equations can be solved by a Newton-Raphson algorithm [18] or incremental techniques [19].

The Kirchhoff transform technique can also be applied to transient problems governed by the diffusion equation

$$\frac{\partial}{\partial x}\left(k\frac{\partial T}{\partial x}\right) + \frac{\partial}{\partial y}\left(k\frac{\partial T}{\partial y}\right) + \frac{\partial}{\partial z}\left(k\frac{\partial T}{\partial z}\right) = \rho c\frac{\partial T}{\partial t} \qquad (32)$$

in which the density ρ and specific heat c may also be temperature-dependent. However, contrary to steady-state problems, this transformation is not sufficient in the sense that the equation for the transform variable still contains a temperature-dependent diffusivity coefficient, in the form

$$\nabla^2 U = \frac{1}{\alpha} \frac{\partial U}{\partial t} \tag{33}$$

where $\alpha = \alpha(T) = k/\rho c$. Brebbia and Skerget[20],[21], Kikuta *et al.*[22], and Pasquetti and Caruso[23] all assumed that the variation of k with T is usually not strong and employed some kind of mean value of k to linearize Eq. (33). A more elegant formulation has been derived by Wrobel and Brebbia[24]. This involves the definition of a modified time variable in the form

$$\tau = \int_0^{t'} \alpha(x, y, z, t) dt \tag{34}$$

which leads to

$$\frac{\partial \tau}{\partial t} = \alpha \tag{35}$$

Substituting the above into Eq. (33), one finally obtains

$$\nabla^2 U = \frac{\partial U}{\partial \tau} \tag{36}$$

Equation (36) can now be readily converted into a boundary integral equation. It is interesting to notice that Eq. (36) is still nonlinear in that the modified time variable τ is a function of position, and an iterative solution process is again necessary. Wrobel and Brebbia[24] developed an efficient Newton-Raphson scheme and implemented it in conjunction with a dual reciprocity approximation which resulted in a boundary-only formulation.

The efficiency of the above formulation was displayed in several numerical tests. The results for the temperature distribution along the thickness of a wall with nonlinear material properties are reproduced in Tables 1 and 2. The wall is 20 cm long, 1 cm high, and is initially at $100°C$. The temperature of the left-end surface is suddenly raised to $200°C$ and kept at this value for $10s$, after which it is decreased to $100°C$ again; the temperature of the right-end surface is kept at $100°C$, while the other surfaces are insulated. The thermal conductivity is $k = 2 + 0.01T\,W/(cm°C)$ and the heat capacity $\rho c = 8\,J/(cm^3 °C)$.

The tables compare results for boundary element discretizations of 22 equal quadratic and 44 equal constant elements, taking into account symmetry with respect to the $x-$axis, with a finite-element solution obtained with 20 linear elements [25]. The time step value was $1\,s$ in all cases, and the average number of iterations was 4 for the BEM and 3 for the FEM. The agreement of the results can be seen to be very good.

Table 1   Results for $t = 10s$.

$x[cm]$	FEM	BEM-Q	BEM-C
0	200.00	200.00	200.00
1	176.16	174.86	175.29
2	153.21	151.03	151.48
3	134.47	131.33	131.74
4	118.60	117.32	117.63
5	108.98	108.74	108.94
6	103.72	104.14	104.27
7	101.29	101.91	102.01
8	100.37	100.87	100.97
9	100.08	100.39	100.50
10	100.01	100.14	100.27

Table 2   Results for $t = 13s$.

$x[cm]$	FEM	BEM-Q	BEM-C
0	100.00	100.00	100.00
1	128.53	130.46	130.15
2	139.97	138.70	138.95
3	136.95	132.01	132.47
4	124.72	121.29	121.71
5	114.40	112.37	112.64
6	107.18	106.56	106.71
7	103.24	103.27	103.36
8	101.29	101.56	101.62
9	100.45	100.72	100.77
10	100.13	100.33	100.36

4.2 Nonlinear Boundary Conditions

The most common cause of nonlinear boundary conditions, other than the boundary condition arising from convection over conducting media with temperature-dependent thermal conductivity, is due to thermal radiative heat transfer. The complexity introduced by radiation can be quite severe as in the case of a concave portion of the boundary where it produces strong interactions between the temperature at points which can "see" each other. No serious difficulties arise from the treatment of nonlinear boundary conditions by the BEM, and several formulations have been presented with different iterative algorithms. These include Newton-Raphson solvers [18],[19], Brown's iterative method [20],[21], incremental techniques [17],[18], "local" linearization [23], and others. The choice of a robust iterative solver is of importance as some simple iterative schemes of Gauss-Seidel type can diverge in the presence of the severe nonlinearities caused by thermal radiation, as pointed out in [17].

The previous papers all dealt with nonlinear boundary conditions considering that the heat flux at a given boundary point depends on the temperature at that point only. In the case when the boundaries are concave, the heat flux at a certain position depends on the temperature at all points which can be seen from that position (providing the fluid filling the enclosure is transparent to thermal radiation). The problem of coupling BEM radiative analysis with the conductive part of the mathematical model of a body having concave boundaries has been discussed by Bialecki[26]-[28]. Since radiation is one of the few phenomena directly governed by integral equations, the idea of applying the BEM to solve radiative problems arises naturally. Bialecki's formulation has been applied to gray models of both radiating walls and participating isothermal medium [26], monochromatic analysis [27], non-isothermal, participating non-gray gas radiation [28] and gray, absorbing, emitting and isotropic scattering media [29].

The cross-section in Fig. 1, reproduced from [28], shows the radiative heat fluxes at the base of an open rectangular cavity of dimensions 0.7x0.7x0.4 m filled by a participating non-gray gas having known non-uniform temperature. The walls of the enclosure were subdivided into 210 square elements of 0.1x0.1 m with specified temperature. The base of the cavity is maintained at temperature 700K whereas its top is open to an environment at temperature 300K. The temperature of the other walls varies linearly from 700K at the bottom to 300K at the top. A five-band radiating gas is considered with constant emissivity and absorption coefficient within each band. The volume of the enclosure has been divided into 27 cubic isothermal cells consisting of three layers of nine cells each.

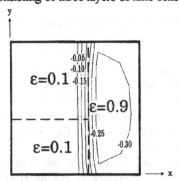

Figure 1 Total incoming radiative heat flux (kW/m^2) at base of rectangular cavity.

4.3 Nonlinear Sources

This situation is characteristic of some sort of chemical reaction taking place within the solid medium. Generally, problems with distributed nonlinear sources can be solved by the BEM using domain discretization and iteration. However, approximate techniques have been developed to allow the domain integral to be transformed into equivalent boundary integrals, thus retaining the boundary-only character which is the main feature of the BEM [30],[31].

A recent example of application of one such technique, namely the dual reciprocity method (DRM), to a heat conduction problem with a nonlinear source has been presented by Partridge and Wrobel[32] who studied the case of spontaneous ignition of a reactive solid. The partial-differential equation describing the problem is the following diffusion-type equation [33]

$$k\nabla^2 T + \rho Q z e^{-E/RT} = \rho c \frac{\partial T}{\partial t} \qquad (37)$$

in which Q is the heat of decomposition of the solid, z the collision number, E is the Arrhenius activation energy and R the universal gas constant. The form of Eq. (37) can be simplified by a change of variables. Defining

$$u = E \frac{(T - T_a)}{RT_a^2} \qquad (38)$$

where T_a is the ambient temperature, the equation in terms of the new variable becomes

$$\nabla^2 u + \gamma e^u = \frac{1}{\alpha} \frac{\partial u}{\partial t} \qquad (39)$$

with

$$\gamma = \frac{\rho Q E z}{k R T_a^2} e^{-E/RT_a} \qquad (40)$$

in which the Frank-Kamenetskii approximation [34] was employed, *i.e.*

$$e^{-E/RT} = e^{-E/RT_a} e^u \qquad (41)$$

It is usual to define the non-dimensional Frank-Kamenetskii parameter as $\delta = l^2 \gamma$, where l is a characteristic problem dimension. The physical significance of δ is discussed next. For this, it is assumed that a given reactive material has an ignition temperature T_m, the ambient temperature is considered to be T_a and the initial temperature is T_o at time $t = 0$. From Eq. (38) the corresponding values of u_m, u_a and u_o can be obtained.

For $\delta < \delta_c$, where δ_c is a critical value which must be exceeded to initiate ignition, a reactive solid at temperature T_o can be placed in an environment at temperature $T_a (T_a < T_m)$, and the temperature field within the solid will change according to Eq. (39) until all points are at a temperature $T_a \le T < T_m$ (where T is not uniform), thus achieving a steady-state situation where ignition does not occur. For $\delta > \delta_c$, the temperature within the solid will continue to increase until some point reaches the ignition temperature T_m and spontaneous ignition then occurs. Thus, to model the process of spontaneous ignition, a knowledge of the critical value, δ_c, is essential. This value can be obtained for a given geometric shape considering $\partial u/\partial t = 0$ in Eq. (39) [34], i.e.

$$\nabla^2 u = -\gamma e^u \tag{42}$$

Partridge and Wrobel[32] developed a dual reciprocity boundary element formulation and applied it to the calculation of critical values through Eq. (42) and to simulate the full process of spontaneous ignition using Eq. (39). Zhu and Satravaha[35] used a similar formulation, but in the Laplace transform space, to solve problems of spontaneous ignition and microwave heating. The problem of finding the critical value, δ_c, was solved for a series of common geometrical shapes for which solutions are available in the literature, including a square of unit side, a circle of unit radius and an ellipse with major axis $= 2$ and minor axis $= 1$. Thus in all cases the characteristic dimension of the problem is $l = 1$, such that $\gamma = \delta$.

Numerical results are presented in Table 3 for different boundary discretizations with fixed numbers of regularly spaced internal nodes (9 for the square, 17 for the circle and the ellipse). It can be seen that, in this case, the use of curved quadratic elements produces little difference in the results if the same number of boundary and internal nodes is used. The reason for this behavior has to do with the boundary conditions of the problem, which impose a constant value of u everywhere. For the circle, this produces a constant boundary flux q. For the ellipse and the square, the flux is variable but its influence on the solution at internal points is of less importance. A finite-element (FEM) solution [34] is also included for comparison. The FEM mesh for each problem consisted respectively of 8, 20 and 34 quadratic isoparametric elements with 37, 75 and 125 node points for half of the region.

The results given in Table 4 are for the same geometric shapes with a fixed boundary discretization and different numbers of regularly spaced DRM internal nodes. It can be seen that, by increasing the number of internal nodes, the solution converges towards the exact value; however, solutions within the usual engineering accuracy may be obtained with a reduced number of internal nodes, with little computer time and data preparation effort.

Table 3 DRM results for γ_c for different discretizations.

shape	32 linear BE	16 quadratic BE	FEM	Exact
square	1.763	1.770	1.703	1.7 [31]
circle	2.032	2.031	2.001	2.0 [34]
ellipse	1.251	1.252	1.234	

Table 4 Results for 16 quadratic boundary elements and different numbers of DRM internal nodes.

shape	5 nodes	9 nodes	17 nodes	80 nodes
square	-	1.770	-	1.707
circle	2.080	-	2.031	2.004
ellipse	1.276	-	1.252	1.235

The dual reciprocity boundary element formulation of Partridge and Wrobel[32] was also applied to the problem of spontaneous ignition of a long cylinder of unit radius, initially at temperature T_o at all points, exposed at time $t = 0$ to an environment at temperature T_a. The cylinder is of a uniform isotropic reactive material, the ignition temperature of which is T_m. An essential boundary condition $T = T_a$ is imposed at all boundary nodes. The critical value of γ is $\gamma_c = 2.0$ in this case (see Table 3). The cylinder was discretized with 16 linear boundary elements, and 19 internal nodes were spaced at intervals of $10\,cm$ along a diagonal. The numerical values of the physical parameters for RDX taken from [34] are $\alpha = 7.778 \times 10^{-4}\,cm^2/s$, $T_m = 425K$, $T_o = 25°C = 298K$, $T_a = 400K$, $E = 198740\,J/mole$ and $R = 8.315\,J/(mole\,K)$, such that $T = (u + 59.76)/$ 0.1494. A variable time step was used because this process has varying stability characteristics. Initially a large Δt may be used, but this must be rapidly reduced as ignition nears. The value of Δt was altered at each time step to maintain the average temperature change at all interior nodes between 5° and 20°.

Results are shown in Fig. 2 for the case $\gamma = 4$. It can be noticed that ignition is first reached at the center of the cylinder in this case. For cases with higher values of γ, the ignition point moves out from the center towards the outer surface, and the elapsed time before ignition reduces. This situation is depicted in Fig. 3 for $\gamma = 50$. The results presented are very similar to those obtained using theoretical considerations in [36].

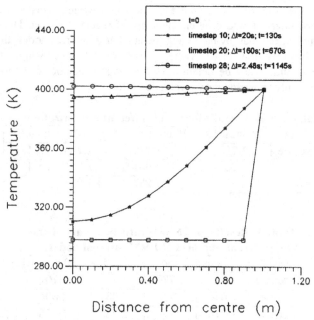

Figure 2 Temperature distribution for γ=4.

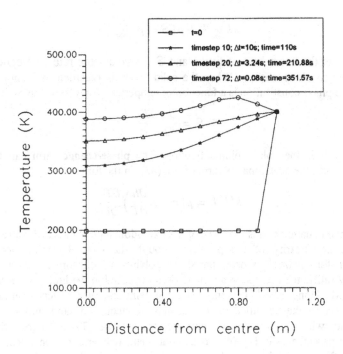

Figure 3 Temperature distribution for $\gamma = 50$.

4.4 Moving Boundary Problems

Moving boundaries are an inherent part of phase change phenomena such as solidification, melting, ablation, welding, electrochemical machining, etc. The physics of these phenomena is so complex that no consistent theory exists. The main reason for this is the lack of thermodynamic equilibrium often encountered in phase change phenomena.

The simplest possible mathematical model is the case of a pure substance with a single temperature of phase change, the so-called Stefan problem [37]. Even such a simple model can become complicated if mushy regions, *i.e.* transition zones filled with a mixture of liquid and crystals, are included in the analysis.

The algorithms employed by the numerous discrete approximation methods for phase-change problems can be broadly divided into fixed-domain and front-tracking schemes. Fixed-domain methods have the great advantage of simple numerical implementation and are generally most useful in problems where the temperature distribution is more important than the interface position. Front-tracking methods, on the other hand, are more complex due to the need of re-meshing during each time step and are mainly used when the primary unknown is the motion of the interface.

The heat transfer equation to be treated here is of the form

$$k\nabla^2 T + \rho L S = \rho c \frac{\partial T}{\partial t} \tag{43}$$

where L is the latent heat of fusion and S the volumetric rate of production of solid due to phase change. When the solid phase is regarded as immobile, and when the phase constitution is a function of temperature only, one may write

$$S = \frac{\partial h}{\partial T}\frac{\partial T}{\partial t} \tag{44}$$

in which h is the solid volume fraction. The phase-change term can then be absorbed into the heat capacity term of Eq. (43), in the form

$$k\nabla^2 T = \rho\left(c - L\frac{\partial h}{\partial T}\right)\frac{\partial T}{\partial t} \tag{45}$$

The above equation forms the basis of the fixed-domain methods and expresses the inherent nonlinearity of the problem through the thermal coefficients. Many different formulations, mostly based on solving a diffusion equation for the enthalpy rather than temperature, have been proposed in the literature. Regarding the BEM, Hong et al.[38] employed the temperature recovery method with application to the solidification of castings in metal and sand molds. Another possibility which is common in finite-element analysis [39] is to treat the term between parentheses in Eq. (45) as an equivalent, temperature-dependent specific heat coefficient. Similar formulations can in principle be developed for the BEM using the algorithms previously discussed for nonlinear materials [24].

A different BEM enthalpy formulation which allows for the treatment of each phase separately was derived by Sarler et al.[40]. The resulting decoupled all-solid and all-liquid geometrically fixed problem in each phase were solved by a standard, direct BEM for the diffusion equation using the alternating phase truncation method. Recently, Sarler[41] presented a more general formulation for coupled mass, momentum and energy transfer based on the dual reciprocity BEM.

Generally, phase change is often active only within a temperature and space range which is narrow relative to the overall range considered in a problem. In particular, the phase-change zone is often infinitesimally thin, for all practical purposes. Thus the problem features a zone of intense activity, usually more concentrated than the mesh spacing, which moves about in a manner which is unknown "a priori". The problem can then be treated as a moving boundary problem in which the phase change is confined to an infinitesimally thin surface across which the discontinuity of flux of sensible heat and the evolution of latent heat must balance [42]. In this approach, the governing equation is the linear diffusion equation within each mutually exclusive phase domain, with the condition

$$\rho L \frac{\partial X_i}{\partial t}n_{si} = k_s\frac{\partial T_s}{\partial n_s} - k_l\frac{\partial T_l}{\partial n_l} \tag{46}$$

attached to the interface Γ_I between the solid (s) and liquid (l) phases. Thus, all latent heat effects are expressed through Eq. (46) which is also used to locate the interface position (X_i are the components of a vector X defining the location of a point on Γ_l). The above formulation, which gives rise to front-tracking algorithms, was initially developed by Chuang and Szekely[43] for one-dimensional, one-phase problems (only the solid phase was considered in the analysis). This formulation was later extended and improved by Shaw[44] and Wrobel[45]. Coleman[46] presented a two-dimensional formulation but, similarly to the previous cases, it involved a domain integral originated through the integration over a variable domain in which the limits of integration are a function of time. O'Neill[42] gave a general integral equation formulation in which the domain integral was analytically transformed into an equivalent boundary integral, using the Leibnitz rule. The formulation was generalized to multi-dimensional problems, but results were only shown for the simpler quasi-static approximation, which is valid when the boundary temperature does not change too rapidly. Heinlein et al.[47] numerically implemented O'Neill's formulation in an application to one-dimensional solidification problems. Zabaras and Mukherjee[48] extended this work for two-dimensional problems, studying the solidification of a square region and a rectangular prism for two distinct situations, assuming that the liquid initially filling the domain was at the melting temperature or at a temperature higher than the freezing temperature. Another implementation of O'Neill's formulation has been presented by Hsieh et al.[49]. DeLima-Silva Jnr and Wrobel[50],[51] extended further the works of O'Neill[42] and Zabaras and Mukherjee[48] with applications to the ablation of solids and the melting of a steel slab in molten steel. Zerroukat and Wrobel[52] derived a general front-tracking technique which can be used for multiple moving-boundary problems. Applications included a three-phases problem, i.e. solid, liquid and vapor, with two simultaneous interfaces. Takahashi et al.[53] presented the first BEM formulation to include a complete interaction between heat transport and mass diffusion in which the latent heat effects behaved as a temperature-dependent heat source. They solved the two-dimensional problem of a steel slab with a series of circular holes with concentration-dependent melting temperature in conjunction with the diffusion of the alloy components. Zerroukat et al.[54] also studied alloy phase change using a front-tracking approach by solving two coupled equations for both heat and solute transfer. Figures 4 and 5, reproduced from [52], show results for interface velocities and position for the problem of a solid wall of thickness $1m$ subjected to a constant heat flux $F = 2500\,W/m^2$ at $x = 0$ and thermally insulated at $x = 1m$. The results are compared with finite-difference[55] and finite-element[56] solutions. The values for the thermophysical properties used in the problem are given in [52] and [56].

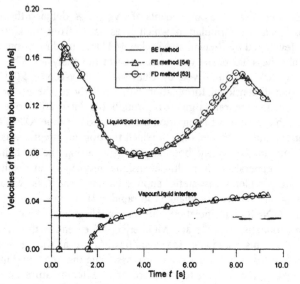

Figure 4 Moving boundaries velocity.

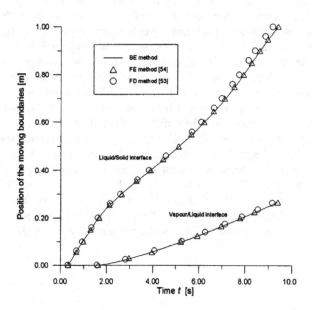

Figure 5 Moving boundaries position.

5 BEM SOLUTION OF HYPERBOLIC HEAT CONDUCTION

In this section, we present BEM models for the non-Fourier mode of heat conduction modeled by the Hyperbolic Heat Conduction Equation (HHCE). This damped wave equation has been proposed by many researchers (see Catteno[57] and Morse and Feshbach[10]) to model a finite speed of propagation of energy in heat conducting media. An excellent review of the subject is found in the text by Tzou[58]. Numerical solutions of the 1-D HHCE were undertaken by Glass *et al.*[59] using the MacCormak finite difference method, by Tamma and Railkar[60] using specially tailored transfinite element methods, and by Yang[61] (who presents a comparative study of various schemes) using characteristics based finite-volume methods. Yang[62] also solved the 2-D HHCE using highly accurate finite-volume based TVD schemes. Tzou[63] presents a review of recent research on the HHCE and moving heat sources and their effects on crack propagation. Kaminski[64] experimentally determined the relaxation time, τ, for several materials such as sand and glass ballotini. There is, however, a dearth of thermophysical data for τ. Vedarvarz *et al.*[65] present a regime map indicating the characteristic length and time scales for which the HHCE is applicable and indicate some applications in which the Fourier model for the heat conduction equation may be in error. Kassab and Nordlund[66] use the Laplace transform BEM to solve two-dimensional HHCE problems, while Priedeman and Kassab[67] and Vicki and West[68] use time-dependent fundamental solutions and a convolution marching scheme to resolve the 1-D damped wave propagation. The first authors consider application to heat transfer, while the second consider application to damped propagating waves in a rod. Lu *et al.*[69] develop a dual reciprocity BEM to solve the HHCE with applications to thermal wave propagation in biological tissue. In this section, the BEM solution of the HHCE is considered using two approaches: time-dependent fundamental solutions and Laplace transform method.

5.1 Laplace Transform BEM for the HHCE

Although arrived at by different means such as the kinetic theory of gases, irreversible thermodynamics, and heuristic arguments, the formulation which leads to the HHCE first modifies the Fourier Law of heat conduction to account for a thermal lag between the heat flux and the temperature gradient such that

$$\tau \frac{\partial \vec{q}}{\partial t} + \vec{q} = - k\nabla T \qquad (47)$$

where, $\tau = \alpha/c^2$ is the ratio of the thermal diffusivity α to the square of the thermal propagation speed, c, and is commonly referred to as the thermal relaxation time. The additional term $\tau \partial q/\partial t$ in Eq. (47) accounts for the thermal lag. Indeed, Tzou[58] proposes that this equation arises by assuming that, in general, the heat flux lags the imposed thermal gradient, e.g. $\vec{q}(\mathbf{r}, t+\tau) = - k\nabla T(\mathbf{r}, t)$, and upon expanding the left-hand side in a Taylor series in time, assuming τ is very small, and retaining only the linear terms, Eq (47) is obtained. Introduction of Eq. (47) in the energy equation leads to the HHCE

$$\frac{1}{c^2}\frac{\partial^2 T}{\partial t^2} + \frac{1}{\alpha}\frac{\partial T}{\partial t} = \nabla^2 T \tag{48}$$

which is also known as the damped-wave or telegrapher's equation, and it was studied by Lord Kelvin in his modeling studies of the propagation of electric signals along the first trans-Atlantic cable. Indeed, with a slight modification, this equation models the electric field and current along a transmission line. The second derivative in time renders Eq. (48) hyperbolic in nature, while the first-order derivative accounts for the diffusive nature of thermal propagation and introduces damping. Thus the HHCE models the propagation of a sharp thermal wave front damped in time.

Following a method first proposed by Rizzo and Shippy[70] to solve the PHCE, the Laplace transform of the HHCE is first taken, resulting in

$$\nabla^2 \bar{T}(x,s) - \left[\frac{s}{\alpha} + \left(\frac{s}{c}\right)^2\right]\bar{T}(x,s) = -\left[\frac{1}{\alpha} + \frac{s}{c^2}\right]T(x,0) - \frac{1}{c^2}\frac{\partial T}{\partial t}(x,0) \tag{49}$$

where s is the Laplace parameter, $\bar{T}(x,s)$ is the Laplace transform of the temperature, and x generically refers to space. Assuming initial thermal equilibrium, $T(x,y)=0$ and $\partial T/\partial t(x,y,0)=0$, a BIE for the Laplace transform temperature is derived as

$$C(\xi)\bar{T}(\xi,s) + \int_S \bar{T}(y,s)\frac{\partial \bar{T}^*(\xi,y,s)}{\partial n_y}dS_y = \int_S \frac{\partial \bar{T}(y,s)}{\partial n_y}\bar{T}^*(\xi,y,s)dS_y \tag{50}$$

where $C(\xi)$ is the usual free term, which is one for ξ at an interior point and 1/2 for ξ on a smooth boundary, see Eq. (2). The above integral equation is discretized following the methods described in Section 1. However, in this case, the free-space Green's function (fundamental solution) is, in 2-D,

$$\bar{T}^*(x,y,s) = -\frac{1}{2\pi}K_0\left(\sqrt{\frac{s}{\alpha} + \left(\frac{s}{c}\right)^2}\,r\right) \tag{51}$$

where K_0 is the modified Bessel function of the second kind of order zero, and $r=|x-y|$. As $c \to \infty$, the HHCE fundamental solution degenerates into the PHCE fundamental solution. Once the Laplace domain temperature is found, it is numerically inverted to retrieve its corresponding time history. Real variable-based numerical inversion methods, such as Schapery's collocation method and the Gaver-Stehfest method, tend to smear out the solution at the wave front and are unable to capture discontinuities of traveling thermal waves. The complex variable method of Honing and Hirdes[71] provides sharp front resolution. The technique is a variant of the Fourier series inversion method of Durbin that provides accelerated series convergence and an optimal choice for the free parameter, v. The inversion formula is

$$f(t) = \frac{e^{vt}}{T}\left[-\frac{1}{2}Re\{F(v)\} + \sum_{k=0}^{N}Re\left\{F\left(v + i\frac{k\pi}{T}\right)\right\}\cos(\frac{k\pi}{T}t)\right.$$

$$\left. - \sum_{k=0}^{N}Im\left\{F\left(v + i\frac{k\pi}{T}\right)\right\}\sin(\frac{k\pi}{T}t)\right] - F1(v,t,T)$$

(52)

Here, $0 < t < 2T$, where t is time, s is the Laplace parameter equal to $v + iw$, F is the Laplace transform of the function $f(t)$, N is the finite number of terms taken in the Fourier series, and $F1(v,t,T)$ is the discretization error. It is obvious from Eq. (51) that the Laplace parameter, s, takes on complex values. This requires modification of standard BEM codes to accommodate complex variable formulation. Care must be given to evaluating the modified Bessel function of order zero and one for large complex arguments, and asymptotic expressions found in Abramowitz and Stegun[72] should be used.

In the two examples presented below, the following non-dimensionalized variables for space and time are used: $\bar{x} = c\,x/2\alpha$ and $\bar{t} = c^2 t/2\alpha$. First, a 1-D problem in an insulated rectangular solid with one side heated is used to demonstrate the propagating shock front characteristic of the HHCE (see Fig. 6a). Three hundred evenly-spaced linear isoparametric elements are used, and the BEM predicted temperature at $\bar{t} = 0.5$ is shown in Fig. 6b. Comparison of Honing and Hirdes inversion with Stehfest inversion reveals that the shock front is highly smeared using the latter method in contrast with the sharp resolution of the Fourier-based technique. In Fig. 6c, the temperatures at $\bar{t} = 0.5$ and $\bar{t} = 1.5$ are plotted illustrating thermal wave reflection form the right side wall and diffusion of the shock front magnitude.

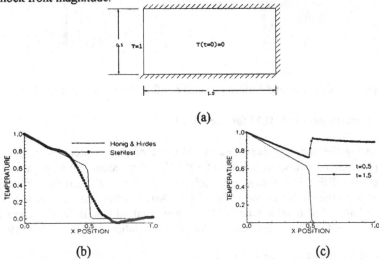

Figure 6 (a) Geometry, boundary and initial conditions, (b) Comparison of Stehfest and Honing and Hirdes BEM inversions at $\bar{t} = 0.5$ and (c) BEM-computed temperatures at $\bar{t} = 0.5, 1.5$ using Honing and Hirdes inversion.

The second example is a 2-D problem of localized heating of an insulated rectangular solid, as illustrated below. The BEM model again uses 300 linear elements. Solutions at three representative times are provided in Fig. 7. The propagation of thermal shock front as well as multiple reflections can be easily traced. Multiple reflections and interactions of oblique thermal shocks from insulated walls are clearly resolved. These intricate interior solutions are obtained from boundary data only, and results compare exceptionally well with those obtained using high resolution TVD finite volume schemes [62].

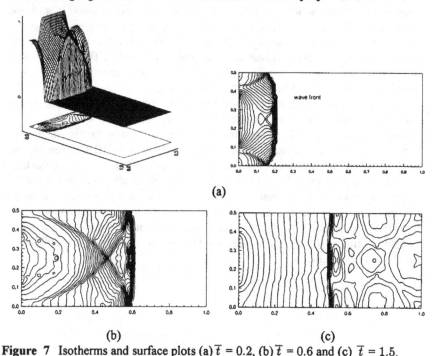

(a)

(b)   (c)

Figure 7   Isotherms and surface plots (a) $\bar{t}$ = 0.2, (b) $\bar{t}$ = 0.6 and (c) $\bar{t}$ = 1.5.

5.2   Time-Dependent BEM for the HHCE

Following the approach presented in Section 1 of this chapter, the fundamental solution (Green free-space solution) for the HHCE is required to formulate a time-dependent BIE. The fundamental solution, $T^*(\xi, x; \tau, t)$, satisfies the adjoint equation to the HHCE perturbed by a Dirac Delta function shifted to some point in space ξ and some point in time t_F. The fundamental solution is derived using a full Fourier transform and subsequently inverting the solution, and in 1-D it is [10]

$$T^*(\xi, \bar{x}; \tau, \bar{t}) = \frac{1}{2}e^{-(\bar{t}-\tau)}I_o[(\bar{t}-\tau)^2 - (\bar{x}-\xi)^2]^{\frac{1}{2}} H[(\bar{t}-\tau) - |\bar{x}-\xi|] \quad (53)$$

where non-dimensional time and space are used, I_o is the modified Bessel function of the first kind of order zero, and H is the Heaviside step function. Note that in

Eq. (53) non-dimensional time, $\bar{t}$, and space, $\bar{x}$, are used. The overbar superscript is dropped onwards with the understanding that space and time are dimensionless. Assuming equilibrium initial conditions, the two domain integrals due to initial conditions go to zero, and, in 1-D, the boundary integral equation is

$$C(\xi)T(\xi,t_F) + \int_{t_0}^{t_F} T(a,t)q^*(\xi,a;t_F,t)dt - \int_{t_0}^{t_F} T(b,t)q^*(\xi,b;t_F,t)dt$$

$$= \int_{t_0}^{t_F} q(a,t)T^*(\xi,a;t_F,t)dt - \int_{t_0}^{t_F} q(b,t)T^*(\xi,b;t_F,t)dt \tag{54}$$

where a and b are the left-most and right-most endpoints of the 1-D domain, q denotes the normal derivative of the temperature, and q^* denotes the normal derivative of T^* which is given in [67]. This equation is discretized in time and collocated at each endpoint, leading to two equations

$$\frac{1}{2}T(a,t_F) = \sum_{f=1}^{N}[T(a,t_f)H_{Ff}(a,a) - T(b,t_f)H_{Ff}(a,b)]$$

$$+ \sum_{f=1}^{N}[q(b,t_f)G_{Ff}(a,b) - q(a,t_j)G_{Ff}(a,a)] \tag{55}$$

$$\frac{1}{2}T(b,t_F) = \sum_{f=1}^{N}[T(a,t_f)H_{Ff}(b,a) - T(b,t_f)H_{Ff}(b,b)]$$

$$+ \sum_{f=1}^{N}[q(b,t_f)G_{Ff}(b,b) - q(a,t_f)G_{Ff}(b,a)] \tag{56}$$

where t_F is the time of interest, $t_f = f\Delta t$, Δt is the time step, and, using a constant in time model, the temporal influence coefficients are

$$H_{F,f}(x,\xi) = \int_{t_{f-1}}^{t_f} q^*(\xi,x;t_F,t_f)dt \,;\, G_{F,f}(x,\xi) = \int_{t_{f-1}}^{t_f} T^*(\xi,x;t_F,t_f)dt \tag{57}$$

These can be evaluated with adaptive Gauss-Kronrod-based quadratures, and it is noted that whenever $x = \xi$, H_{Ff} is zero for all times. It should be noted that even when a constant heat flux condition is imposed, care must be taken at start-up, since the Catteno equation relates the temperature gradient at the wall to the heat flux as well as the time rate of change of the heat flux. Consequently, when starting from equilibrium conditions and imposing a constant heat flux, the boundary condition at the wall can be approximated at the first time step as

$$\tau \frac{\Delta \vec{q}_w}{\Delta t} + \vec{q}_w = -k\nabla T \cdot \hat{n}|_w \tag{58}$$

where the subscript w denotes the location to be taken at the wall. Subsequently, the usual definition of the wall flux is retrieved if $\vec{q}_w$ is constant. Care must also be taken to account for the behavior of the Dirac Delta function which appears in $H_{F,f}(x,\xi)$ and in the Heaviside step function which appears both in $H_{F,f}(x,\xi)$ and $G_{F,f}(x,\xi)$ (see[67]). Upon imposing boundary conditions at each endpoint, Eqs. (55) and (56) lead to two simultaneous equations which must be solved at each new time for the current value of the endpoint temperature or wall gradient, whichever is not specified as a boundary condition. Once the time histories of the temperature and wall gradients are resolved, the interior temperature distribution is determined by locating the source point ξ at the point of interest, discretizing Eq. (54), and reconstructing the time history of the point of interest. An example of temperature distributions predicted using the method discussed in this section is provided in Fig. 8. Here, in Fig. 8a, one hundred sixty-two time steps are taken to advance the solution from $t = 0$ to $t = 2.4$ for a medium originally in thermal equilibrium and whose left wall is imposed with a temperature of one and the right maintained at zero temperature. Results compare well with the analytical solution. In Fig. 8b, the same problem is considered with the right wall insulated to illustrate the reflection of the thermal shock front. Again BEM results compare well with the exact solution. This completes the review of the treatment of the HHCE using the BEM. In the next section, BEM applications in inverse problems are considered.

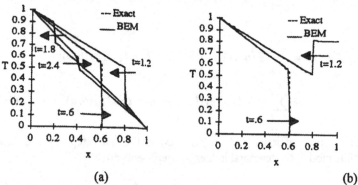

(a)   (b)

Figure 8 BEM-computed temperature profiles (a) 1-D problem with $T(0,t) = 1$ and $T(1,t) = 0$ and (b) 1-D problem with $T(0,t) = 1$ and $\partial T/\partial x|_{(1,t)} = 0$.

6 BEM APPLICATION TO INVERSE PROBLEMS IN HEAT CONDUCTION

In contrast to a well-posed forward problem, in an inverse problem, one of the following is not completely specified: the governing equation for the temperature, thermo-physical properties, boundary conditions, initial condition, or system geometry. The purpose of the inverse problem is to find the unknown with aid of additional (overspecified) information. Typically, the overspecified condition is provided by measuring the temperature at an exposed boundary or by internal measurements via embedded sensors. Noise becomes an important concern in the solution of most inverse problems as the overspecified condition is usually provided by an experiment. General review and developments of inverse problems in heat transfer can be found in monographs by Beck *et al.*[73], Alifanov[74], and Kurpisz and Nowak[75], while BEM application in inverse thermal problems can be found in Ingham and Wrobel[76] and Ingham and Yuan[77]. This section will restrict itself to BEM application to two inverse heat conduction problems: inverse geometric problem (IGP) and identification of unknown convective heat transfer coefficient.

6.1 Inverse Geometric Problem

In the IGP problem, with application to nondestructive evaluation of flaws and cavities, the portion of the geometry hidden from view is unknown and to be identified with the aid of a Cauchy (overspecified) condition at the exposed surface. This can be accomplished by measuring transient or steady-state thermal responses of a conducting medium subjected to a thermal load. In this section we consider the steady-state-based method termed IR-CAT (infrared computerized axial tomography), (see Kassab and Hsieh[78]). This technique relies on surface temperature measurement using an IR-scanner to reconstruct cross-sectional views of the scanned object by solving the inverse geometric problem, see Fig. 9.   A cavity Γ_c is detected using the surface temperature distribution T_s provided by the IR-scanner. The exposed surface Γ_e, whose temperature is scanned, is exchanging heat with the surroundings at T_∞. The convective coefficient h at the exposed surface is known, and, thus, both temperature and heat flux are available on Γ_e. This provides a Cauchy condition at this surface. An inverse problem is solved reconstructing the subsurface cavity geometry iteratively by matching the measured surface temperatures with the BEM-predicted surface temperature distribution under the current geometric configuration. The BEM offers a distinct advantage in solving this type of problem as only boundary information at a

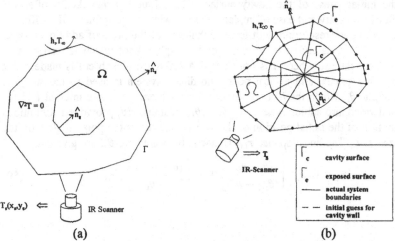

(a)   (b)

Figure 9 IGP problem (a) inverse geometric problem configuration and (b) star-anchored grid pattern along with BEM discretization and a circular initial guess.

continually evolving boundary is required, and, as such, complex domain re-meshing of each new geometry required by FDM or FEM is avoided.

Hsieh *et al.*[79] were the first to use BEM in a pattern search method for the solution of the inverse geometric problem. Other researchers apply the BEM to solve the inverse geometric problem[80,81] but limit detectable geometries to elliptic or super-elliptic shapes. Essentially, these approach the inverse problem by searching for a restricted set of unknown parameters, the location of the center (x_o, y_o), the semi-major axes values, (a, b), and the exponent n (in case of a super-elliptic trial) of the function $[(x - x_o)/a]^n + [(y - y_o)/b]^n - 1 = 0$. Kassab and Pollard[82,83] developed flexible BEM-based anchored grid pattern (AGP) algorithms which are free of geometric constraints. We now describe this method which is based on either linear or cubic spline representations of the inner cavity.

The AGP algorithm requires solution of a series of forward heat conduction problems. The body under investigation is modeled under steady-state conditions with no heat sources and sinks and with constant thermal conductivity. If important due to large expected temperature differences, variation of the thermal conductivity with temperature can be accounted for by use of the Kirchhoff transform as described in Section 1 of this chapter. Thus, regardless of whether the problem is linear or nonlinear, a Laplace equation in the dependent variable (temperature or Kirchhoff transform) is solved by the BEM. The AGP iterative method starts with an initial guess for the sought cavity geometry. Typically, the shape of a circle or an ellipse may be used, and the exposed surface temperature distribution often provides clues to locate the initial cavity. A pattern of JE nodes is used to discretize the exposed surface Γ_e; see Fig. 9. The initial guess for the cavity is discretized using JC nodes. An AGP is laid out in the interior of the body. In the star-like AGP illustrated in the figure, each ray is associated with an exposed surface node and an interior cavity node, and the base points of all the rays are located at the same point (anchor point). Both linear and cubic splines can be used for the initial guess of the cavity surface (Γ_c). Linear spline AGP's offer the benefit of simplicity and ease of implementation, while cubic spline AGP's offer the benefit of higher resolution with fewer unknowns at the price of additional logic. In the AGP, a cavity node corresponds to a spline knot which is restricted to move along the ray where it is positioned. A reference branch cut is made along the positive x-axis, and a local polar coordinate system is used to measure the polar angle, θ_i, which the i-th ray makes with the branch cut. The interval $I = [0, 2\pi]$ is divided into subintervals $I_i = [\theta_{i-1}, \theta_i]$. A spline $r(\theta)$ provides a continuous distribution of the radial position of the cavity wall versus the polar angle θ on the subinterval I_i. The linear spline, $r_{ls}(\theta)$, and cubic spline, $r_{ls}(\theta)$, are given as

$$r_{ls}(\theta) = r_{i-1}\left[\frac{(\theta - \theta_i)}{(\theta_{i-1} - \theta_i)}\right] + r_i\left[\frac{(\theta - \theta_{i-1})}{(\theta_i - \theta_{i-1})}\right], r_{ls} \in I_i \qquad (59)$$

$$r_{cs}(\theta) = \frac{(\theta_i - \theta)^3}{6\Delta\theta_i} M_{i-1} + \frac{(\theta - \theta_{i-1})^3}{6\Delta\theta_i} M_i$$
$$+ \frac{\theta_i - \theta}{\Delta\theta_i}\left(r_{i-1} - \frac{M_{i-1}\Delta\theta_i^2}{6}\right) + \frac{\theta - \theta_{i-1}}{\Delta\theta_i}\left(r_i - \frac{M_i\Delta\theta_i^2}{6}\right), r_{cs} \in I_i \tag{60}$$

respectively. Spacing between the knots is $\Delta\theta_i = \theta_i - \theta_{i-1}$. The second derivative of the spline at the i-th knot is M_i, and the position of the i-th knot is r_i. The linear spline is defined by the values of r at the knots, while the cubic spline is defined by the values of its second derivatives M_i at the knots which is found by solving a set of tridiagonal simultaneous equations for the second derivative values M_i. Both splines interpolate the data at the knots, e.g., $r(\theta_i) = r_i$.

With the above cubic spline defining the current estimate for the cavity geometry, the forward heat conduction problem is solved by the BEM. At any iteration, so long as the current estimate of the cavity geometry deviates from the true cavity shape, current BEM-computed surface temperatures will differ from IR-detected surface temperatures. Consequently, residuals are defined at the exposed surface nodes to measure this difference as

$$R_i(r_1, r_2, \ldots, r_{JC}) = \alpha\left[\max\left|T_{j,BEM} - T_{j,IR}\right|\right]_{\theta_j \in I_i}, \quad i = 1, 2, \ldots, JC \tag{61}$$

where α is the sign associated with the difference between the BEM-computed and IR-detected surface nodal temperatures where the absolute difference is a maximum on the I_i subinterval, and θ_j is the polar angle of the j-th node on the exposed surface. The cavity wall is located iteratively by solving the nonlinear system of Eq. (61) for the radial location of the spline knots. This is most effectively accomplished using the Newton-Raphson method to update the current estimate of the cavity geometry $\underline{r}^{n+1}$ from its prior estimate, $\underline{r}^n$, as: $\underline{r}^{n+1} = \underline{r}^n - \underline{r}_c^n$. The correction vector, $\underline{r}_c^n$, is obtained by solving the following linear equations: $[J(\underline{r}^n)]\underline{r}_c^n = \{R(\underline{r}^n)\}$. The Jacobian matrix of partial derivatives, $J_{ij} = \partial R_i(\underline{r}^n)/\partial r_j$, measures the response (sensitivity) of the residuals to a perturbation in the system geometry. The Jacobian matrix can be evaluated at the first iteration and subsequently updated efficiently using the Broyden update, which involves only a sequence of vector multiplications. Iteration proceeds until residuals are within an acceptable tolerance.

The logic of the AGP algorithm is illustrated in Fig. 10. The advantage of the AGP is that it limits the space searched to obtain a solution and readily allows physical constraints to be implemented as the spline knots are confined to move along the AGP. Location updates of the spline knots are prohibited to yield negative values of r_i or locating a knot outside its associated exterior node. The method effectively avoids folding of the geometry in intermediate steps. Pollard and Kassab[82-84] devised a traveling hole algorithm to dynamically relocate the AGP and its anchor point towards the centroid of the current configuration in case of poor or incorrect initial guesses. They also develop a strategy employing various non star-like AGP's for multiple subsurface cavities.

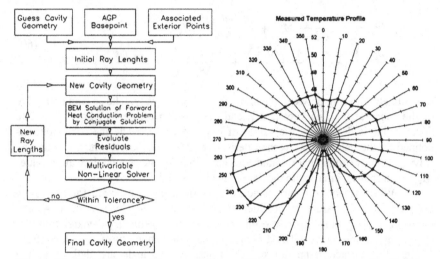

Figure 10 The AGP algorithm.   **Figure 11** Surface temperature distribution.

The flexibility of the AGP technique to detect irregular cavity geometries has been established in [82,83]. It is illustrated here using results from an experiment [83] in which a plaster model with irregular inner and outer geometries was constructed, heated at the cavity wall, and allowed to reach steady state. An AGA Thermovision System 680 (Indium-Antimonide IR-Scanner) was used to retrieve the surface temperature profile nonintrusively, and the profile is displayed as a polar plot in Fig. 11. The inner-lobe temperature was measured as 63°C, room temperature was 27.8°C, and a value of $h/k = 0.28 \, cm^{-1}$ was used. It is assumed that this value is applicable to the entire model surface. A cross section of the exterior geometry and the detected and actual cavity geometries are shown below, along with the initial guess used to start the iteration. It is noted that the cubic spline AGP locates the cavity using six unknowns (one of the rays corresponds to the vertical y-axis).

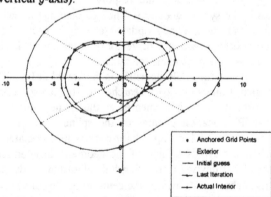

Figure 12 Comparison of actual and detected cavity.

We now discuss another approach based on the adjoint method which was proposed by Meric[88,89] in a series of papers in which the author couples the technique with the BEM and the material derivative method to address the related problem of shape optimization of heat conducting systems. Huang and Chao[86] and Huang and Chiang[87] recently used this approach to solve the IGP using BEM with measurements taken at the surface[86] and at the interior of the body[87]. The adjoint method is useful in gradient-based iterative processes which minimize a quadratic functional (cost function) measuring the error incurred with the current approximation of the cavity. In this technique, the gradient vector is evaluated explicitly by solving related problems, which themselves are amenable to solution by BEM. Although by no means limited to the IGP, we now describe the technique in the context of the inverse geometric problem.

Here again, the steady case is considered, and the Laplace equation governs the temperature distribution. Denote measured temperatures taken at N-points along the exposed convecting surface Γ_e, see Fig. 9b, as $Y_i = Y(x_i, y_i)$. Since the exposed surface is convecting heat, and the temperature is known, then the heat flux, q, is known there as well. For the sake of discussion, let us assume that the temperature is prescribed at the interior cavity Γ_c. Therefore, the field problem is

$$\nabla^2 T = 0, (x, y) \in \Omega \; ; T = Y \text{ and } \frac{\partial T}{\partial n_y} = q, (x, y) \in \Gamma_e \; ; T = T_o, (x, y) \in \Gamma_c \quad (62)$$

where Ω is the domain bounded by the boundary Γ. As long as the current cavity estimate does not match the actual cavity, then the BEM-computed temperatures, T_i, will not match the measured temperatures, Y_i, and a measure of error is

$$R(\underset{\sim}{r}) = \sum_{i=1}^{N} [T_i(\underset{\sim}{r}) - Y_i]^2 \quad (63)$$

Here, $R(\underset{\sim}{r})$ is a scalar function of $\underset{\sim}{r}$, which generically refers to the parametric representation of the unknown geometry. For example, Meric uses a fixed anchored linear AGP-type discretization to solve the optimization problem. Huang and his colleagues actually solve for the (x, y) nodal coordinate locations of the estimated cavity geometry, but do not mention particular schemes used to avoid the folding of the evolving cavity geometry and physically unrealistic intermediate shapes.

An iterative process is then used to advance the current guess, $\underset{\sim}{r}^n$, for the parametric representation of the cavity wall to obtain a new guess, $\underset{\sim}{r}^{n+1}$, as $\underset{\sim}{r}^{n+1} = \underset{\sim}{r}^n + \Delta \underset{\sim}{r}^n$, and in gradient-based methods, the correction vector $\Delta \underset{\sim}{r}^n$ is related to $R'(\underset{\sim}{r}) = \nabla R = [\partial R/\partial r_1, \partial R/\partial r_2, ... \partial R/\partial r_M]^T$, the gradient of the cost function. For instance, if steepest descent is used, the update is

$$\underset{\sim}{r}^{n+1} = \underset{\sim}{r}^n - \beta^n \underset{\sim}{R}'(\underset{\sim}{r}^n) \quad (64)$$

where β^n is a parameter which solves a minimum problem to be discussed shortly. The gradient of the cost function is evaluated by solving the adjoint problem, which is efficiently carried out by BEM. A Lagrange multiplier is introduced to adjoin the governing equation to the residual as

$$R(\underline{r}) = \int_{\Gamma_e} [T(\underline{r}) - Y]^2 D(x - x_i)(y - y_i) d\Gamma + \lambda \int_{\Omega} \nabla^2 T d\Omega \qquad (65)$$

where $D(x - x_i)(y - y_i)$ is the sum of Dirac Delta functions shifted at the measuring points (x_i, y_i), $i = 1, 2 ... N$. Taking the variation of the above, δR, due to a variation in the geometry of the inner cavity parameterization, $\delta \underline{r}$, leads to

$$\delta R = \int_{\Gamma_e} 2[T(\underline{r}) - Y] D(x - x_i)(y - y_i) \delta T d\Gamma$$
$$+ \int_{\Omega} [\delta \lambda \nabla^2 T + \lambda \nabla^2 (\delta T)] d\Omega \qquad (66)$$

Applying Green's second identity to the second domain integral and recognizing that the first domain integral is zero, due to the governing equation, results in

$$\delta R = \int_{\Omega} \nabla^2 \lambda \, \delta T d\Omega + + \int_{\Gamma_c} \left\{ \lambda \frac{\partial (\delta T)}{\partial n} - \delta T \frac{\partial \lambda}{\partial n} \right\} d\Gamma \qquad (67)$$
$$+ \int_{\Gamma_e} \left[\left\{ 2[T(\underline{r}) - Y] D(x - x_i)(y - y_i) - \frac{\partial \lambda}{\partial n} \right\} \delta T - \lambda \frac{\partial (\delta T)}{\partial n} \right] d\Gamma$$

Taking into account the known boundary conditions for the direct problem in Eq. (62), and taking into account the contour integrals above, the following adjoint problem is constructed to eliminate all but one contour integral. First, the integrand of the domain integral is set to zero, then setting $\lambda = 0$ at the cavity and setting the term in the braces in the integrand of the integral over the exterior boundary to zero, and recognizing that at the external boundary $\delta T = 0$, the following boundary value problem for the adjoint variable arises

$$\begin{array}{ll}
\nabla^2 \lambda = 0 & (x, y) \in \Omega \\
\lambda = 0 & (x, y) \in \Gamma_c \\
\dfrac{\partial \lambda}{\partial n} = 2[T(\underline{r}) - Y] D(x - x_i)(y - y_i) & (x, y) \in \Gamma_e
\end{array} \qquad (68)$$

where the boundary condition at the exposed boundary is evaluated at measuring points. This leaves the following contour integral

$$\delta R = \int_{\Gamma_c} -\frac{\partial T}{\partial n} \frac{\partial \lambda}{\partial n} \delta \underline{r} \, d\Gamma \qquad (69)$$

where the relation $\delta T = [\partial T / \partial n] \delta \underline{r}$ on Γ_c has been used. Upon recognizing that

$\delta R = \int_{\Gamma_c} \nabla R(\underset{\sim}{r}) \, \delta \underset{\sim}{r} \, d\Gamma$ [74], the expression for the gradient is established as

$$\nabla R(\underset{\sim}{r}) = -\frac{\partial T}{\partial n} \frac{\partial \lambda}{\partial n}\bigg|_{\Gamma_c} \tag{70}$$

It is noted that the boundary value problem for λ is not only readily solved using the BEM, but, furthermore, the normal derivative of λ is itself a nodal unknown which is determined upon the solution of the BEM equations, see Eq.(3), without having resort to numerical differentiation, as would be the case with finite difference or finite elements.

The stepping parameter β^n is found by implicitly expressing $T_i(\underset{\sim}{r})$ in Eq. (64) as $T_i(\underset{\sim}{r}^n - \beta^n \underset{\sim}{R}'(\underset{\sim}{r}^n))$, expanding in a Taylor series and retaining only linear terms, introducing the results into the quadratic functional in Eq. (63), and minimizing with respect to β^n. There results [74]

$$\beta^n = \sum_{i=1}^{N} \left[T_i - Y_i\right]^2 \delta T_i \bigg/ \sum_{i=1}^{N} \delta T_i^2 \tag{71}$$

Thus, β^n is evaluated by solving another auxiliary problem for the temperature sensitivity, δT. This problem is derived by perturbing T by δT due to a variation of $\underset{\sim}{r}$ by $\delta \underset{\sim}{r}$ in the original problem and, upon neglecting second order terms, there results

$$\nabla^2 \delta T = 0 \qquad\qquad (x, y) \in \Omega \tag{72}$$
$$\delta T = 0 \qquad\qquad (x, y) \in \Gamma_c$$
$$\delta T = \frac{\partial T}{\partial n} \delta \underset{\sim}{r} \qquad\qquad (x, y) \in \Gamma_e$$

which is also efficiently solved by BEM. Thus, solving the inverse problem by this approach requires the following sequence:

1.' For the current guess $\underset{\sim}{r}^n$, solve direct problem in Eq. (62) for T by BEM using the flux boundary condition.
2. Solve adjoint problem, Eq. (68), for λ using BEM, and compute $\underset{\sim}{R}'(\underset{\sim}{r}^n)$.
3. Set $\delta \underset{\sim}{r} = \underset{\sim}{R}'(\underset{\sim}{r}^n)$ and solve sensitivity problem, Eq. (72), for δT by BEM.
4. Compute β^n using Eq. (71).
5. Check convergence criterion and continue if needed.

The convergence criterion is taken following Alifanov[74] as $R(\underset{\sim}{r}^{n+1}) \leq N\sigma^2$, where N is the number of measuring points and σ is the standard deviation of the error in measurement of the temperature which is assumed to be constant over all measurements. The steepest descent method is not necessarily the most efficient means of minimizing the functional; however, the basic principles and the use of the BEM in solving this problem are well illustrated with this example. Faster converging techniques, such as the conjugate gradient method, require an additional level of computation to find the conjugate directions. Regularization of the functional has not been addressed so far, and this topic is now taken up in the next application of BEM to inverse problems, the identification of the heat transfer coefficient from surface temperature measurements.

6.2 Identification of the Convective Heat Transfer Coefficient

Identifying the convective heat transfer coefficient, h, from surface temperature measurements is routinely carried out in industry and academic research labs. Thin-film gauges used to measure surface heat fluxes or surface temperature measurements are taken using liquid crystals or temperature-sensitive paints. Typically, measurements are transient, and an inverse analysis is subsequently carried out to determine heat transfer coefficients. It is current standard practice to use a one-dimensional conduction model for a semi-infinite slab

$$\frac{T(t) - T_i}{T_\infty - T_i} = 1 - e^{\left(\frac{h^2 t}{k\rho c}\right)} erfc\left[h\sqrt{t/k\rho c}\right] \tag{73}$$

as the basis of the analysis. It is obvious that this approach leads to erroneous values of h, in particular in regions such as corners and stagnation points where heat transfer cannot be modeled as 1-D for any duration of time. Consequently, improvement of the analysis has been sought to produce accurate h values [90].

The BEM is ideally suited to solve this type of inverse problem as the surface temperature and heat flux appear as nodal unknowns, and these are precisely the variables required in the inverse analysis. Further, the method affords ease of multi-dimensional modeling of temperature and heat flux with a minimal effort of meshing the surface. This feature has been capitalized in the work of Maillet et al.[91], who formulate an inverse BEM-based approach to retrieve the angular distribution of the heat transfer coefficient from a cylinder in a numerical study. The authors propose a second-order regularization technique to stabilize results. Hsieh and Farid[92] also used an inverse BEM approach to retrieve the angular variation of the natural convective heat transfer coefficient over a rough, heated horizontal cylinder in an experiment in which the steady-state surface temperature of the cylinder is measured non-intrusively by infrared scanning. Recently, Martin and Dulikravich[93] also use a steady-state, BEM-based approach in a numerical study to retrieve the heat transfer coefficient and use singular value decomposition to stabilize results. However, transient surface temperature measurements are more commonly used in experiments. Divo et al.[94] recently developed an inverse transient-BEM-based technique to retrieve multi-dimensional heat transfer coefficients. This technique is now presented.

The method relies on the transient BEM to solve the diffusion problem; see Section 1. Let NB denote the number of surface temperature measurements taken at the convective boundary of interest, $\widehat{T}_i^F$ denote these measured temperatures, and T_i^F denote the temperatures computed by the BEM at these nodes. The superscript F identifies the current time level. Film coefficient retrieval is actually a direct problem due to the fact that the time history of temperature measurements is used as imposed first-kind boundary conditions at convective boundaries. However, small measurement errors at these boundaries translate into large deviations of the computed heat fluxes and, consequently, erroneous film coefficients are modeled. To overcome this sensitivity problem, convective coefficient retrieval is treated as an inverse problem in which the following least-square functional is minimized at the current time level F:

$$Z^F = \sum_i^{NB} \left(\frac{\widehat{T}_i^F - T_i^F}{\Delta T^F} \right)^2 + \beta \sum_i^{NB} \left(\frac{q_i^F - \overline{q}_i^F}{\Delta q^F} \right)^2 \tag{74}$$

Here, ΔT^F is the maximum range of temperatures used to normalize the quadratic functional, and the second term is a regularization term modulated by the regularization parameter β. The terms q_i^F are the BEM-computed fluxes at the measuring points, $\overline{q}_i^F$ is an interpolation of currently computed heat flux distribution at the convective boundary, and Δq^F is the maximum range of heat fluxes used to normalize the regularization term. With addition of the regularization term to the quadratic functional, minimization of Z^F allows the solution to adjust temperatures at the convective nodes to "best fit" the input data. The regularization parameter β plays an important role by providing a means to tune the smoothing of the solution while controlling deviations from measured temperatures.

The term $\overline{q}_i^F$ is not commonly used in standard regularization. Rather, heat fluxes are usually regularized with respect to zero. Subtracting this term allows the optimized heat fluxes q_i^F to approach $\overline{q}_i^F$ as the regularization parameter β approaches infinity, and the optimized temperatures T_i^F to approach the measurements $\widehat{T}_i^F$ as the regularization parameter β approaches zero. The term $\overline{q}_i^F$ is the current interpolation of computed heat fluxes at the convective nodes and is represented through a linear combination of nodal fluxes as

$$\overline{q}_i^F = A_{ik} q_k^F \tag{75}$$

where the matrix A_{ik} is derived from an orthogonal polynomial fit of the current BEM-computed surface fluxes at the convective nodes (explained later). Indicial notation is used hereon. Introducing the above in the functional, there results

$$Z^F = \frac{1}{(\Delta T^F)^2} \sum_i^{NB} \left(\widehat{T}_i^F - T_i^F \right)^2 + \frac{\beta}{(\Delta q^F)^2} \sum_i^{NB} \left(B_{ik} q_k^F \right)^2 \tag{76}$$

where the matrix $B_{ik} = \delta_{ik} - A_{ik}$, and δ_{ik} is the Krönecker delta. Minimization is accomplished by differentiating Z^F with respect to q_j^F and setting to zero

$$\frac{\partial Z^F}{\partial q_j^F} = \frac{2}{(\Delta T^F)^2} \left(T_i^F - \widehat{T}_i^F \right) \frac{\partial T_i^F}{\partial q_j^F} + \frac{2\beta}{(\Delta q^F)^2} B_{ik} q_k^F B_{ik} \delta_{kj} = 0 \tag{77}$$

and furthermore,

$$X_{ij}\left(\widehat{T}_i^F - T_i^F\right) = m^F \beta C_{jk} q_k^F \tag{78}$$

where $X_{ij} = \partial T_i^F/\partial q_j^F$ is the sensitivity matrix, $m^F = (\Delta T^F/\Delta q^F)^2$, and $C_{jk} = B_{ik}B_{ij}$. The relationship between temperatures and heat fluxes can be established rewriting the BEM equations (see Section 1) as follows:

$$T_i^F = \bar{H}_{ik}^{FF} G_{kj}^{FF} q_j^F + d_i^F \tag{79}$$

where $\bar{H}_{ik}^{FF}$ is the inverse of $\widehat{H}_{ik}^{FF}$ and $d_i^F = \bar{H}_{ik}^{FF} c_k^F$. From this expression,

$$X_{ij} = \partial T_i^F/\partial q_j^F = \bar{H}_{ik}^{FF} G_{kj}^{FF} \tag{80}$$

or

$$X_{ij}\left(\widehat{T}_i^F - X_{ik}q_k^F - d_i^F\right) = m^F \beta C_{jk} q_k^F \tag{81}$$

and finally, rearranging, we arrive at

$$\left(X_{ij}X_{ik} + m^F \beta C_{jk}\right)q_k^F = X_{ij}\left(\widehat{T}_i^F - d_i^F\right) \tag{82}$$

This algebraic set of equations can be written in the simple form $[A]\{x\} = \{b\}$ and can be solved by direct or indirect methods at each time level to obtain the optimized boundary heat fluxes. Once heat fluxes are obtained, Eq. (79) is used to calculate the optimized boundary temperatures at non-convective nodes, if this is of interest. Finally, Newton's cooling law is used to retrieve the heat transfer coefficient at each node as $h_i^F = q_i^F/(T_i^F - T_{\infty,i}^F)$, where $T_{\infty,i}^F$ is the reference ambient temperature at node i.

It is important to point out that when performing the matrix product $X_{ij}X_{ik}$, the indices i and j are valid only for those elements at which inverse convective conditions are imposed, while the index k is valid for all elements ($k = 1, ..., N$). The rest of the indices in Eq. (82) are valid just for those elements at which the inverse convective condition is applied. Therefore, the algebraic system resulting from Eq. (82) should be augmented to the size of N (number of boundary elements) in order to obtain a square system. Moreover, it should be noted that the above method is readily applied to the steady state case by setting $d_i^F = 0$ in Eq. (82) and using the steady state H and G coefficients, described in Section 2, to evaluate the sensitivity coefficients X_{ij}.

Interpolation of convective boundary heat fluxes used in the functional regularization term is carried out with the aid of Gram-Schmidt orthogonalization to determine the interpolant matrix A_{ik}. A homogeneous parameter η is used to provide a 1-D parameterization of the axis along the convective boundary, and it is used as the interpolation variable. However, sharp corners are usually encountered along convective boundaries precluding a smooth transformation into the new dimension. To resolve this problem, independent interpolations are performed along smooth boundaries and then arranged together to form the final matrix A_{ik}, as illustrated in Fig. 12. The process of determining the interpolation matrix for the boundary heat fluxes can be readily automated. When an angle γ is detected between two adjoining boundary elements that differs from 180° by a fixed

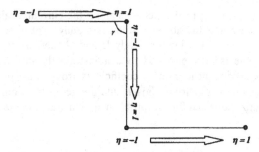

Figure 12 Interpolation process of the boundary heat fluxes.

amount $\Delta\gamma$ (in this case fixed to 30°), a new independent curve-fit is performed. The polynomial representation of the heat fluxes along the straight path is

$$\bar{q}\,(\eta) = \sum_{n=0}^{N} c_n\, p_n(\eta) \tag{83}$$

where the Gram-Schmidt polynomials $p_n(x)$ have the following form

$$p_{j+1}(\eta) = (\eta - \alpha_{j+1})p_j(\eta) - \beta_j\, p_{j-1}(\eta) \tag{84}$$

where $p_0(\eta) = 1$ and $p_{-1}(\eta) = 0$. The coefficients α and β are

$$\alpha_{k+1} = \frac{\sum_{i=1}^{n}\eta_i\,[p_k(\eta_i)]^2}{\sum_{i=1}^{n}[p_k(\eta_i)]^2} \quad \text{and} \quad \beta_k = \frac{\sum_{i=1}^{n}[p_k(\eta_i)]^2}{\sum_{i=1}^{n}[p_{k-1}(\eta_i)]^2} \tag{85}$$

where the η_i's correspond to the boundary nodes along the interpolation path $(-1 < \eta < 1)$. The order of the polynomials used in the code varies between three and five depending directly on the number of boundary nodes along the interpolation path. A least-squares cofactor matrix is constructed as

$$d_{lm} = \sum_{i=1}^{n} p_l(\eta_i) p_m(\eta_i) \tag{86}$$

which will turn to be diagonal due to the orthogonality of the polynomials p_l at the points η_i. Therefore, the interpolated heat fluxes along the path are

$$\bar{q}_i = p_l(\eta_i) d_{lm}^{-1} p_m(\eta_k)\, q_k \tag{87}$$

From the previous expression, the interpolation matrix A_{ik} can be easily identified as

$$A_{ik} = p_l(\eta_i)d_{lm}^{-1}p_m(\eta_k) \tag{88}$$

for i and k corresponding to the elements along the current straight interpolation path. It should be noted the above method is readily applied to the steady-state case by setting $d_i^F = 0$ and using the steady H and G coefficients.

An example is now provided to demonstrate the efficacy of the inverse approach in identifying heat transfer coefficients from transient surface data. A backward-facing step geometry is chosen, and the geometry, boundary conditions, and BEM discretization used for this problem are illustrated in the figure below.

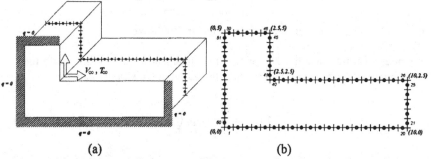

(a)   (b)

Figure 13 Backward facing step (a) Geometry and boundary conditions and (b) BEM discretization.

Thermophysical properties are taken as those of aluminum: $k = 2.37W/cmK$, $\rho = 0.002702\ g/cm^3$, $c = 903\ J/Kg$. The initial condition is set to $T_{ini} = 30°C$, and the ambient convective temperature is $T_{amb} = 800°C$. Due to the irregular geometry, this problem precludes an analytical solution. For this purpose, the direct problem is solved using the BEM with sixty equally spaced quadratic subparametric boundary elements. The temperature history at the convective nodes provided by the direct BEM solution is used as input data or measurements for the optimization problem in order to retrieve the film coefficients.

The imposed film coefficient distribution for the direct BEM solution grows exponentially between a value of $h = 0.01W/cm^2K$ at node 26 to a value of $h = 0.074W/cm^2K$ at node 40 and decreases exponentially from a value of $h = 0.055W/cm^2K$ at node 41 to a value of $h = 0.0075W/cm^2K$ at node 45. Twenty time levels are equally distributed between 0 and 20 seconds for the convolution scheme. Examination of a contour plot of BEM-computed isotherms after 20 seconds in Fig. 14 reveals that the temperature is not well modeled as 1-D and, therefore, the 2-D analysis presented above will be required to obtain accurate values for h.

The boundary temperature solution given by the BEM is now used as measurements to simulate an experiment. The temperature history at each boundary node where convective conditions are imposed is used to force time-dependent first-kind boundary conditions and retrieve the film coefficient distribution from the 1-D and the BEM approximations. In Fig. 15, the 1-D approximation of the film coefficient is compared with the exact distribution along nodes 26 to 45 after 1 second and 20 seconds.

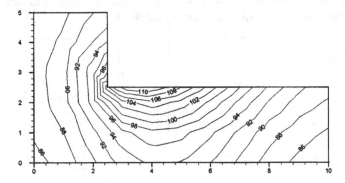

Figure 14 BEM computed isotherms after 20 seconds.

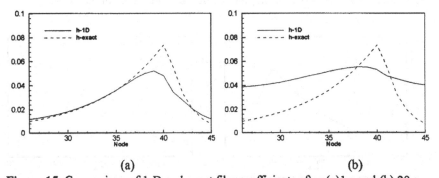

Figure 15 Comparison of 1-D and exact film coefficients after (a)1 s and (b) 20 s.

Next, the temperature history is used as measurement data for the retrieval of film coefficients is slightly modified as $T_{input} = T_{exact} + \epsilon$, with maximum normally distributed random errors of $\epsilon_{max} = \pm 0.5°C$ and $\pm 1°C$ to simulate noise in the input data. First, the film coefficient is retrieved using a non-effective regularization parameter of $\beta = 0$. A comparison of the BEM-retrieved film coefficient distribution for different measurement errors after 1 second and 20 seconds is shown in Fig. 16. Next, the effect of the regularization parameter is shown below. The value of the regularization parameter is found iteratively through experimentation to yield a smooth representation of the film coefficient distribution when a value of $\beta = 5\sigma$ was chosen, where σ is the standard deviation of the simulated input error in measured temperatures which is directly related to the simulated input random error by $\sigma = |\epsilon_{max}|/2.576$ for a level of confidence of 99% that the maximum random error lies within one σ.

Results plotted in Fig. 17 reveal a considerable improvement with respect to the non-regularized solution. The oscillations of the film coefficient distribution apparent in the non-regularized results have been effectively damped down, yielding a smooth prediction for the convective coefficient. This concludes the treatment of inverse problem applications using the BEM.

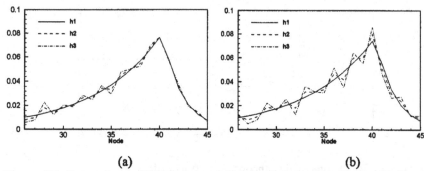

(a)   (b)

Figure 16 Comparison of BEM-retrieved film coefficients after (a) 1 s. (b) 20 s. Measurement errors: $\pm 0°C$ for h1, $\pm 0.5°C$ for h2, and $\pm 1°C$ for h3. $\beta = 0$.

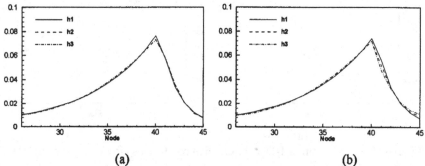

(a)   (b)

Figure 17 Comparison of BEM-retrieved film coefficients after (a) 1 s. (b) 20 s. Measurement errors: $\pm 0°C$ for h1, $\pm 0.5°C$ for h2, and $\pm 1°C$ for h3. $\beta = 5\sigma$.

7 CONCLUSION

In this chapter, the basics of BEM for steady and transient heat conduction have been reviewed. It is shown that the BEM can be applied to most nonlinear problems of practical heat conduction, and this includes nonlinearities due to material properties, boundary conditions, sources, and phase change. Applications of the BEM to modeling hyperbolic heat conduction, to solving the inverse geometric problems, and to solving the inverse heat transfer coefficient identification problem are also presented. Due to the boundary-only discretization retained in all these applications, the BEM offers definite advantages over domain-meshing methods. This feature can be capitalized upon in formulating efficient and accurate numerical schemes for a variety of heat conduction problems.

8 REFERENCES

1. Brebbia, C.A., Telles, J.C.F. and Wrobel, L.C., *Boundary Element Techniques*, Springer-Verlag, Berlin, 1984.
2. Brebbia, C.A. and Dominguez, J., *Boundary Elements: An Introductory Course*, Computational Mechanics Pub., Southampton and McGraw-Hill, New York,1989.
3. Kane, J.H., *Boundary Element Analysis in Engineering Continuum Mechanics*, Prentice-Hall, New Jersey, 1993.
4. Kellogg, O.D., *Foundations of Potential Theory*, Dover, New York, 1953.
5. Chang, Y.P., Kang, C.S. and Chen, D.J., The Use of Fundamental Green's Functions for the Solution of Problems of Heat Conduction in Anisotropic Media, *Int. J. Heat Mass Transfer*, Vol. 16, pp. 1905-1918, 1973.
6. Kassab, A.J. and Divo, E., A Generalized Boundary Integral Equation for Isotropic Heat Conduction with Spatially Varying Thermal Conductivity, *Eng. Analysis with Boundary Elements*, Vol. 18, pp. 273-286, 1996.
7. Divo, E. and Kassab, A.J., A Boundary Integral Equation for Steady Heat Conduction in Anisotropic and Heterogeneous Media, *Num. Heat Transfer, Part B*, Vol. 32, pp. 37-61, 1997.
8. Divo, E. and Kassab, A.J., A Generalized BIE for Transient Heat Conduction in Heterogeneous Media, *J. Thermophysics Heat Trans.*, Vol. 12, No.3, pp. 364-373, 1998.
9. Divo, E. and Kassab, A.J., A Generalized Boundary Integral Equation for Axisymmetric Heat Conduction in Non-homogeneous Media, in *BEM XIX*, *Proc. of the 19th Int. Conf. on Boundary Elements*, (ed. Brebbia, C.A., Santini, P., Aliabadi, M.H., and Orlandi, P.), Computational Mechanics Pub., Southampton, pp. 453-464, 1997.
10. Morse, P.M. and Feshbach, H., *Methods of Theoretical Physics*, McGraw-Hill, New York, 1953.
11. Carslaw, H.S. and Jaeger, J.C., *Conduction of Heat in Solids*, 2nd ed., Clarendon Press, Oxford, 1959.
12. Shaw, R.P., An Integral Equation Approach to Diffusion, *Int. J. Heat Mass Trans.*, Vol. 17, pp. 693-699, 1974.
13. Wrobel, L.C. and Brebbia, C.A., The Boundary Element Method for Steady-State and Transient Heat Conduction, in *Numerical Methods in Thermal Problems*, Vol. 1, (ed. R.W. Lewis and K. Morgan), Pineridge Press, Swansea, 1979.
14. Davey, K. and Hinduja, S., An Improved Procedure for Solving Transient Heat Conduction Problems Using the Boundary Element Method, *Int. J. Num. Meth. Eng.*, Vol. 28, pp. 2293-2306, 1989.

15. Greengard, L. and Strain, J., A Fast Algorithm for the Evaluation of Heat Potentials, *Comm. Pure Appl. Math.*, Vol. 43, pp. 949-963, 1990.
16. Zerroukat, M., A Fast Boundary Element Algorithm for Time-Dependent Potential Problems, *Appl. Math. Modelling*, Vol. 22, pp. 183-196, 1998.
17. Bialecki, R. and Nowak, A.J., Boundary Value Problems in Heat Conduction with Nonlinear Material and Nonlinear Boundary Conditions, *Appl. Math. Modelling*, Vol. 5, pp. 417-421, 1981.
18. Azevedo, J.P.S. and Wrobel, L.C., Non-Linear Heat Conduction in Composite Bodies: A Boundary Element Formulation, *Int. J. Numer. Meth. Eng.*, Vol. 26, pp. 19-38, 1988.
19. Bialecki, R. and Nhalik, R., Solving Nonlinear Steady State Potential Problems in Inhomogeneous Bodies Using the Boundary Element Method, *Num. Heat Trans., Part B*, Vol. 15, pp. 79-96, 1989.
20. Brebbia, C.A. and Skerget, P., Nonlinear Time Dependent Potential Problems Using BEM, in *Boundary Elements VI, Proc. of the 6th Int. Conf. on Boundary Elements*, (ed. Brebbia.,C.A.), Springer-Verlag, Berlin, pp. 2.9-2.23, 1984.
21. Skerget, P. and Brebbia, C.A., Time Dependent Nonlinear Potential Problems, in *Topics in Boundary Element Research*, Vol. 2, (ed. Brebbia, C.A.), Springer-Verlag, Berlin, pp. 63-86, 1985.
22. Kikuta, M., Togoh, H. and Tanaka, M., Boundary Element Analysis of Nonlinear Transient Heat Conduction Problems, *Comp. Methods Appl. Mech. Eng.*, Vol. 62, pp. 321-329, 1987.
23. Pasquetti, R. and Caruso, A., Boundary Element Approach for Transient and Nonlinear Thermal Diffusion, *Num. Heat Transfer, Part B*, Vol. 17, pp. 83-99, 1990.
24. Wrobel, L.C. and Brebbia, C.A., The Dual Reciprocity Boundary Element Formulation for Nonlinear Diffusion Problems, *Comp. Methods Appl. Mech. Eng.*, Vol. 65, pp. 147-164, 1987.
25. Orivuori, S., Efficient Method for Solution of Nonlinear Heat Conduction Problems, *Int. J. Num. Meth. Eng.*, Vol.14, pp. 1461-1476, 1979.
26. Bialecki, R., Applying BEM to Calculation of Temperature Fields in Bodies Containing Radiating Enclosures, in *BEM VII, Proc. of the 7th Int. Conf. on Boundary Elements*, (ed. Brebbia, C.A. and Maier, G.), Springer-Verlag, Berlin, Vol. 1, pp.2.35-2.50, 1985.
27. Bialecki, R., Radiative Heat Transfer in Cavities: BEM Solution, in *BEM X, Proc. of the 10th Int. Conf. on Boundary Elements*, (ed. Brebbia, C.A.) Vol. 2, Springer-Verlag, Berlin, pp. 246-256, 1988.
28. Bialecki, R., Applying the Boundary Element Technique to the Solution of Heat Radiation Problems in Cavities Filled by a Nongray Emitting-Absorbing Medium, *Num. Heat Transfer, Part A*, Vol.20, pp. 41-64, 1991.
29. Sun, B., Zheng, D., Klimpke, B. and Yildir, B., Modified Boundary Element Method for Radiative Heat Transfer Analyses in Emitting, Absorbing and Scattering Media, *Eng. Analysis with Boundary Elements*, Vol. 21, pp. 93-104, 1998.

30. Partridge, P.W., Brebbia, C.A. and Wrobel, L.C., *The Dual Reciprocity Boundary Element Method*, Computational Mechanics Pub., Southampton, 1991.
31. Nowak, A.J., Solution of Transient Heat Conduction Problems Using Boundary-Only Formulation, in *BEM IX, Proc. the 9th Int. Conf. on Boundary Elements*, (ed. Brebbia, C.A., Wendland, W.L., and Khun, G.), Springer-Verlag, Berlin, Vol. 3, pp., 265-276, 1987.
32. Partridge, P.W. and Wrobel, L.C., The Reciprocity Boundary Element Method for Spontaneous Ignition, *Int. J. Numer. Meth. Eng.*, Vol. 30, pp. 953-963, 1990.
33. Boddington, T., Gray, P. and Harvey, D.I., Thermal Theory of Spontaneous Ignition: Criticality in Bodies of Arbitrary Shape, *Phil. Trans. Royal Soc. London*, Vol. 270, pp. 467-506, 1970.
34. Anderson, C.A. and Zienkiewicz, O.C., Spontaneous Ignition: Finite Element Solution for Steady and Transient Conditions, *Trans. ASME J. Heat Transfer*, Vol. 96, pp. 398-404, 1974.
35. Zhu, S. and Satravaha, P., An Efficient Computational Method for Modelling Transient Heat Conduction with Nonlinear Source Terms, *Appl. Math. Modelling*, Vol. 20, pp. 513-522, 1996.
36. Zinn, J. and Mader, C.L., Thermal Initiation of Explosives, *J. Appl. Physics*, Vol. 31, pp. 323-328, 1960.
37. Rubinstein, L.I., *The Stefan Problem*, AMS Translations of Mathematical Monographs, Vol. 27, American Mathematical Society, Providence, R.I., 1971.
38. Hong, C.P., Umeda, T. and Kimura, Y., Application of the Boundary Element Method in Two and Three Dimensional Unsteady Heat Transfer Involving Phase Change: Solidification Problems, in *BEM V, Proc. of the 19th Int. Conf. on Boundary Elements*, (ed. Brebbia, C.A., Futogami, T., and Tanaka, M.), Springer-Verlag, Berlin, pp. 153-162, 1983.
39. Comini, G., Del Giudice, R., Lewis, R.W. and Zienkiewicz, O.C., Finite Element Solution of Non-Linear Heat Conduction Problems with Special Reference to Phase Change, *Int. J. Numer. Meth. Eng.*, Vol. 8, pp. 613-624, 1974.
40. Sarler, B., Alujevic, A. and Kuhn, G., Treatment of Multi-Dimensional Stefan Problems by the Alternating Phase Truncation Boundary Element Method, in *Boundary Element Methods in Mechanical and Electrical Engineering*, (ed. Brebbia, C.A. and Chaudouet-Miranda), Springer-Verlag, Berlin, pp. 371-391, 1990.
41. Sarler, B., Dual Reciprocity Boundary Element Method Formulation for Coupled Energy, Mass and Momentum Transport in Solid-Liquid Phase Change Systems, in *Computational Methods for Free and Moving Boundary Problems IV*, (ed. Brebbia, C.A. and Van Keer), Computational Mechanics Pub., Southampton, pp. 13-25, 1997.

42.   O'Neill, K., Boundary Integral Equation Solution for Moving Boundary Phase Change Problems, *Int. J. Numer. Meth. Eng.*, Vol. 19, pp. 1825-1859, 1983.

43.   Chuang, K. and Szekely, J., On the Use of Green's Functions for Solving Melting and Solidification Problems, *Int. J. Heat Mass Transfer*, Vol. 14, pp. 1285-1294, 1971.

44.   Shaw, R.P., A Boundary Integral Approach to One-Dimensional Ablation Problems, in *Boundary Element Methods in Engineering*, (ed. Brebbia, C.A.), Springer-Verlag, Berlin, pp. 127-140, 1982.

45.   Wrobel, L.C., A Boundary Element Solution to Stefan's Problem, in *BEM V, Proc. the 5th Int. Conf. on Boundary Elements*, (ed. Brebbia, C.A., Futogami, T., and Tanaka, M.), Springer-Verlag, Berlin, pp. 173-182 ,1983.

46.   Coleman, C.J., A Boundary Integral Formulation of the Stefan Problem, *Appl. Math. Modelling*, Vol. 10, pp. 445-449, 1986.

47.   Heinlein, M., Mukherjee, S. and Richmond, O., A Boundary Element Method Analysis of Temperature Fields and Stresses During Solidification, *Acta Mech.*, Vol. 59, pp. 59-81, 1986.

48.   Zabaras, N. and Mukherjee, S., An Analysis of Solidification Problems by the Boundary Element Method, *Int. J. Numer. Meth. Eng.*, Vol. 24, pp.1879-1900, 1987.

49.   Hsieh, C.K., Choi, C-Y. and Kassab, A.J., Solution of Stefan Problems by a Boundary Element Method, in *BETECH VII, Proc. of the 7th Int. Conf. on Boundary Element Technology*, (ed. Brebbia, C.A. and Ingber, M.S.), Computational Mechanics Pub., Southampton, pp. 473-493, 1992.

50.   De Lima-Silva Jnr, W. and Wrobel, L.C., A Boundary Element Formulation of Multi-Dimensional Ablation Problems, in *Boundary Elements XIV, Proc. of 14th Int. Conf. on Boundary Element Methods*, (ed. Brebbia, C.A, Dominguez, J.J., and Paris,F.), Computational Mechanics Pub., Southampton, pp. 391-406, 1992.

51.   De Lima-Silva Jnr, W. and Wrobel, L.C., A Front-Tracking BEM Formulation for One-Phase Solidification/Melting Problems, *Eng. Analysis with Boundary Elements*, Vol. 16, pp. 171-182, 1995.

52.   Zerroukat, M. and Wrobel, L.C., A Boundary Element Method for Multiple Moving Boundary Problems, *J. Comput. Phys.*, Vol. 138, pp. 501-519, 1997.

53.   Takahashi, S., Onishi, K., Kuroki, T. and Hayashi, K., Boundary Elements to Phase Change Problems, in *BEM V, Proc. of the 5th Int. Conf. on Boundary Elements*, (ed. Brebbia, C.A., Futogami, T., and Tanaka, M.), Springer-Verlag, Berlin, pp. 163-172, 1983.

54.   Zerroukat, M., Power, H. and Wrobel, L.C., Heat and Solute Diffusion with a Moving Interface: A Boundary Element Approach, *Int. J. Heat Mass Transfer*, Vol. 41, pp. 2429-2436, 1998.

55.   Zerroukat, M. and Chatwin, C.R., A Finite Difference Algorithm for Multiple Moving Boundary Problems Using Real and Virtual Grid Networks, *J. Comput. Phys.*, Vol. 112, pp. 298-307, 1994.

56.   Bonnerot, R. and Jamet, P., A Conservative Finite Element Method for One-Dimensional Stefan Problems with Appearing and Disappearing Phases, *J. Comp. Physics*, Vol. 41, pp. 357-388, 1981.

57.   Catteno, M.C., Sur Une Forme de L'Equation de la Chaleur Eliminant le Paradoxe d'une Propagation Instantanee, *Comptes Rendus Hebdomadaire Seances de L'Academie des Sciences*, Vol. 247, No. 4, pp. 431-433, 1958.

58.   Tzou, D.Y., *Macro- to Microscale Heat Transfer*, Taylor and Francis,1997.

59.   Glass, D.E., Öziṣik, N.M., McRae, D.S. and Vick, B. On the Numerical Solution of Hyperbolic Heat Conduction, *Num. Heat Transfer*, Vol. 8, pp.497-504, 1985.

60.   Tamma,K. and Railkar, S. B., Specially Tailored Transfinite-Element Formulations for Hyperbolic Heat Conduction Involving Non-Fourier Effects, *Num. Heat Transfer, Part B*, Vol. 15, pp. 221-226, 1989.

61.   Yang, H.Q., Characteristics Based High-Order Accurate and NonOscillatory Numerical Method for Hyperbolic Heat Conduction Equation, *Num. Heat Transfer, Part B*, Vol. 18, pp. 221-241, 1990.

62.   Yang, H.Q., Solution of Two-Dimensional Hyperbolic Heat Conduction Equation by High Resolution Numerical Methods, *Num. Heat Transfer, Part A*, Vol. 21, pp. 33-349, 1992.

63.   Tzou, D.Y., Thermal Shock Phenomenon Under High Rate Response in Solids, in *Annual Review of Heat Transfer*, (ed. Tien, C.L.), Chapt. 3, Hemisphere Publishing Inc., Washington, D.C., pp. 111-185, 1992.

64.   Kaminski, W., Hyperbolic Heat Conduction Equation for Materials with a Nonhomogenous Inner Structure, *J. Heat Transfer*, Vol.112, pp.555-560, 1990.

65.   Vedavaraz, A., Kumar, S., and Moallemi, M.K., Significance of Non-Fourier Heat Waves in Conduction, *ASME J. Heat Transfer*, Vol. 116, pp. 221-224, 1994.

66.   Nordlund, R.S. and Kassab, A.J., Non-Fourier Heat Conduction: a Boundary Element Solution of the Hyperbolic Heat Conduction Equation, *BEM17: Proc. of the 17th Int. Conf. on Boundary Elements*, (ed. Brebbia, C.A., Kim, S., Osswald, T., and Power, H.), Computational Mechanics Pub., Southampton, pp. 279-286, 1995.

67.   Preideman, D. and Kassab, A.J., A Boundary Element Solution of the Hyperbolic Heat Conduction Equation Using Time Dependent Fundamental Solutions, in *BEMXVII, Proc. of 17th Int. Conf. on Boundary Elements*, (ed. Brebbia, C.A., Kim, S., Osswald, T., and Power, H.), Computational Mechanics Pub., Southampton, pp. 305-314, 1995.

68.   Vick, B. and West, R. L., Analysis of Damped Waves Using the Boundary Element Methods, in *BETECH XII, Proc. of 12th Int. Conf. on Boundary Element Technology*, (eds. Frankel, J., Brebbia, C.A., and Aliabadi, M.H.), Computational Mechanics Pub.,Southampton, pp. 264-278, 1996.

69. Lu, W.Q., Zeng, Y., and Liu, J., Extension of the Dual Reciprocity Boundary Element Method to the Problems of Phase-Change and Thermal Wave Propagation, in *BEMXX, Proc. of 20th Int. Conf. on Boundary Elements*, (eds. Kassab, A.J., Brebbia, C.A., and Chopra, M.B.), Computational Mechanics Pub, Southampton, pp. 649-659, 1998.

70. Rizzo, F. and Shippy, D.J., A Method of Solution for Certain Problems of Transient Heat Conduction, *AIAA Journal*, Vol. 8, No. 11, pp. 2004-2009, 1970.

71. Honing, G. and Hirdes, A Method for the Numerical Inversion of Laplace Transforms, *J. Comp. Appl. Math.*, Vol. 10, pp. 113-132, 1984.

72. Abramowitz, M. and Stegun, I., *Handbook of Mathematical Functions*, Dover Publications, Inc., New York, 1972.

73. Beck, J.V., Blackwell, B., and St. Clair, C.R., *Inverse Heat Conduction: Ill-Posed Problems*, John Wiley and Sons, New York, 1985.

74. Alifanov, O.M., *Inverse Heat Transfer Problems*, Springer Verlag, New York, 1994.

75. Kurpisz, K. and Nowak, A.J., *Inverse Thermal Problems*, Computational Mechanics Pub., Southampton, 1995.

76. Ingham, D.B. and Wrobel, L.C., *Boundary Integral Formulations for Inverse Analysis*, Computational Mechanics Pub., Southampton, 1997.

77. Ingham, D.B. and Yuan, Y., *The Boundary Element Method for Solving Improperly Posed Problems*, Computational Mechanics Pub., Southampton, 1994.

78. Kassab, A.J. and Hsieh, C.K., Application of Infrared Scanners and Inverse Heat Conduction Methods to Infrared Computerized Axial Tomography, *Rev. Sci. Inst.*, Vol. 58, pp. 89-95, 1987.

79. Hsieh, C.K., Choi, C.Y. and Liu, K.M., A Domain Extension Method for Quantitative Detection of Cavities by Infrared Scanning, *J. Nondes. Eval.*, Vol. 8, pp. 195-211, 1989.

80. Das, S. and Mitra, A.K., An Algorithm for the Solution of Inverse Laplace Problems and Its Application in Flaw Identification in Materials, *J. Comp. Physics*, Vol. 99, pp. 99-105, 1992.

81. Dulikravich, G.S. and Martin, Geometrical Inverse Problems in Three-Dimensional Non-linear Steady State Heat Conduction, *Eng. Analysis with Boundary Elements*, Vol. 15, pp. 161-169, 1995.

82. Pollard, J. and Kassab, A.J., Automated Algorithm for the Nondestructive Detection of Subsurface Cavities by the IR-CAT Method, *J. Nondes. Eval.*, Vol. 12, No.3, pp. 175-186, 1993.

83. Kassab, A.J. and Pollard, J., Automated Cubic Spline Anchored Grid Pattern Algorithm for the High Resolution Detection of Subsurface Cavities by the IR-CAT Method, *Num. Heat Transfer, Part B: Fund.*, Vol. 26, No.1, pp. 63-78, 1994.

84. Kassab, A.J. and Pollard, J., An Inverse Heat Conduction Algorithm for the Thermal Detection of Arbitrarily Shaped Cavities, *Inverse Prob. Eng.*, Vol. 1, No.3, pp. 231-245, 1995.

85. Kassab, A.J., Hsieh, C.K., and Pollard, J., The Inverse Geometric Problem Applied to the IR-CAT Method for Detection of an Irregular Subsurface Cavity, *Inverse Problems in Engineering Mechanics, Proc. ISIP98, 1998 Int. Symp. Inverse Prob. Eng. Mech.*, (eds. Tanaka, M. and Dulikravich, G.S.), Elsevier Press, New York, pp. 111-120, 1998.

86. Huang, C.H., and Chao, B.R., An Inverse Geometry Problem in Identifying Irregular Boundary Configurations, *Int. J. Heat Mass Transfer*, Vol. 40, No. 9, pp. 2045-2053, 1997.

87. Huang, C.H., and Chiang, C.C., Shape Indentification Problem in Estimating Geometry of Multiple Cavities, *J. Thermophysics Heat Transfer*, Vol. 12, No. 2, pp. 270-277, 1998.

88. Meric, A., Boundary Integral Equation and Conjugate Heat Gradient Method for Optimal Boundary Heating of Solids, *Int. J. Heat Mass Transfer*, Vol. 26, No.2 , pp. 261-2267, 1983.

89. Meric, A., Differential and Integral Sensitivity Formulations and Shape Optimization by BEM, *Eng. Analysis Boundary Elements*, Vol. 15, No.2, pp.181-188, 1995.

90. Walker, D.G. and Scott, E.P., A Method for Improving Two-Dimensional High Heat Flux Estimates from Surface Temperature Measurements, *AIAA Paper 97-2574*.

91. Maillet, D., DeGiovanni, and Pasquetti, R., Inverse Heat Conduction Applied to the Measurement of Heat Transfer Coefficient on a Cylinder: Comparison Between an Analytical and a Boundary Element Technique, *ASME J. Heat Transfer*, Vol. 113, pp. 549-557, 1991.

92. Farid, M.S. and Hsieh, C.K., Measurement of the Free Convection Heat Transfer Coefficient for a Rough Horizontal Non-Isothermal Cylinder in Ambient Air by Infrared Scanning, *ASME J. Heat Transfer*, Vol. 114, pp. 1054-1056, 1992.

93. Martin, T.J. and Dulikravich, G.S., Inverse Determination of Steady Convective Coefficient Distributions, *ASME J. Heat Transfer*, Vol. 120, pp. 119-334, 1998.

94. Divo, E., Kassab, A.J., and Chyu, M.K., A BEM-Based Inverse Approach to Retrieve Multi-dimensional Heat Transfer Coefficients from Transient Temperature Measurements, in *BETECH99, Proc. of the 13th Int. Conf. on Boundary Element Technology*, (ed. Chen, C.S., Brebbia, C.A., and Pepper, D.), Computational Mechanics, Southampton, pp. 65-74, 1999.

SIX

MOLECULAR DYNAMICS METHOD FOR MICROSCALE HEAT TRANSFER

S. Maruyama

1 INTRODUCTION

Molecular level understandings and treatments have been recognized to be more and more important in heat and mass transfer research. A new field, "Molecular Thermophysical Engineering," has a variety of applications in further development of macroscopic heat transfer theory and in handling the extreme heat transfer situations related to advanced technologies.

For example, studies of basic mechanisms of heat transfer such as in phase change heat transfer demand the microscopic understanding of liquid-solid contact phenomena. The nucleation theory of liquid droplet in vapor or of vapor bubble in liquid sometimes needs to take account of nuclei in size of molecular clusters. The efficient heat transfer in three-phase interface (evaporation and condensation of liquid on the solid surface) becomes the singular problem in the macroscopic treatment. Some modeling of the heat transfer based on the correct understandings of molecular level phenomena seems to be necessary. The effect of the surfactant on the heat and mass transfer through liquid-vapor interface is also an example of the direct effect of molecular scale phenomena on the macroscopic problem. The surface treatment of the solid surface has a similar effect.

Even though there has been much effort of extending our macroscopic analysis to extremely microscopic conditions in space (micrometer scale and nanometer scale system), time (microsecond, nanosecond and picosecond technology), and rate (extremely high heat flux), there is a certain limitation in the extrapolations. Here, the development of the molecular dynamics (MD) computer simulation technique has shown the possibility of taking care of such microscale phenomena from the other direction. The MD methods have long been used and are well developed as a tool in statistical mechanics and chemistry. However, it is a new challenge to extend the method to the spatial and temporal scale of

189

macroscopic heat transfer phenomena. On the other hand, by developments of high energy-flux devices such as laser beam and electron beam, more physically reasonable treatment of heat transfer is being required. The thin film technology developed in the semiconductor industry demands the prediction of heat transfer characteristics of nanometer scale materials.

In this chapter, one of the promising numerical techniques, the classical molecular dynamics method, is first overviewed with a special emphasis on applications to heat transfer problems in section 2 in order to give the minimum knowledge of the method to a reader not familiar with it. The van der Waals interaction potential for rare gas, effective pair potential for water and many-body potential for silicon and carbon are discussed in detail. Then, the molecular scale representation of the liquid-vapor interface is discussed in section 3. The surface tension, Young-Laplace equation, and condensation coefficient are discussed from the viewpoint of molecular scale phenomena. Section 4 deals with the solid-liquid-vapor interactions. MD simulations of liquid droplet in contact with solid surface and a vapor bubble on solid surface are introduced. The validity of Young's equation of contact angle is also discussed. Then, demonstrations of real heat transfer phenomena are discussed in section 4. Since heat transfer is intrinsically a non-equilibrium phenomenon, the non-equilibrium MD simulations for constant heat flux system and the homogeneous nucleation of liquid droplet in supersaturated vapor and nucleation of vapor bubble in liquid are discussed. Then, the heterogeneous nucleation of vapor bubble on the surface is also discussed. Some interesting non-equilibrium MD simulations dealing with the formation of molecular structures are introduced in section 5.4. Finally, in section 6, future developments of molecular scale heat transfer are discussed.

2 MOLECULAR DYNAMICS METHOD

Knowledge of statistical mechanical gas dynamics has been helpful to understand the relationship between molecular motion and macroscopic gas dynamics phenomena [1]. Recently, a direct simulation method using the Monte Carlo technique (DSMC) developed by Bird [2] has been widely used for the practical simulations of rarefied gas dynamics. In the other extreme, statistical mechanical treatment of solid-state matters has been well developed as solid state physics [e.g. 3]. For example, the direct simulation of the Boltzmann equation of phonon is being developed and applied to the heat conduction analysis of thin film [4] for example. However, when we need to take care of liquid or inter-phase phenomenon, which is inevitable for phase-change heat transfer, the statistical mechanics approach is not as much developed as for the gas-dynamics statistics and the solid-state statistics. The most powerful tool for the investigation of the microscopic phenomena in heat transfer is the MD method [e.g. 5]. In principal, the MD method can be applied to all phases of gas, liquid and solid and to interfaces of these three phases.

2.1 Equation of Motion and Potential Function

In the MD method, the classical equations of motion (Newton's equations) are solved for atoms and molecules as

$$m_i \frac{d^2 r_i}{dt^2} = F_i = -\frac{\partial \Phi}{\partial r_i},$$ (1)

where m_i, r_i, F_i are mass, position vector, force vector of molecule i, respectively, and Φ is the potential of the system. This classical form of equation of motion is known to be a good approximation of the Schrödinger equation when the mass of atom is not too small and the system temperature is not too low. Equation (1) itself should be questioned when applied to light molecules such as hydrogen and helium and/or at very low temperature. Once the potential of a system is obtained, it is straightforward to numerically solve Eq. (1). In principal, any of gas, liquid, solid states, and inter-phase phenomena can be solved without the knowledge of "thermo-physical properties" such as thermal conductivity, viscosity, latent heat, saturation temperature and surface tension.

The potential of a system $\Phi(r_1, r_2, ... r_N)$ can often be reasonably assumed to be the sum of the effective pair potential $\phi(r_{ij})$ as

$$\Phi = \sum_i \sum_{j>i} \phi(r_{ij}),$$ (2)

where r_{ij} is the distance between molecules i and j. It should be noted that the assumption of Eq. (2) is often employed for simplicity even though the validity is questionable. The covalent system such as carbon and silicon cannot accept the pair-potential approximation.

2.2 Examples of Potential Forms

In order to simulate practical molecules, the determination of the suitable potential function is very important. Here, the well-known Lennard-Jones potential for inert gas and for a statistical mechanical model system is introduced; also introduced are potential forms for water and many-body potential for silicon and carbon. The interaction potential forms between metal atoms are intentionally excluded because the luck of the effective technique of handling free electron for heat conduction prevents from the reasonable treatment of heat conduction through solid metal.

2.2.1 Lennard-Jones potential. An example of the pair potential is the well-known Lennard-Jones (12-6) potential function expressed as

$$\phi(r) = 4\varepsilon \left[\left(\frac{\sigma}{r} \right)^{12} - \left(\frac{\sigma}{r} \right)^6 \right],$$ (3)

3

Table 1 Parameters for Lennard-Jones potential for inert molecules.

	σ [nm]	ε [J]	ε/k_B [K]
Ne	0.274	0.50×10^{-21}	36.2
Ar	0.340	1.67×10^{-21}	121
Kr	0.365	2.25×10^{-21}	163
Xe	0.398	3.20×10^{-21}	232

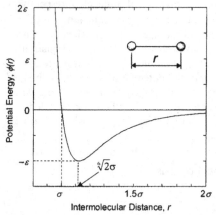

Figure 1 Lennard-Jones (12-6) potential.

where ε and σ are energy and length scales, respectively, and r is the intermolecular distance as shown in Fig. 1. The intermolecular potential of inert monatomic molecules such as Ne, Ar, Kr and Xe is known to be reasonably expressed by this function. Typical values of σ and ε for each molecule are listed in Table 1. Moreover, many computational and statistical mechanical studies have been performed with this potential as the model pair potential. Here, the equation of motion can be non-dimensionalized by choosing σ, ε and m as length, energy and mass scale, respectively. The reduced formulas for typical physical properties are listed in Table 2. When a simulation system consists of only Lennard-Jones molecules, the non-dimensional analysis has an advantage in order not to repeat practically the same simulation. Then, molecules are called Lennard-Jones molecules, and argon parameters $\sigma = 0.34$ nm, $\varepsilon = 1.67\times10^{-21}$ J, and $\tau = 2.2 \times10^{-12}$ s are used to describe dimensional values in order to illustrate the physical meaning. The phase-diagram of Lennard-Jones system [6] is useful for a design of a simulation. The critical and triplet temperatures are $T_c^* = 1.35$ and $T_t^* = 0.68$, or $T_c = 163$ K and $T_t = 82$ K with argon property [7].

For the efficient calculation of potential, which is the most CPU demanding, Lennard-Jones function in Eq. (3) is often cutoff at the intermolecular distance $r_C = 2.5\ \sigma$ to $5.5\ \sigma$. However, for pressure or stress calculations, the contribution to potential from far-away molecules can result in a considerable error as demonstrated for surface tension [8]. In order to reduce this discrepancy,

Table 2 Reduced properties for Lennard-Jones system.

Property		Reduced Form
Length	$r^* =$	r/σ
Time	$t^* =$	$t/\tau = t(\varepsilon/m\sigma^2)^{1/2}$
Temperature	$T^* =$	$k_B T/\varepsilon$
Force	$f^* =$	$f\sigma/\varepsilon$
Energy	$\phi^* =$	ϕ/ε
Pressure	$P^* =$	$P\sigma^3/\varepsilon$
Number density	$N^* =$	$N\sigma^3$
Density	$\rho^* =$	$\sigma^3\rho/m$
Surface tension	$\gamma^* =$	$\gamma\sigma^2/\varepsilon$

several forms of smooth connection of cutoff have been proposed such as in Eq. (4) by Stoddard & Ford [9].

$$\phi(r) = 4\varepsilon\left[\left\{\left(\frac{\sigma}{r}\right)^{12} - \left(\frac{\sigma}{r}\right)^6\right\} + \left(6r_C^{*-12} - 3r_C^{*-6}\right)\left(\frac{r}{r_C}\right)^2 - \left(7r_C^{*-12} - 4r_C^{*-6}\right)\right] \quad (4)$$

2.2.2 Effective pair potential for water. The effective pair potential form for liquid water has been intensively studied. The classical ST2 potential proposed in 1974 by Stillinger and Rahman [10] based on BNS model [11] was widely used in the 1980s. The rigid water molecule was modeled as Fig. 2a, with the distance of OH just 0.1 nm and the angle of HOH the tetrahedral angle $\theta = 2\cos^{-1}\left(1/\sqrt{3}\right) \cong$ 109.47°. Point charges at four sites shown in Fig. 2a were assumed: positive charge of 0.235 7 e each on hydrogen sites and two negative charges at positions of lone electron pairs (tetrahedral directions). They modeled the potential function as the summation of Coulomb potential between charges and the Lennard-Jones potential between oxygen atoms. Hence, the effective pair potential of molecules at R_1 and R_2 are expressed as

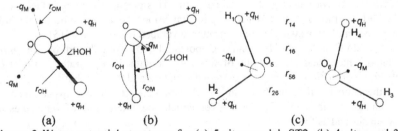

(a)   (b)   (c)

Figure 2 Water potential structures for (a) 5 sites model, ST2, (b) 4 sites and 3 sites models, TIP4P, CC, SPC, SPC/E, (c) definition of interatomic length of MCY and CC potential.

Table 3 Potential parameters for water.

		ST2	SPC/E	TIP4P	CC
r_{OH}	[nm]	0.100	0.100	0.095 72	0.095 72
$\angle HOH$	[°]	109.47	109.47	104.52	104.52
σ_{OO}	[nm]	0.310	0.316 6	0.315 4	N/A
ε_{OO}	$\times 10^{-21}$ [J]	0.526 05	1.079 7	1.077 2	N/A
r_{OM}	[nm]	0.08	0	0.015	0.024 994
q_H[a]	[C]	0.235 7 e	0.423 8 e	0.52 e	0.185 59 e
q_M	[C]	-0.235 7 e	-0.847 6 e	-1.04 e	-0.371 18 e

[a]Charge of electron $e = 1.60219 \times 10^{-19}$ C

$$\phi_{12}(\boldsymbol{R}_1, \boldsymbol{R}_2) = 4\varepsilon_{OO}\left[\left(\frac{\sigma_{OO}}{R_{12}}\right)^{12} - \left(\frac{\sigma_{OO}}{R_{12}}\right)^{6}\right] + S(R_{12})\sum_i\sum_j\frac{q_iq_j}{4\pi\varepsilon_0 r_{ij}}, \qquad (5)$$

where R_{12} represents the distance of oxygen atoms, and σ_{OO} and ε_{OO} are Lennard-Jones parameters. The Coulombic interaction is the sum of 16 pairs of point charges. $S(R_{12})$ is the modulation function to reduce the Coulombic force when two molecules are very close.

Later, much simpler forms of SPC (Simple Point Charge) [12] and SPC/E (Extended SPC) [13] potentials were introduced by Berendsen *et al.* SPC/E potential employed the configuration in Fig. 2b, with charges on oxygen and hydrogen equal to −0.8476 and +0.4238 e, respectively. Lennard-Jones function of oxygen-oxygen interaction was used as ST2 as in Eq. (5) but without the modulation function $S(R_{12})$.

TIP4P potential proposed by Jorgensen *et al.* [14] employed the structure of water molecule as $r_{OH} = 0.09572$ nm and $\angle HOH = 104.52°$ based on the experimentally assigned value for the isolated molecule. The positive point charges q were on hydrogen atoms, and the negative charge $-2q$ was set at r_{OM} from the oxygen atom on the bisector of the HOH angle, as in Fig. 2b. The function can be written as Eq. (5) without $S(R_{12})$ function. The parameters listed in Table 3 were optimized for thermodynamic data such as density, potential energy, specific heat, evaporation energy, self-diffusion coefficient and thermal conductivity, and structure data such as the radial distribution function and neutron diffraction results at 25 °C and 1atm. This potential is regarded as one of the OPLS (optimized potential for liquid simulations) set covering liquid alcohols and other molecules with hydroxyl groups developed by Jorgensen [15].

MYC potential [16] and CC potential [17] were based on *ab initio* quantum molecular calculations of water dimer with the elaborate treatment of electron correlation energy. The assumed structure and the distribution of charges are the same as TIP4P as shown in Fig. 2b with a different length r_{OM} and amount of charge as in Table 3. For CC potential, the interaction of molecules is parameterized as follows.

$$\phi_{12}(R_1, R_2) = \sum_i \sum_j \frac{q_i q_j}{4\pi\varepsilon_0 r_{ij}} + a_1 \exp(-b_1 r_{56})$$
$$+ a_2 \left[\exp(-b_2 r_{13}) + \exp(-b_2 r_{14}) + \exp(-b_2 r_{23}) + \exp(-b_2 r_{24})\right] \quad (6)$$
$$+ a_3 \left[\exp(-b_3 r_{16}) + \exp(-b_3 r_{26}) + \exp(-b_3 r_{35}) + \exp(-b_3 r_{45})\right]$$
$$- a_4 \left[\exp(-b_4 r_{16}) + \exp(-b_4 r_{26}) + \exp(-b_4 r_{35}) + \exp(-b_4 r_{45})\right]$$

$$
\begin{array}{llll}
a_1 = & 315.708 & \times 10^{-17} [J], & b_1 = & 47.555 & [1/nm], \\
a_2 = & 2.4873 & \times 10^{-17} [J], & b_2 = & 38.446 & [1/nm], \\
a_3 = & 1.4694 & \times 10^{-17} [J], & b_3 = & 31.763 & [1/nm], \\
a_4 = & 0.3181 & \times 10^{-17} [J], & b_4 = & 24.806 & [1/nm].
\end{array}
$$

Among these rigid water models, SPC/E, TIP4P and CC potentials are well accepted in recent simulations of liquid water such as the demonstration of the excellent agreement of surface tension with experimental results using SPC/E potential [18]. Because all of these rigid water models are "effective" pair potential optimized for liquid water, it must be always questioned if these are applicable to small clusters, wider range of thermodynamics condition, or liquid-vapor interface. Even though the experimental permanent dipole moment of isolated water is 1.85 D[1], most rigid models employ higher value such as 2.351 D for SPC/E to effectively model the induced dipole moment at liquid phase. The direct inclusion of the polarizability to the water models results in the many-body potential, which requires the iterative calculation of polarization depending on surrounding molecules. The polarizable potential based on TIP4P [19], MCY [20] and SPC [21] are used to simulate the structure of small clusters and transition of monomer to bulk properties. On the other hand, flexible water models with spring [22] or Morse type [23] intramolecular potential are examined seeking for the demonstration of vibrational spectrum shift and for the reasonable prediction of dielectric constant.

2.2.3 Many-body potential for carbon and silicon. The approximation of pair potential cannot be applied for atoms with covalent chemical bond such as silicon and carbon. SW potential for silicon proposed by Stillinger and Weber in 1985 [24] was made of two-body term and three-body term that stabilize the diamond structure of silicon. Tersoff [25, 26] proposed a many-body potential function for silicon, carbon, germanium and combinations of these atoms. For simulations of solid silicon, this potential [26] is widely used. Brenner modified the Tersoff potential for carbon and extended it for a hydrocarbon system [28]. A simplified form of Brenner potential removing rather complicated 'conjugate terms' is widely used for studies of fullerene [29, 30] and carbon-nanotube. Both Tersoff potential and the simplified Brenner potential can be expressed as following in a unified form. The total potential energy of a system is expressed as the sum of every chemical bond as

---

[1] 1 D = 3.3357×10^{-30} Cm in SI unit.

$$\Phi = \sum_i \sum_{j(i<j)} f_C(r_{ij}) \{V_R(r_{ij}) - b^*_{ij} V_A(r_{ij})\}, \tag{7}$$

where the summation is for every chemical bond. $V_R(r)$ and $V_A(r)$ are repulsive and attractive parts of the Morse type potential, respectively.

$$V_R(r) = f_C(r) \frac{D_e}{S-1} \exp\{-\beta\sqrt{2S}(r - R_e)\} \tag{8}$$

$$V_A(r) = f_C(r) \frac{D_e S}{S-1} \exp\{-\beta\sqrt{2/S}(r - R_e)\} \tag{9}$$

The cutoff function $f_C(r)$ is a simple decaying function centered at $r = R$ with the half width of D.

$$f_C(r) = \begin{cases} 1 & (r < R - D) \\ \frac{1}{2} - \frac{1}{2}\sin\left[\frac{\pi}{2}(r - R)/D\right] & (R - D < r < R + D) \\ 0 & (r > R + D) \end{cases} \tag{10}$$

Finally, b^*_{ij} term expresses the modification of the attractive force $V_A(r)$ depending on θ_{ijk}, the bond angle between bonds i-j and i-k.

$$b^*_{ij} = \frac{b_{ij} + b_{ji}}{2}, \quad b_{ij} = \left(1 + a^n \left\{\sum_{k(\neq i,j)} f_C(r_{ik}) g(\theta_{ijk})\right\}^n\right)^{-\delta} \tag{11}$$

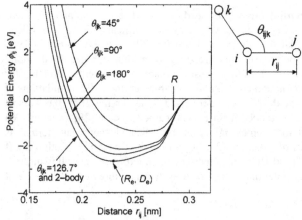

Figure 3 Many-body characteristics of Tersoff potential for silicon.

Table 4 Parameters for Tersoff potential and Brenner potential.

	Tersoff (Si)	Tersoff (C)	Brenner (C)
D_e [eV]	2.6660	5.1644	6.325
R_e [nm]	0.2295	0.1447	0.1315
S	1.4316	1.5769	1.29
β [nm^{-1}]	14.656	19.640	1.5
A	1.1000×10^{-6}	1.5724×10^{-7}	1.1304×10^{-2}
N	7.8734×10^{-1}	7.2751×10^{-1}	1
δ	$1/(2n)$	$1/(2n)$	0.80469
C	1.0039×10^{5}	3.8049×10^{4}	19
D	1.6217×10^{1}	4.384	2.5
H	-5.9825×10^{-1}	-5.7058×10^{-1}	-1
R [nm]	0.285	0.195	0.185
D [nm]	0.015	0.015	0.015

$$g(\theta) = 1 + \frac{c^2}{d^2} - \frac{c^2}{d^2 + (h - \cos\theta)^2} \tag{12}$$

Parameter constants for Tersoff potential for silicon (improved elastic properties) [26] and carbon and Brenner potential for carbon are listed in Table 4. In order to illustrate the characteristic of Tersoff and Brenner potential function, a potential energy contribution from a bond is expressed in Fig. 3. The Tersoff parameters for silicon are assumed and the energy of i-j bond under the influence of the third atom k, $\phi' = f_C(r_{ij})\{V_R(r_{ij}) - b_{ij}V_A(r_{ij})\}$ is drawn. The effect of the third atom k is negligible only when the angle θ_{ijk} is 126.7°.

2.3 Integration of the Newtonian Equation

The integration of the equation of motion is straightforward. Unlike the simulation of fluid dynamics, simpler integration scheme is usually preferred [5]. Verlet's integration scheme, as follows, can be simply derived by the Taylor series expansion of the equation of motion.

$$r_i(t + \Delta t) = 2r_i(t) - r_i(t - \Delta t) + (\Delta t)^2 F_i(t)/m_i \tag{13}$$

$$v_i(t) = \{r_i(t + \Delta t) - r_i(t - \Delta t)\}/2\Delta t \tag{14}$$

where Δt is the time step. A bit modified leap-frog method, as follows, is widely used in practical simulations [5]. After the velocity of each molecule is calculated as Eq. (15), the position is calculated as Eq. (16).

$$v_i\left(t + \frac{\Delta t}{2}\right) = v_i\left(t - \frac{\Delta t}{2}\right) + \Delta t \frac{F_i(t)}{m_i} \tag{15}$$

$$r_i(t + \Delta t) = r_i(t) + \Delta t v_i\left(t + \frac{\Delta t}{2}\right) \tag{16}$$

Typical time step Δt is about 0.005 τ or 10 fs with argon property of Lennard-Jones potential. More elaborate integration schemes such as Gear's predictor-corrector method [5] are sometimes employed depending on the complexity of the potential function and the demand of the accuracy of motion in each time step.

2.4 Boundary Condition: Spatial and Temporal Scale

Since the spatial and temporal scale handled with the MD method is extremely small compared to the scale of macroscopic heat transfer phenomena, the most important point of the design of a MD simulation applied to the macroscopic problem is the boundary condition. Many problems in chemistry, where the reaction process in the macroscopic chamber can be described with simple chemical reaction formulas, can be simulated in a relatively small equilibrium system. This situation can be understood by noting that the energy scale of chemical reaction is much higher than the energy scale of interaction with ambient molecules. Then, the interaction with other molecules can all be included with the thermodynamic properties such as temperature and pressure. On the other hand, because most problems in heat transfer deal with the temperature itself, interaction with 'ambient' molecules is usually very important.

Many MD simulations in chemistry and statistical mechanics have used the fully periodic boundary condition, which assumes that the system is simply homogeneous for an infinite length scale. The implementation of the periodic boundary condition is very simple. Any information beyond a boundary can be calculated with the replica of molecules, as in Fig. 4. This boundary condition is

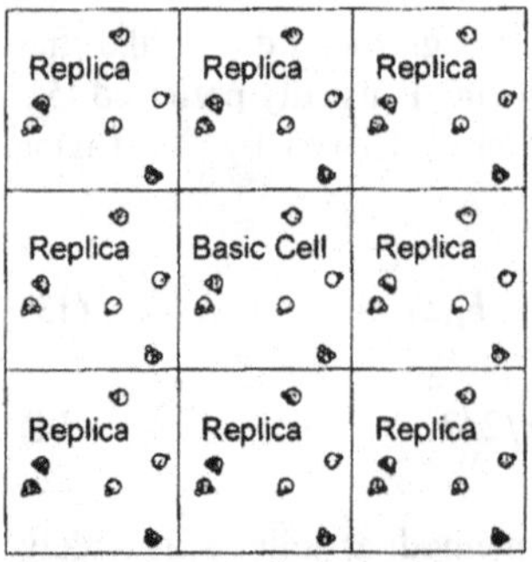

Figure 4 Periodic boundary condition.

used for two or four directions even for non-equilibrium calculations. The interaction of molecules is calculated beyond the periodic boundary with replica molecules. In order to avoid the calculation of potential between a molecule and its own replica, the potential must be cutoff to smaller than half the width of the base-cell scale. This cannot be a big problem for the short-range force such as Lennard-Jones potential, which decays as r^{-6}. Since Coulombic force decays only with r^{-1}, the simple cutoff does not give a good result. Usually, the well-known Ewald sum method [5] is employed, where the contribution from molecules in replica cells is approximated by a sophisticated manner. This is also somewhat of a problem for a system without the fully periodic conditions. The calculation of pressure using the virial theorem in Eq. (23) is also not straightforward. The sum of the potential terms in principal should be for the molecules inside the control volume V. However, for the fully periodic condition, the treatment of the pairs of potential as others gives a good result. For a spatially non-equilibrium situation, measurements of pressure and stress tensor are very complicated. The stress tensor defined in a surface rather than the volume as in Eq. (23) is demonstrated to be better [31, 32].

Many problems in heat transfer may include a phenomenon with a larger scale than the calculation domain, such as instability or a large modulation of properties. The temperature and specific volume condition where the phase separation happens in a macroscopic condition may be simulated as formation of the cluster in the small-scale calculation. Furthermore, for the non-equilibrium simulations, the establishment of the proper boundary condition is very difficult. In addition, the time scale that a MD simulation can handle might be too short to simulate the dynamic process. Examples of non-equilibrium systems are discussed in section 5.

The difficulty in the boundary condition is less for gas-phase molecules because the contribution of potential energy compared to kinetic energy is small. If the potential contribution is ignored, some simple boundary condition such as mirror reflection boundary can be used. Simply changing the velocity component as if a molecule makes an inelastic reflection. There is no good boundary condition for a liquid system. When it is impossible to use the periodic boundary condition, a solid wall or a vapor layer should be connected. Several different levels of the solid boundary conditions can be used. By locating an array of stationary molecules, the 0 K solid boundary can be constructed. Since the stationary molecules do not exchange the kinetic energy, they can be regarded as thermally adiabatic.

A one-dimensional potential function equivalent to the integration of the solid molecules can be used to represent an adiabatic wall. For example, the integration of a layer of fcc (111) surface of Lennard-Jones molecules can be expressed as

$$\Phi(z) = \frac{4\sqrt{3}\pi}{15} \frac{\varepsilon_{INT}\sigma_{INT}^{2}}{R_0^{2}} \left\{ 2\left(\frac{\sigma_{INT}}{z}\right)^{10} - 5\left(\frac{\sigma_{INT}}{z}\right)^{4} \right\} \qquad (17)$$

where ε_{INT} and σ_{INT} are Lennard-Jones energy and length parameters between the solid molecule and the liquid molecule. R_0 and z are the nearest neighbor distance of solid molecules and the coordinate normal to the surface, respectively. On the other hand, the volume integral is possible by imagining as if solid molecules are a continuum.

$$\Phi(z) = \frac{2\pi}{45} \frac{\rho_S}{m_S} \varepsilon_{INT} \sigma_{INT}{}^3 \left\{ 2\left(\frac{\sigma_{INT}}{z} \right)^9 - 15\left(\frac{\sigma_{INT}}{z} \right)^3 \right\} \qquad (18)$$

where ρ_S/m_S is the number density of solid.

However, most heat transfer simulations prefer to use the constant temperature solid wall. The simple velocity scaling in section 2.5 is often applied to three crystal layers of harmonic molecules. Since the velocity scaling is too artificial, the following phantom technique [33-36] is recommended. Phantom molecules model the infinitely wide bulk solid kept at a constant temperature T with the proper heat conduction characteristics. An example of the configuration of phantom molecules for a harmonic fcc solid system is shown in Fig. 5. A phantom molecule is connected to each molecule of the solid layer through a spring of $2k$ and a damper of $\alpha = m\omega_D \pi / 6$ in the vertical direction and with springs of $3.5k$ and dampers of α in two horizontal directions. Here, ω_D is the Debye frequency. Each phantom molecule is excited by the random force of Gaussian distribution with the standard deviation $\sigma_F = \sqrt{2\alpha k_B T / \Delta t}$. The energy flux to the calculation system can be accurately calculated by integrating the exciting force and the damping force applied to phantom molecules [36].

Through the careful matching of the boundary conditions, the MD simulation can find a way to connect to statistical techniques for gas and solid, which can easily handle much larger spatial and temporal scales. Some indirect examples are modeling the collision dynamics [37, 38] or the gas-surface interaction [39] for DSMC simulations through MD simulations. Furthermore, the boundary condition of phonon dynamics should be handled by the MD method.

2.5 Initial Condition and Control of Temperature and/or Pressure

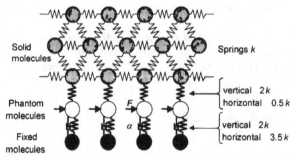

Figure 5 Constant temperature solid wall with phantom molecules.

The initial condition for each molecule is usually assigned by giving the velocity $v = \sqrt{3k_B T_C / m}$ with random directions for all molecules. The Maxwell-Boltzmann velocity distribution can be obtained after some equilibration calculations. The equilibrium system can often be calculated for constant temperature and constant pressure conditions. The simple temperature control of the equilibrium system can be realized by just scaling the velocity of molecules as $v_i' = v_i \sqrt{T/T_C}$ with the current temperature T and the desired temperature T_C. More elaborate techniques to realize the constant temperature system are known as the Anderson method [40] and the Nosé-Hoover method [41, 42].

Anderson method mimics random collisions with an imaginary heat bath particle. At intervals, the velocity of a randomly selected molecule is changed to a value chosen from the Maxwell-Boltzmann distribution. The choice of this interval is an important parameter. The Nosé-Hoover thermostat method involve the modification of the equation of motion as

$$m_i \frac{d^2 r_i}{dt^2} = F_i - \zeta m_i \frac{dr_i}{dt}, \quad \frac{d\zeta}{dt} = \frac{2\left(E_k - E_k^{\,0}\right)}{Q} \tag{19}$$

where ζ is the friction coefficient, E_k kinetic energy of the system, $E_k^{\,0}$ kinetic energy corresponding to the desired temperature T_C, and Q the thermal inertia parameter. All of these constant temperature techniques have been probed to give a statistically correct canonical ensemble, but the physical meaning of velocity re-scaling is not clear.

Andersen [40] described a technique to realize the constant pressure simulation. The simulation box size V is altered similar to the action of a piston with a mass. Parrinello and Rahman [43, 44] extended this technique to change the simulation box shape as well as size for solid crystal. Berendsen proposed a very simple "pressure bath" technique for the constant pressure simulation. The system pressure obeys $dP/dt = (P_C - P)/t_p$ by scaling the box size and position of molecules by a factor $\chi^{1/3}$ for each step.

$$r' = \chi^{1/3} r, \quad \chi = 1 - \beta_T \frac{\Delta t}{t_p}\left(P_C - P\right) \tag{20}$$

where β_T, t_p are the isothermal compressibility and time constant parameter, respectively.

Finally, it should be stressed again that all of these techniques of constant temperature or constant pressure are artificial to reproduce the statistical ensemble, and the physical meaning of the modification of position or velocity of each molecule is not clear.

2.6 Thermophysical and Dynamic Properties

According to statistical mechanics, thermodynamics properties such as temperature, internal energy and pressure can be defined as

Temperature
$$T = \frac{1}{3Nk_B} \left\langle \sum_{i=1}^{N} m_i v_i^2 \right\rangle \tag{21}$$

Internal energy
$$U = \frac{3}{2} Nk_B T + \left\langle \sum_i \sum_{j>i} \phi(r_{ij}) \right\rangle \tag{22}$$

Pressure
$$P = \frac{N}{V} k_B T - \frac{1}{3V} \left\langle \sum_i \sum_{j>i} \frac{\partial \phi}{\partial r_{ij}} \cdot r_{ij} \right\rangle \tag{23}$$

Here, temperature is simply the kinetic energy of molecules, and the internal energy is the total energy of kinetic and potential energies. Calculations of these properties are trivial. The pressure is defined through the virial theorem. There is no established technique to measure entropy and free energy by the MD method. These properties can be obtained by the statistical Monte Carlo method briefly discussed in section 2.7.

Some dynamics properties such as self-diffusivity, thermal conductivity and viscosity can be calculated by the equilibrium simulations though the fluctuations of properties, assuming that the macroscopic concepts of the linear equations such as Fick's law or Fourier's law are valid. The definitions of the equations, statistical mechanical Green-Kubo formula and the practical formulas derived using the Einstein relations are listed in Table 5.

The far-infrared and infrared absorption spectrum related to the radiative heat transfer can be calculated [46, 47] by employing the quantum mechanical perturbation theory. The absorption cross-section per molecule $\alpha(\omega)$ for light of frequency ω is derived as follows: Assuming that light interferes only to the permanent electric dipole moment, $\alpha(\omega)$ is expressed as

Table 5 Calculations of dynamic properties.

Property	Definition	Statistical Mechanical Green-Kubo Formula	With Einstein Relation For large t
Diffusion coefficient	$\dot{n} = -D \dfrac{\partial n}{\partial x}$	$\dfrac{1}{3} \int_0^\infty \langle v_i(t) \cdot v_i(0) \rangle dt$	$\dfrac{1}{6t} \langle |r_i(t) - r_i(0)|^2 \rangle$
Thermal conductivity[1]	$q = -\lambda \dfrac{\partial T}{\partial x}$	$\dfrac{V}{k_B T^2} \int_0^\infty \langle \tilde{q}_\alpha(t) \cdot \tilde{q}_\alpha(0) \rangle dt$	$\dfrac{V}{k_B T^2 2t} \langle (\delta \varepsilon_\alpha(t) - \delta \varepsilon_\alpha(0))^2 \rangle$
Shear viscosity[2]	$F = \mu \dfrac{\partial U}{\partial y}$	$\dfrac{V}{k_B T} \int_0^\infty \langle \tilde{p}_{\alpha\beta}(t) \cdot \tilde{p}_{\alpha\beta}(0) \rangle dt$	$\dfrac{V}{k_B T 2t} \langle (\tilde{D}_{\alpha\beta}(t) - \tilde{D}_{\alpha\beta}(0))^2 \rangle$

1. $\tilde{q}_\alpha = \dfrac{d\delta\varepsilon_\alpha}{dt}$, $\delta\varepsilon_\alpha = \dfrac{1}{V} \sum_i r_{i\alpha}(\varepsilon_i - \langle \varepsilon_i \rangle)$, $\varepsilon_i = \dfrac{m_i v_i^2}{2} + \dfrac{1}{2} \sum_{j \neq i} \phi(r_{ij})$, $\alpha = x, y, z$

2. NVE only. $\tilde{p}_{\alpha\beta} = \dfrac{1}{V} \left(\sum_i m_i v_{i\alpha} v_{i\beta} + \sum_i \sum_{j>i} r_{ij\alpha} f_{ij\beta} \right)$, $\tilde{D}_{\alpha\beta} = \dfrac{1}{V} \sum_i m_i r_{i\alpha} v_{i\beta}$, $\alpha\beta = xy, yz, zx$

$$\alpha(\omega) = \frac{\pi\omega\{1 - \exp(-\hbar\omega/k_{\mathrm{B}}T)\}}{3\varepsilon_0\hbar ncN} I(\omega),$$ (24)

$$I(\omega) = \frac{1}{2\pi} \int_{-\infty}^{\infty} \exp(-i\omega t)dt\langle\mu(0)\cdot\mu(t)\rangle_0$$ (25)

where c and n are the speed of light and the refractive index, which is often assumed to be unity, respectively. $I(\omega)$ and $\mu(t)$ are the transition rate and the electric dipole moment, respectively. The ensemble average $\langle\mu(0)\cdot\mu(t)\rangle_0$ is equivalent to the autocorrelation, and $I(\omega)$ reduces to the power spectrum of the electric dipole moment of the system. For the classical equilibrium system, the absorption cross section tends to the following equation as the classical limit of $\hbar\omega/k_{\mathrm{B}}T \to 0$.

$$\alpha(\omega) = \frac{\pi\omega^2}{3\varepsilon_0 k_{\mathrm{B}}TncN} I(\omega)$$ (26)

2.7 Monte Carlo Simulation

The Monte Carlo (MC) method or Metropolis method is often compared to the MD method. With the MC method, the same potential function as MD can be used. Instead of propagating positions of molecules based on the equation of motion in Eq. (1), configurations of molecules are generated with random numbers so that the probability of a configuration is proportional to the statistical probability for the ensemble considered. For example, the configuration should obey the Boltzmann distribution for a constant NVT (number, volume and temperature: canonical) ensemble. After generating such molecular configurations, the average value of any physical property can be obtained as a weighted integral of the configurations. The MC method has an advantage compared with MD when a physical property for a statistical ensemble is calculated. The MC method is established for constant NVE (number, volume and energy: microcanonical), constant NVT (number, volume and temperature: canonical), constant NPT (number, pressure and temperature), and even constant μVT (chemical potential, volume and temperature: grand-canonical) ensembles. However, the dynamic properties such as diffusivity or viscosity cannot be calculated by the statistical MC method. Furthermore, the non-equilibrium system such as the system with heat flux cannot be handled with the MC method.

3 LIQUID-VAPOR INTERFACE

3.1 Surface Tension

Surface tension is one of the benchmark properties to examine the applicability of a potential function to the liquid-vapor interface. Figure 6 shows an example of liquid-vapor interfaces of liquid slab [48, 49]. The calculation region had periodic boundary conditions for all six boundaries. Starting from a crystal of argon continuing over side boundaries, the liquid slab with flat liquid-vapor interface in Fig. 6 was realized after 2 ns MD simulation. Considering the periodic boundary conditions, this liquid slab can be regarded as an infinitely wide thin liquid film. During the simulation, the number of molecules, volume and total energy of the system were conserved except for the early temperature control period. When the liquid layer is thick enough, the bulk property of liquid can be obtained at the central region, and two liquid-vapor interfaces can be realized. The vapor (open), interfacial (gray), and liquid (solid) molecules are distinguished by the potential felt by each molecule. By taking a time average, the density profile in Fig. 6b, pressure tensor, and surface tension can be reasonably predicted. This is the typical molecular configuration for the measurement of surface tension. The quite accurate prediction of surface tension has been demonstrated for Lennard-Jones fluid [8] and water [18] by integrating the difference of normal $P_N(z)$ and tangential $P_T(z)$ components of pressure tensor across the surface as

$$\gamma_{LG} = \int_{z_L}^{z_G} [P_N(z) - P_T(z)]dz, \tag{27}$$

where z is the coordinate perpendicular to the interface. Here, P_N and P_T are equal to the thermodynamic pressure P in bulk vapor position z_G and bulk liquid position z_L. In the case of liquid slab as shown in Fig. 6, the integration between two vapor regions results in $2\gamma_{LG}$ since there are two liquid-vapor interfaces. In principle, the normal pressure $P_N(z)$ should be completely constant through the

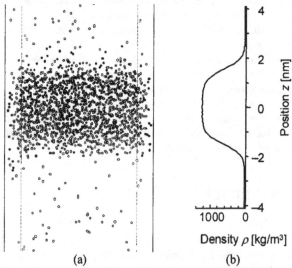

(a)   (b)

Figure 6 A flat interface of liquid and vapor (1944 argon molecules saturated at 100 K in 5.5×5.5×20 nm box), (a) a snapshot, (b) density distribution.

interface for the mechanical balance required at equilibrium. The reason for integrating the pressure difference in Eq. (27) is believed to reduce the numerical fluctuations by canceling the common kinetic term of the pressure expression [the first term of Eq. (23)]. However, it seems to also cancel the problem of the pressure definition in locally non-uniform density variation across the interface [32].

3.2 Liquid Droplet and Young-Laplace Equation

Figure 7 shows examples of argon and water liquid droplet surrounded by its vapor [48, 49]. This configuration can be obtained when the initial argon crystal is placed at the center of the fully periodic cubic region. This is regarded as an isolated liquid droplet floating in its vapor. When the size of the droplet is large enough, the bulk property of liquid is expected at the central region. The well-known Young-Laplace equation relates the curvature of a liquid-vapor interface and surface tension to the pressure difference. For a liquid droplet, the Young-Laplace equation is described as

$$\gamma_{LG} = \frac{(P_L - P_G)R}{2}.$$ (28)

The microscopic representation of the Young-Laplace equation can be used for the evaluation of the surface tension itself, which should be a kinetic property derived from the molecular parameters. It is necessary to obtain the pressure variation across the liquid and vapor interface in order to obtain P_L and P_G as asymptotic values. The estimation of the pressure profile is quite difficult and results in a considerable error. Thompson $et\ al.$ [50] used the following spherical extension of Irving-Kirkwood's formula to calculate the normal pressure profile:

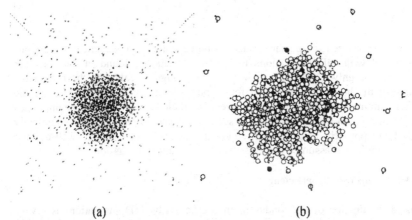

(a)   (b)

Figure 7 Liquid droplet and surrounding vapor. (a) 2 048 argon molecules saturated at 95 K in a 12 nm cubic box, (b) Water droplet at 380 K.

$$P_N(r) = k_B T \frac{\rho(r)}{m} - \frac{1}{4\pi r^2} \sum_k f_k \tag{29}$$

$$P_N(r) = k_B T \frac{\rho(r)}{m} - \frac{1}{4\pi r^3} \sum_k |r \cdot r_{ij}| \frac{1}{r_{ij}} \frac{d\phi(r_{ij})}{dr_{ij}} \tag{30}$$

The normal pressure in Eq. (29) is measured as the force across the control spherical surface of radius r from the center of the droplet. The sum over k is over the normal component f_k of forces acting across the control surface between a pair of molecules i and j. The sign of f_k is defined as positive for repulsive forces and negative for attractive forces. Using the vector $r_{ij} = r_j - r_i$ and the potential $\phi(r_{ij})$, the pressure is expressed as in Eq. (30).

The definition of the radius of a droplet is not straightforward, since the size of the droplet is normally very small and the liquid-vapor interface has a certain width as shown in Fig. 6b (planar surface). The equimolar dividing radius R_e defined as follows is a convenient choice.

$$mN = \frac{4\pi}{3} R_e^3 \rho_L + \left\{ L^3 - \frac{4\pi}{3} R_e^3 \right\} \rho_G \tag{31}$$

where ρ_L, ρ_G, N, and L are liquid density, vapor density, number of molecules, and unit length of the cell, respectively. This R_e means the radius of a hypothetical sphere of uniform density ρ_L in a cubic cell of density ρ_G. However, the statistical mechanical choice of the radius is so-called surface of tension R_S. The first-order correction of the surface tension of a curved surface compared to that for a planar surface $\gamma_{LG\infty}$ is expressed by the Tolman length δ.

$$\gamma_{LG} = \gamma_{LG\infty} \left(1 - \frac{2\delta}{R_e} \right) + O\left(R_e^{-2} \right), \tag{32}$$

where $\delta = z_e - z_S$ for a planar surface. Detailed statistical mechanical discussions compared with MD simulations for small droplets are found in the literature [50-53]. Roughly a thousand molecules are enough to calculate the reasonable value of the bulk surface tension for argon without this correction [49]. In the other extreme, the surface tension for very small clusters, which may be important in the nucleation theory discussed in section 5.2, should require some completely different approach, because such small cluster does not have the well-defined central liquid part assumed in the statistical mechanical discussions.

3.3 Condensation Coefficient

The determination of the condensation coefficient by MD simulations is a very fascinating task, as demonstrated in the review by Tanasawa [54]. The condensation coefficient has been simply defined as the ratio of rates of the number of condensation molecules to incident molecules. Through the detailed studies of the liquid-vapor inter-phase phenomena of argon, water, and methanol, Matsumoto et al. [55-57] pointed out that this macroscopic concept couldn't be directly converted to the molecular scale concept. They calculated the condensation coefficient, for the first time, through MD simulations, and stressed the importance of the 'molecular exchange' process: a molecule condensed into the liquid phase lets another liquid molecule vaporize. By excluding those molecules from the number of condensing molecules, they had shown a good agreement with experiments at least for the equilibrium condition [58, 59]. In fact, the apparent 'self-reflection' of condensing molecules was always about 10 % regardless of molecular species. On the other hand, Tsuruta et al. [60] had reported a significant dependence of the trapping rate on the normal velocity of incident molecules. They seek the connection to the DSMC method for the calculation of the condensation process. Since there are significant differences in these two approaches, it appears that a new microscopic definition of the condensation coefficient may be necessary which is physically plausible and also useful for the further connection to the macroscopic theories.

Most of the studies with MD simulations have dealt with the equilibrium system of liquid and vapor, assuming that the condensation coefficient is a "coefficient" independent of supersaturating pressure or temperature. Recent experiment [61] has shown, however, a considerable dependence of the "coefficient" on supersaturation conditions. It seems that it is not easy to handle the non-equilibrium MD simulation [62] to explain these experimental results. On the other hand, according to the DSMC calculation [63] of the condensation phenomena, there is a quite thick layer where the vapor temperature varies from the liquid-vapor interface. Since the direct simulation of such a wide scale with the non-equilibrium MD method seems to be impossible, a connection of these two methods through a reasonable boundary treatment is desired.

4 SOLID-LIQUID-VAPOR INTERACTIONS

4.1 Liquid Droplet on Solid Surface

Solid-liquid-vapor interaction phenomena or simply contact phenomena of liquid to the solid surface have a very important role in phase-change heat transfer. Except for the direct contact heat transfer, most practical phase-change heat-transfer problems involve the solid surface as a heater or a condenser. The importance of the liquid wettability to the surface is apparent in a dropwise condensation, high-heat-flux boiling heat transfer and capillary liquid film evaporators. The mechanical and thermodynamic treatments of the traditional macroscopic approach had difficulty in the treatment of the line of three-phase contact. The contact line is the singular point in the macroscopic sense, since the non-slip condition of fluid dynamics, i.e. $U = V = 0$ at the surface, simply denies the movement of the contact line. The curious "monolayer liquid film" considered

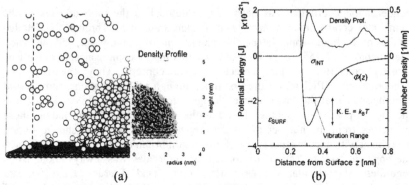

(a)   (b)

Figure 8 A liquid droplet in contact with solid surface. (a) A snapshot compared with the two-dimensional density profile, (b) integrated potential profile and the density profile.

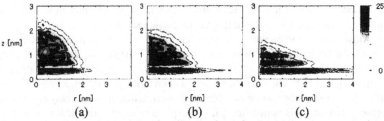

(a)   (b)   (c)

Figure 9 Two-dimensional density distribution of a liquid droplet on a surface for (a) E2, (b) E3, (c) E4. See Table 6 for potential parameters.

Table 6 Calculation conditions of the solid-liquid interaction.

Label	ε_{INT} [$\times 10^{-21}$ J]	σ_{INT} [nm]	ε^*_{SURF}
E2	0.575	0.308 5	1.86
E3	0.750	0.308 5	2.42
E4	0.925	0.308 5	2.99

$\varepsilon^*_{SURF} = \varepsilon_{SURF} / \varepsilon_{AR}$

in some macroscopic theories of heat transfer should be examined.

There are good reviews of the connection between microscopic and macroscopic views of the wetting phenomena by Dussan [64], and from a slightly more microscopic point of view by Koplik and Banavar [65]. Figure 8a compares a snapshot of the liquid droplet in contact with a solid surface with a two-dimensional density distribution. Simulation conditions are similar to our previous reports [66, 67], but 1 944 argon molecules are included and about 1 600 molecules constitute the liquid droplet surrounded by saturated vapor. Solid molecules are located as three layers of fcc (111) surfaces with harmonic potential (only one layer is shown in Fig. 8a for simplicity). The interaction potential between argon and solid molecule expressed by the L-J potential is chosen so that the apparent contact angle becomes about 90°.

The effect of the interaction potential on the shape of the liquid droplet is

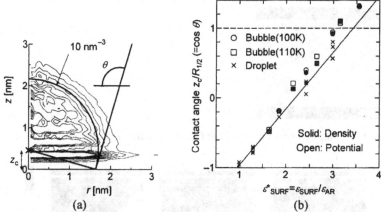

(a)   (b)

Figure 10 Contact angle measured for liquid droplet and vapor bubble. (a) Definition of contact angle for liquid droplet, (b) dependence of contact angle on the integrated depth of surface potential ε_{SURF}.

apparent in Fig. 9 [66].   With increase in the strength of the interaction potential between the surface molecule and argon, the flatter shape is observed. Furthermore, with stronger interaction potential, the spread of the first layer of liquid film is much more pronounced [66]. The layered structure is commonly observed for liquid-solid interfaces and explained as due to the solvation force [68]. Figure 8b explains the reason for this layered structure more clearly. For a liquid molecule, the effect of the solid molecules can be integrated to the one-dimensional function $\Phi(z)$ in Eq. (17). This potential function is compared with the density profile in Fig. 8b. The similarity of these is remarkable, and the temperature level correlates the sharpness of the density profile. It should noticed that the integrated function $\Phi(z)$ in Eq. (17) has a minimum $\varepsilon_{SURF} = (4\sqrt{3}\pi/5)(\sigma_{INT}^2/R_0^2)\varepsilon_{INT}$ at $z = \sigma_{INT}$. The peak of the second layer of the density appears around $z = \sigma_{INT} + \sigma_{AR}$ because the second layer is trapped by the integrated potential of argon molecules layered at $z = \sigma_{INT}$.

Except for the two or three liquid layers near the surface, the averaged shape of the liquid droplet is close to the semi-spherical. In order to measure the "contact angle," we can fit a circle to a density contour disregarding the two layers of liquid near the solid surface as in Fig. 10a [66]. Controversially enough, the cosine of measured contact angle or the average shape of the droplet far from the surface was linearly dependent on the strength of the surface potential [Fig. 10b]. Comparing the simulation changing the different parameter of the interaction σ_{INT} and different configuration of the solid surface of one-dimensional function in Eq. (17), one layer of fixed molecules, three layers of harmonic molecules, the contact angle was determined by the effective integrated potential energy ε_{SURF} [66, 67].

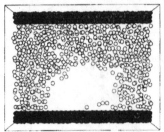

Figure 11 A snapshot of a vapor bubble (1 nm thick slice).

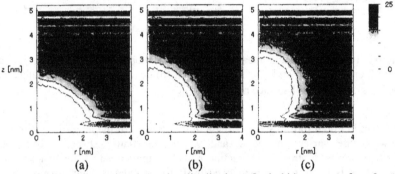

Figure 12 Two-dimensional density distribution of a bubble on a surface for (a) E2, (b) E3, (c) E4.

4.2 Vapor Bubble on Solid Surface

The opposite configuration of liquid and vapor, i.e. a vapor bubble in liquid, is realized for negative pressure as in Fig. 11 [35, 69]. Here, a sliced view through the center of the vapor is shown to visualize the vapor bubble in the liquid. Two dimensional density distributions for three different interaction potentials compatible to Fig. 9 are summarized in Fig. 12. The completely opposite situation of liquid and vapor is apparent, except for the layered liquid structure, which is always extending from liquid to vapor area. The contact angle measured in the same manner compared well to the liquid droplet case in Fig. 10b. The contact angle measured for the vapor bubble is slightly smaller in Fig. 10. This may be due to the effect of the surface tension on the contact line because the curvature of the contact line is opposite in two systems. One interesting point about the vapor bubble is that the first liquid layer completely covers the surface for the very wettable case of E4 in Fig. 12c. It was revealed that the $\cos\theta$ could be generalized to be $z_c/R_{1/2}$, to continuously express the dependency of the contact angle for the extremely wettable surface where z_C and $R_{1/2}$ are the center height and radius of the fitting circle (see Fig. 10a).

4.3 Contact Angle and Young's Equation

The contact angle is introduced to represent the degree of the partial wettability of the solid surface in macroscopic studies. The well-known Young's equation relates the contact angle to the balance of surface energies.

$$\cos\theta = \frac{\gamma_{SG} - \gamma_{SL}}{\gamma_{LG}}, \tag{33}$$

where γ_{SG}, γ_{SL} and γ_{LG} are surface energies between solid-vapor, solid-liquid, and liquid-vapor, respectively. This equation can be understood from the mechanical balance of forces or from the thermodynamic concept of minimizing the Helmholtz free energy. Since it is usually impossible to independently measure the surface energies except for the surface tension γ_{LG}, the well-known and useful Young's equation is still somewhat conceptual. Furthermore, the definition of the contact angle seems to be controversial if the thin liquid film exists over the 'dry' surface.

In 1977, Saville [70] claimed that the Young's equation is not satisfied from his MD results. He enclosed a liquid slab and coexisting vapor between two parallel surfaces represented by the one-dimensional potential function (Eq. (18)). Using 255 to 1205 L-J molecules at about the triplet temperature, he measured the meniscus of the liquid-vapor interface and compared it with the calculated surface tensions γ_{LG} and γ_{SL} - γ_{SG}. However, Nijmeijer et al. [71] showed good agreement of the observed contact angle and the contact angle calculated from Young's equation. Sikken et al. [72] and Nijmeijer et al. [71] used a little different configuration with 8500 fluid molecules and 2904 solid molecules, and the difficulty of the calculation of the surface tension term γ_{SL} - γ_{SG} was also overcome. Later, Thompson et al. [73] further supported the soundness of Young's equation and even discussed the dynamic contact angle. Furthermore, the contact angle measurement by the MD simulation can be useful to predict the wettability of realistic molecules on a realistic surface [74].

It seems that all these arguments and discrepancies exist not only because of the difficulties in measuring the surface energies but because the definition of the observed contact angle is not clear. As in the case of the surface tension of a droplet, a certain dividing surface of liquid-vapor must be defined to measure the contact angle. Macroscopic definition of the contact angle is valid only when the number of molecules is so large that the thickness of the interfaces is negligible. Finally, it should be noticed that the effect of the gravity is completely negligible for such a small-scale droplet. Those readers familiar with the macroscopic system should compare this system size of order of 5 nm to the capillary length.

4.4 Dynamic Process of Contact

It is well known that the measured macroscopic contact angle is a function of the velocity of the contact line U. When the contact line is moving from the liquid to vapor direction ($U > 0$), it is called advancing condition. And, the opposite (U < 0) is called receding condition. It is very interesting that the limit of $U = 0$ for advancing conditions (called advancing contact angle) and that for receding conditions (receding contact angle) do not coincide. The contact angle remembers its moving history called contact angle hysteresis. From extensive macroscopic studies, it is believed that the dynamics contact angle shows the range of angles between advancing and receding due to the metastable contact directly related to the surface conditions such as roughness and chemical heterogeneity. On the other hand, there are reports of MD simulations [65, 73] that claim the reproduction of the dynamic contact angle, even though the surface is perfectly smooth and chemically homogeneous. This contradiction is still open question. It is likely to be simply that the system size of MD simulations is too small so that the crystal of solid molecules may be felt as the periodically rough potential field.

When the contact-line speed is increased for advancing conditions, the dynamic contact angle generally increases until it finally reaches 180°. Further increase in the advancing speed beyond this critical speed induces a macroscopic saw-tooth instability of the contact line. It seems that the shape of the contact line is adjusted so that the velocity component normal to the curved contact line is kept at the critical speed. Such instability cannot be reproduced in the small system used in the MD simulations. If such an extreme condition is applied to the small molecular system, probably a new instability will be induced, which is not corresponding to any macroscopic phenomena.

5 NON-EQUILIBRIUM SIMULATIONS

Heat transfer is a non-equilibrium phenomenon. Even though the thermophysical properties and inter-phase dynamics discussed in previous sections are useful for heat transfer analysis, the direct simulation of the heat transfer problem is much more desired. Here, a spatial non-equilibrium simulation refers to the system with spatial temperature gradient or heat flux. On the other hand, a temporal non-equilibrium simulation refers to the system with the temporal evolution of temperature, internal energy or other properties. Certain phenomena inherent to small scale can be studied in such a technique. On the other hand, it is not easy to extend to the macroscopic scale phenomena since the scale in the non-equilibrium direction is very small, such as the thickness in the spatially non-equilibrium system and the simulation time in temporally non-equilibrium systems. Then, the gradient of non-equilibrium is extremely large such as large heat flux and large supersaturation rate. As a typical example, the melting process seems to be easily reproduced, but the solidification process that involves the considerable ordering of molecule structure is far more difficult.

5.1 Spatially Non-Equilibrium Simulation

The thermal conductivity can be calculated with the equilibrium MD by the statistical formula in Table 5. However, the validity of Fourier's law in an extremely microscopic system such as thin film can only be examined by the direct non-equilibrium heat conduction calculation. The mechanism of heat

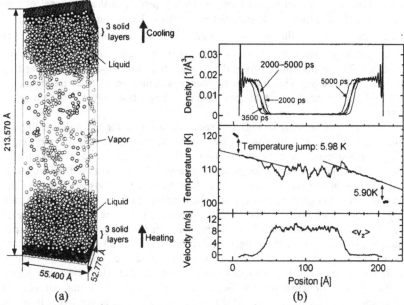

Figure 13 Non-equilibrium MD for inter-phase heat transfer. (a) A snapshot, (b) density, temperature, and velocity distributions.

conduction itself is also interesting [75, 76, 77]. The heat flux through a volume is calculated as

$$q = \frac{1}{2V} \left[\sum_{i}^{N} m_i v_i^2 v_i + \sum_{i}^{N} \sum_{j \neq i}^{N} \phi_{ij} v_i - \sum_{i}^{N} \sum_{j \neq i}^{N} \left(r_{ij} f_{ij} \right) v_i \right]$$ (34)

where the first and second terms related to summations of kinetic and potential energy carried by a molecule i. The third term, the tensor product of vectors r_{ij} and f_{ij}, represents the energy transfer by the pressure work. Because of the third term, the calculation of heat flux is not trivial at all [32].

An example of the spatial non-equilibrium simulation is shown in Fig. 13 [36]. The purpose of this simulation was to measure the thermal resistance in the interface of liquid and solid. A vapor region was sandwiched between liquid layers, which were in contact with two solid walls. While independently controlling temperatures at ends of walls by the phantom method described in section 2.5, energy flux through the system was accurately calculated. The heat flux and vapor pressure became almost constant after about 2 ns after suddenly enforcing the temperature difference between surfaces. The measured temperature distribution normal to interfaces in this quasi-steady condition shown in Fig. 13b revealed a distinctive temperature jump near the solid-liquid interface, which could be regarded as the thermal resistance over the interface. The temperature distribution in the liquid region (see the density profile in the top panel of Fig. 13b) can be fit to a linear line, and the heat conductivity λ_L can be calculated from this gradient and heat flux q_W as $\lambda_L = q_W / (\partial T / \partial z)$. This value was actually in good agreement with the macroscopic value of liquid argon. The thermal resistance R_T was determined from the temperature jump T_{JUMP} and the heat flux q_W as $R_T = T_{JUMP} / q_W$. This thermal resistance is equivalent to 5~20 nm thickness of liquid heat conduction layer, and hence, is important only for such a small system.

The configuration in Fig. 13a seems to be used for the non-equilibrium condensation and evaporation studies since the condensation in the upper liquid-vapor interface and the evaporation in the lower interface are quasi-steady. The heat flux through higher temperature side $q_W{}^{evap}$ was consumed for the latent heat of the evaporation, and the residual heat flux q_V was mostly carried by the net mass flux through the vapor region. The latent heat of condensation was added to q_V to reproduce $q_W{}^{cond}$ at the lower temperature side. The measured value of heat flux q_V was about 1/3 of q_W. Then, the temperature gradient in the vapor phase is too small for the vapor heat conduction. It was revealed that the dominant carrier of energy in the vapor region was the net velocity component $<v_z>$ shown in the bottom panel of Fig. 13b. It should be noticed that when the vaporization coefficient or condensation coefficient is considered for a non-equilibrium liquid-vapor interface, the effect of this net mass flux must be removed. It seems that a considerably large vapor region should be necessary to simplify the calculation.

5.2 Homogeneous Nucleation

The homogeneous nucleation is one of the typical macroscopic phenomena directly affected by the molecular scale dynamics. Recently, Yasuoka and

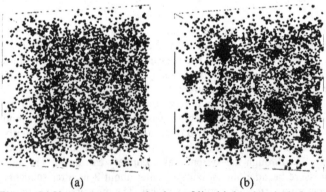

(a)   (b)

Figure 14 Homogeneous nucleation of liquid droplets in L-J system by Yasuoka and Matsumoto [78], (a) after quenched: t=0, (b) at t = 600τ (1.29 ns for argon). [Reprinted from [78] by permission from Journal of Chemical Physics, copyright 1999, American Institute of Physics]

Matsumoto have demonstrated the non-equilibrium MD simulations of the nucleation process for Lennard-Jones [78] and for water [79, 80]. For the Lennard-Jones (argon) fluid, homogeneous nucleation at the triple-point temperature under supersaturation ratio of 6.8 was simulated. Snapshots of the nucleation of argon droplets are shown in Fig. 14 [78]. The appearance of several large liquid droplets is clearly observed in Fig. 14b. The key technique for such a calculation is the temperature control. After quenching to the supersaturation condition, the condensation latent heat must be removed for the successive condensation. They used 5 000 Lennard-Jones molecules for the simulation mixed with 5 000 soft-core carrier gas molecules connected to the Nosé-Hoover thermostat for the cooling agent. This cooling method mimicked the carrier gas of supersonic jet experiments. Through the detailed study of growth and delay of nuclei size distribution, they have estimated the nucleation rate and the critical size of nucleus. The nucleation rate was seven orders of magnitude larger than the prediction of classical nucleation theory, whereas the critical nucleus size was 30-40 atoms compared to 25.4 by the theory. The free energy of the formation was estimated to explain this quick nucleation. They have performed the similar simulation [79] for water of TIP4P potential at 350 K under supersaturation ratio 7.3. The calculated nucleation rate was two orders of magnitude smaller than the classical nucleation theory, just in good agreement with the "pulse expansion chamber" experimental results [81]. The estimated critical nucleus size was 30-40 compared to the prediction of classical theory of order of one.

Ikeshoji et al. [82, 83] have simulated the similar nucleation process of Lennard-Jones molecules with special attention to the magic number clusters of 13, 19 and 23, which are abundantly observed in experimental mass spectra. By their large-scale simulation using 65 526 molecules, the importance of the temperature control method was stressed. The Nosé-Hoover thermostat (see section 2.5) did not reproduce the magic-number clusters because the internal energy (rotation and vibration) and the translational energy decreased at almost the same rate. They introduced a special temperature control that should give a similar effect to that of Yasuoka and Matsumoto [78]. Furthermore, it was suggested that the long-time evaporation process was essential for the reproduction of the magic number clusters. Their simulation time for the evaporation process extended to 26.4 ns (argon) compared to 3.9 ns by Yasuoka and Matsumoto [78].

A MD simulation of homogeneous nucleation of a vapor bubble is much more difficult compared to the nucleation of a liquid droplet. Kinjo and Matsumoto [84] expanded a Lennard-Jones liquid to demonstrate the cavitation in negative pressure. A single cavity was formed at the thermodynamic condition near the spinodal line. Since the generation of a bubble considerably alters the system pressure of liquid, only the inception of the cavity can be studied in such a numerical system. They have roughly estimated the nucleation rate as eight orders of magnitude larger than that of the classical nucleation theory.

5.3 Heterogeneous Nucleation

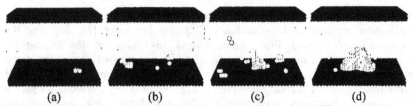

Figure 15 Nucleation of a vapor bubble on a solid surface (E3) at (a)1.42 ns, (b)1.48 ns, (c)1.54 ns, (d)1.60 ns.

The heterogeneous nucleation is more practically important than the homogeneous nucleation in most heat transfer problems. Figure 15 shows an example of the heterogeneous nucleation of vapor bubble on a solid surface [35, 69]. Liquid argon between parallel solid surfaces was gradually expanded until a vapor bubble was nucleated. 5 488 Lennard-Jones molecules represented argon liquid, and three layers of harmonic molecules represented each solid surface with the phantom constant-temperature model in section 2.4. In order to visualize the density variations leading to the vapor bubble nucleation, three-dimensional grid points of 0.2 nm intervals were visualized as 'void' when there were no molecules within $1.2\sigma_{AR}$. The interaction potential between a solid molecule and argon was also expressed by the Lennard-Jones potential with the potential parameters σ_{INT} and ε_{INT}. The solid-argon potential parameter was moderately wettable for the bottom surface and very wettable for the top surface. As a result, the cavity nuclei appeared and disappeared randomly on the bottom surface, and at some point they grew to a certain stable size. The calculation of the nucleation rate is not easy, as in the case of homogeneous nucleation of vapor.

5.4 Formation of Structure

The formation of a certain structure of a cluster is important even for an equilibrium case such as the supercritical condition [85], or the hydrogen-bonded

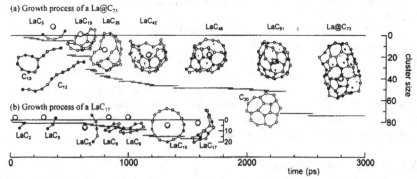

Figure 16 Growth process of La attached clusters: (a) La@C_{73} and (b) La@C_{17}.

cluster of water [86]. On the other hand, the simulation of the formation process of a special molecule structure such as fullerene [29, 30], metal-containing fullerene [87] or clathrate-hydrate [88] is very attractive. Figure 16 shows the formation process of metal (La) containing fullerene simulated by the MD method with the Brenner potential described in section 2.2.3. In addition to the Brenner potential for carbon-carbon interaction, the metal-carbon and metal-metal interaction was constructed [87] by fitting to *ab initio* calculations based on DFT (density functional theory). Since there was considerable charge transfer from the metal atom to carbon cluster, the Coulombic interaction force was included in the potential. Mainly because of this force, the metal atom works as a nucleation site for carbon clusters, as in Fig. 16. The organized clustering process of carbon cage was considerably different from the pure carbon simulation [29, 30]. This simulation is regarded as the nucleation and condensation of a carbon and metal binary mixture. As other nucleation simulations, the time scale of the simulation was compressed about 1000 times. Then the artificial temperature control was enforced to translational, rotational, and vibrational motions of freedom independently. In order to obtain the more realistic structure of C_{60} and C_{70}, the extra annealing calculations for such clusters were necessary. The multi-scale problem is much more severe in this case than in cases of argon or water, because the energy scale of chemical reaction is much higher.

6 FUTURE DIRECTIONS

The sound understanding of molecular level phenomena is required in varieties of phase-change theories such as nucleation of dropwise condensation, atomization, homogeneous and heterogeneous nucleation of vapor bubbles in cavitation and boiling. Moreover, heat transfer right at the three-phase interface, which is a singular point in the macroscopic sense, should be considered for evaporation in a micro-channel and for the micro- and macro-layer of boiling. The upper limit of heat flux of phase change must be clarified since recent advanced technologies such as intense laser light or electron beam easily achieve a very high heat flux. Phase-change phenomena involved in the thin film manufacturing process and laser manufacturing are often out of the range of the conventional approach. Other examples are surfactant effect in liquid-vapor interface and surface treatment effect of a solid surface.

  Even though the MD method is a powerful tool, the reader should notice its shortcomings that the spatial and temporal scale of the system that can be handled is usually too small to directly compare with the macroscopic phenomena. Even with the rapid advances of computer technology in future, most macroscopic problems cannot be handled by directly solving each motion of molecules. Then, the ensemble technique of the molecular motion and the treatment of boundary condition must be improved for the connection to macroscopic phenomena.

  Moreover, the determination of the potential function for molecules in real applications is not straightforward, and the assumption of classical potential fails when the effect of electrons is not confined in the potential form. For example, heat conduction in metals cannot be easily handled due to free electrons. Chemical reaction processes are also difficult to handle with the classical potential. The quantum feature of electrons must be considered when electrons are excited by laser light, by electromagnetic wave or by certain chemical reactions. We encounter these problems in thin-film production and treatment processes such as CVD or plasma etching or new manufacturing techniques utilizing plasma, laser beam, and electron beam. In such processes, many heat transfer problems may be linked to higher energy phenomena than the chemical reaction. Recently, the applicability of the quantum MD method is being explored. The well-known Car-Parrinello method [89] solves the positions of atoms and electronic states at the same time. Since this technique is based on the steady-state Schrödinger equation, the propagation of an electronic state is rather *ad hoc*. On the other hand, the non-adiabatic quantum MD method [90], which is formulated without the Born-Oppenheimer approximation, can currently handle a system with a few atoms. Certain new advances in the quantum MD method for heat transfer must be developed [91, 92].

Finally, the comparison with experiment is often crucial. Since most experiments in heat transfer deal with macroscopic quantities, it is not easy to evaluate simulation results. A few direct comparisons of the MD simulations to experiments have been reported in the heat transfer field such as the EXAFS study of LiBr effect on liquid-vapor interfacial phenomena [93, 94], or FT-ICR mass spectroscopic study on carbon clusters [95, 96].

7 NOMENCLATURE

A	constant in Tersoff/Brenner potential
a	potential parameter in Eq. (11)
a_1-a_4	potential parameters in Eq. (6)
B	constant parameter in Tersoff/Brenner potential
b_1-b_4	potential parameters in Eq. (6)
b_{ij}	function in Tersoff potential
c	constant in Tersoff/Brenner potential, speed of light
D	diffusivity, half width of cutoff function
D_e	potential depth in Tersoff/Brenner potential
$\tilde{D}_{\alpha\beta}$	rotation matrix
d	constant in Tersoff/Brenner potential
E_k	kinetic energy
F	force vector
f	force
f_c	cutoff function
$g(\theta)$	a function in Tersoff/Brenner potential
h	constant in Tersoff/Brenner potential
$\hbar$	Planck's constant: 1.05459×10^{-34} Js
$I(\omega)$	transition rate

k	spring constant of harmonic potential
k_B	Boltzmann constant
L	length of a calculation cell
m	mass
N	number density
n	constant in Tersoff/Brenner potential, density of species
P	pressure
$\tilde{P}_{\alpha\beta}$	instantaneous pressure tensor
q	electric charge, heat flux
$\tilde{q}_a$	α (x, y, z) component of energy flux
R	position of molecule
R	radius, intermolecular distance of oxygen, center of cutoff radius
R_0	nearest neighbor distance of solid crystal
R_e	equilibrium intermolecular distance, equimolar dividing radius
R_S	radius of surface of tension
r	intermolecular distance
r	position vector
S	constant in Tersoff/Brenner potential
$S(R)$	modulation function for ST2 potential
T	temperature
T_c	critical temperature
T_{JUMP}	temperature jump
T_t	triplet temperature
t	time
t_P	time constant
U	internal energy, fluid dynamics velocity
V	volume
$V_A(r)$	an attractive potential function in Tersoff/Brenner potential
$V_R(r)$	a repulsive potential function in Tersoff/Brenner potential
v	velocity vector
z	coordinate perpendicular to the interface

Greek Symbols:

α	absorption cross section, damping factor
β	constant in Tersoff/Brenner potential
β_T	isothermal compressibility
χ	scaling factor
Δt	time step of the finite difference integration
δ	constant in Tersoff/Brenner potential, Tolman length
ε	energy parameter of Lennard-Jones potential
ε_0	permittivity (dielectric constant) in vacuum $= 8.8542\times10^{-12}$ F/m $= 8.8542\times10^{-12}$ C^2/(Nm^2)
ε_i	instantaneous energy of molecule i
Φ	potential
ϕ	pair potential

γ surface tension
λ constant in Tersoff/Brenner potential, thermal conductivity
μ shear viscosity
μ dipole moment
θ angle, contact angle
ρ density
σ length parameter of Lennard-Jones potential
σ_F standard deviation of exiting force
τ time scale of non-dimensional Lennard-Jones system: $\tau = \sigma(m/\varepsilon)^{1/2}$
ω angular frequency
ω_D Debye frequency
ζ friction coefficient

Subscripts:
AR   argon
C   cutoff, control, center of fitting circle
G   gas
H   hydrogen
INT   interaction of solid and liquid
i, j, k   index of molecule
L   liquid
LG   liquid-gas
M   position of negative point charge
N   normal direction
OO   oxygen-oxygen
S   solid
SG   solid-gas
SL   solid-liquid
T   tangential direction
V   vapor
W   solid wall

8 REFERENCES

1. J. O. Hirschfelder, C. F. Curtiss and R. B. Bird, *Molecular Theory of Gases and Liquids*, John Wiley & Sons, New York, 1954.
2. G. A. Bird, *Molecular Gas Dynamics and the Direct Simulation of Gas Flow*, Oxford University Press, New York, 1994.
3. C. Kittel, *Introduction to Solid State Physics*, 7th ed., John Wiley & Sons, New York, 1996.
4. G. Chen, Thermal Conductivity and Ballistic-Phonon Transport in the Cross-Plane Direction of Superlattices, *Phys. Rev. B*, vol. 57, no. 23, pp. 14958-14973, 1998.
5. M. P. Allen and D. J. Tildesley, *Computer Simulation of Liquids*, Oxford University Press, New York, 1987.

6. Nicolas, J. J., Gubbins, K. E., Streett, W. B. and Tildesley, D. J., Equation of State for the Lennard-Jones Fluid, *Molecular Physics*, vol. 37-5, pp. 1429-1454, 1979.

7. J. P. Hansen and L. Verlet, Phase Transitions of the Lennard-Jones System, *Phys. Rev.*, vol. 184, pp. 151-161, 1969.

8. M. J. P. Nijmeijer, A. F. Bakker, C. Bruin, and J. H. Sikkenk, A Molecular Dynamics Simulation of the Lennard-Jones Liquid-Vapor Interface, *J. Chem. Phys.*, vol. 89, no. 6, pp. 3789-3792, 1988.

9. S. D. Stoddard and J. Ford, Numerical Experiments on the Stochastic Behavior of a Lennard-Jones Gas System, *Phys. Rev. A*, vol. 8, pp. 1504-1512, 1973.

10. F. H. Stillinger and A. Rahman, Improved Simulation of Liquid Water by Molecular Dynamics, *J. Chem. Phys.*, vol. 60, no. 4, pp. 1545-1557, 1974.

11. A. Ben-Naim and F. H. Stillinger, Aspects of the Statistical-Mechanical Theory of Water, in *Structure and Transport Processes in Water and Aqueous Solutions* (ed. R. A. Horne), Wiley-Interscience, New York, 1972.

12. H. J. C. Berendsen, J. P. M. Postma, W. F. van Gunsteren, J. Hermans, *in Intermolecular Forces* (ed. B. Pullmann), Reidel, Dordrecht, pp. 331-, 1981.

13. H. J. C. Berendsen, J. R. Grigera, and T. P. Straatsma, The Missing Term in Effective Pair Potentials, *J. Phys. Chem.*, vol. 91, no. 24, pp.6269-6271, 1987.

14. W. L. Jorgensen, J. Chandrasekhar, J. D. Madura, R. W. Impey and M. L. Klein, Comparison of Simple Potential Functions for Simulating Liquid Water, *J. Chem. Phys.*, vol. 79, no. 2, pp. 926-935, 1983.

15. W. L. Jorgensen, Optimized Intermolecular Potential Functions for Liquid Alcohols, *J. Phys. Chem.*, vol. 90, pp. 1276-1284, 1986.

16. O. Matsuoka, E. Clementi and M. Yoshimine, CI Study of the Water Dimer Potential Surface, *J. Chem. Phys.*, vol. 64, no. 4, pp. 1351-1361, 1976.

17. V. Carravetta and E. Clementi, Water-Water Interaction Potential: An Approximation of the Electron Correlation Contribution by a Function of the SCF Density Matrix, *J. Chem. Phys.*, vol. 81, no. 6, pp. 2646-2651, 1984.

18. J. Alejandre, D. J. Tildesley, and G. A. Chapela, Molecular Dynamics Simulation of the Orthobaric Densities and Surface Tension of Water, *J. Chem. Phys.*, vol. 102, no. 11, pp. 4574-4583, 1995.

19. L. X. Dang, and T-M. Chang, Molecular Dynamics Study of Water Clusters, Liquid, and Liquid-Vapor Interface of Water with Many-Body Potentials, *J. Chem. Phys.*, vol. 106, no. 19, pp. 8149-8159, 1997.

20. U. Niesar, G. Corongiu, E. Clementi, G. R. Kneller and D. K. Bhattacharya, Molecular Dynamics Simulations of Liquid Water Using the NCC ab initio Potential, *J. Phys. Chem.*, vol. 94, no. 20, pp. 7949-7956, 1990.

21. D. N. Bernardo, Y. Ding, K. K-Jespersen and R. M. Levy, An Anisotropic Polarizable Water Model: Incorporation of All-Atom Polarizabilities into Molecular Mechanics Force Fields, *J. Phys. Chem.*, vol. 98, pp. 4180-4187, 1994.

22. L. X. Dang and B. M. Pettitt, Simple Intramolecular Model Potentials for Water, *J. Phys. Chem.*, vol. 91, no. 12, pp. 3349-3354, 1987.

23. J. Anderson, J. J. Ullo and S. Yip, Molecular Dynamics Simulation of

Dielectric Properties of Water, *J. Chem. Phys.*, vol. 83, no. 3, pp. 1726-1732, 1987.

24. F. H. Stillinger and T. A. Weber, Computer Simulation of Local Order in Condensed Phase of Silicon, *Phys. Rev. B*, vol. 31, no. 8, pp. 5262-5271, 1985.

25. J. Tersoff, New Empirical Approach for the Structure and Energy of Covalent Systems, *Phys. Rev. B*, vol. 37, no. 12, pp. 6991-7000, 1988.

26. J. Tersoff, Empirical Interatomic Potential for Silicon with Improved Elastic Properties, *Phys. Rev. B*, vol. 38, no. 14, pp. 9902-9905, 1988.

27. J. Tersoff, Modeling Solid-State Chemistry: Interatomic Potentials for Multicomponent Systems, *Phys. Rev. B*, vol. 39, no. 8, pp. 5566-5568, 1989.

28. D. W. Brenner, Empirical Potential for Hydrocarbons for Use in Simulating the Chemical Vapor Deposition of Diamond Films, *Phys. Rev. B*, vol. 42, pp.9458-9471, 1990.

29. Y. Yamaguchi and S. Maruyama, A Molecular Dynamics Simulation of the Fullerene Formation Process, *Chem. Phys. Lett.*, vol. 286-3,4, pp. 336-342, 1998.

30. S. Maruyama and Y. Yamaguchi, A Molecular Dynamics Demonstration of Annealing to a Perfect C_{60} Structure, *Chem. Phys. Lett.*, vol. 286-3,4, pp. 343-349, 1998.

31. D. H. Tsai, The Virial Theorem and Stress Calculation in Molecular Dynamics, *J. Chem. Phys.*, vol. 70, no. 3, pp. 1375-1382, 1979.

32. T. Ikeshoji, B. Hafskjold and H. Furuholt, Molecular-Level Calculation Scheme for Pressure and Energy Flux in Heterogeneous Systems of Planar and Spherical Symmetry, *Molecular Physics*, 1999, submitted.

33. J. C. Tully, Dynamics of Gas-Surface Interactions: 3D Generalized Langevin Model Applied to fcc and bcc Surfaces, *J. Chem. Phys.*, vol. 73, no. 4, pp. 1975-1985, 1980.

34. J. Blömer and A. E. Beylich, MD-Simulation of Inelastic Molecular Collisions with Condensed Matter Surfaces, *Proceedings of 20th International Symposium on Rarefied Gas Dynamics*, Beijing, China, August 19-23, 1996, pp. 392-397, Peking University Press, Beijing, 1997.

35. S. Maruyama, and T. Kimura, A Molecular Dynamics Simulation of a Bubble Nucleation on Solid Surface, *Proceedings of 5th ASME-JSME Thermal Engineering Joint Conference*, San Diego, U.S.A., March 15-19, 1999, AJTE99-6511, 1999.

36. S. Maruyama and T. Kimura, A Study on Thermal Resistance over a Solid-Liquid Interface by the Molecular Dynamics Method, *Thermal Sci. Eng.*, vol. 7, no. 1, pp. 63-68, 1999.

37. T. Tokumasu and Y. Matsumoto, Parallel Computing of Diatomic Molecular Rarefied Gas Flows, *Parallel Computing*, vol. 23, pp. 1249-1260, 1997.

38. T. Tokumasu and Y. Matsumoto, Dynamic Molecular Collision (DMC) Model of Diatomic Molecules for DSMC Calculation, *Proceedings of 21st International Symposium on Rarefied Gas Dynamics*, vol. 2, in print, 1998.

39. N. Yamanishi and Y. Matsumoto, A New Model for Diatomic Molecules Scattering from Solid Surfaces, *Proceedings of 20th International Symposium on Rarefied Gas Dynamics*, Beijing, China, August 19-23, 1996,

pp. 381-386, Peking University Press, Beijing, 1997.
40. H. C. Anderson, Molecular Dynamics Simulations at Constant Pressure and/or Temperature, *J. Chem. Phys.*, vol. 72, no. 4, pp. 2384-2393, 1980.
41. S. Nosé, A Unified Formulation of the Constant Temperature Molecular Dynamics Methods, *J. Chem. Phys.*, vol. 81, no. 1, pp. 511-519, 1984.
42. W. G. Hoover, Canonical Dynamics: Equilibrium Phase-Space Distributions, *Phys. Rev. A*, vol. 31, no. 3, pp. 1695-1697, 1985.
43. M. Parrinello and A. Rahman, Crystal Structure and Pair Potentials: a Molecular Dynamics Study, *Phys. Rev. Lett.*, vol. 45, pp. 1196-1199, 1980.
44. M. Parrinello and A. Rahman, Polymorphic Transitions in Single Crystal: a New Molecular Dynamics Method, *J. Appl. Phys.*, vol. 52, pp. 7182-7190, 1981.
45. H. J. C. Berendsen, J. P. M. Postma, W. F. van Gunsteren, A. DiNola, and J. R. Haak, Molecular dynamics with coupling to an external bath, *J. Chem. Phys.*, vol. 81, no. 8, pp. 3684-3690, 1984.
46. P. H. Berens and K. R. Wilson, Molecular Dynamics and Spectra. I. Diatomic Rotation and Vibration, *J. Chem. Phys.*, vol. 74, no. 9, pp. 4872-4884, 1981.
47. S. Matsumoto and S. Maruyama, Far-Infrared Spectrum of Water by Molecular Dynamics Method, *Proceedings of the Second JSME-KSME Thermal Engineering Conference*, Kitakyushu, Japan, October 19-21, 1992, Vol. 3, pp. 61-64, 1992.
48. S. Maruyama, S. Matsumoto, and A. Ogita, Surface Phenomena of Molecular Clusters by Molecular Dynamics Method, *Thermal Sci. Eng.*, vol. 2, no. 1, pp. 77-84, 1994.
49. S. Maruyama, S. Matsumoto, M. Shoji and A. Ogita, A Molecular Dynamics Study of Interface Phenomena of a Liquid Droplet, *Heat Transfer 1994: Proceedings of the Tenth International Heat Transfer Conference*, Brighton, U.K., August 14-18, 1994, Vol. 3, pp. 409-414, Taylor & Francis, Washington D.C., 1994.
50. S. M. Thompson, K. E. Gubbins, J. P. R. B. Walton, R. A. R. Chantry, and J. S. Rowlinson, A Molecular Dynamics Study of Liquid Drops, *J. Chem. Phys.*, vol. 81, no. 1, pp. 530-542,1984.
51. M. J. P. Nijmeijer, C. Bruin, A. B. van Woerkom, and A. F. Bakker, Molecular Dynamics of the Surface Tension of a Drop, *J. Chem. Phys.*, vol. 96, no. 1, pp. 565-576, 1992.
52. M. J. Haye and C. Bruin, Molecular Dynamics Study of the Curvature Correction to the Surface Tension, *J. Chem. Phys.*, vol. 100, no. 1, pp. 556-559, 1994.
53. R. M. Townsend and S. A. Rice, Molecular Dynamics Studies of the Liquid-Vapor Interface of Water, *J. Chem. Phys.*, vol. 94, no. 3, pp. 2207-2218, 1991.
54. I. Tanasawa, Recent Advances in Condensation Heat Transfer, *Heat Transfer 1994: Proceedings of the Tenth International Heat Transfer Conference*, Brighton, U.K., August 14-18, 1994, Vol. 1, pp. 297-312, Taylor & Francis, Washington D.C., 1994.
55. K. Yasuoka, M. Matsumoto and Y. Kataoka, Evaporation and Condensation

at a Liquid Surface. I. Argon, *J. Chem. Phys.*, vol. 101, no. 9, pp. 7904-7911, 1994.

56. M. Matsumoto, K. Yasuoka and Y. Kataoka, Evaporation and Condensation at a Liquid Surface. II. Methanol, *J. Chem. Phys.*, vol. 101, no. 9, pp. 7911-7917, 1994.

57. M. Matsumoto, K. Yasuoka and Y. Kataoka, Molecular Mechanism of Evaporation and Condensation, *Thermal Sci. Eng.*, vol. 3, no.3, pp. 27-31, 1995.

58. M. Matsumoto and S. Fujikawa, Nonequilibrium Vapor Condensation: Molecular Simulation and Shock-Tube Experiment, *Micro. Thermophys. Eng.*, vol. 1, no. 2, pp. 119-126, 1997.

59. S. Fujikawa, M. Kotani and N. Takasugi, Theory of Film Condensation on Shock-Tube Endwall behind Reflected Shock Wave (Theoretical Basis for Determination of Condensation Coefficient), *JSME Int. J.*, Ser. B, vol. 40, no. 1, pp. 159-165, 1997.

60. T. Tsuruta, H. Tanaka, K. Tamashima and T. Masuoka, Condensation Coefficient and Interphase Mass Transfer, *International Symposium on Molecular and Microscale Heat Transfer in Materials Processing and Other Applications* (ed. I. Tanasawa and S. Nishio), Begell House, New York, pp. International Center Heat Mass Transfer Symposium, Yokohama, pp. 229-240, 1997.

61. M. Kotani, T. Tsuzuyama, Y. Fujii and S. Fujikawa, Nonequilibrium Vapor Condensation in Shock Tube, *JSME Int. J.*, Ser. B, vol. 41, no. 2, pp. 436-440, 1998.

62. M. Matsumoto, Molecular Dynamics of Fluid Phase Change, *Fluid Phase Equilibria*, vol. 144, pp. 307-314, 1998.

63. V. P. Carey and S. M. Oyumi, Condensation Growth of Single and Multiple Water Microdroplets in Supersaturated Steam: Molecular Simulation Predictions, *Micro. Thermophys. Eng.*, vol. 1, no.1, pp. 31-38, 1997.

64. V. E. B. Dussan, On the Spreading of Liquids on Solid Surfaces: Static and Dynamic Contact Lines, *Ann. Rev. Fluid Mech.*, vol. 11, pp. 371-400, 1979.

65. J. Koplik and J. R. Banavar, Continuum Deductions from Molecular Hydrodynamics, *Ann. Rev. Fluid Mech.*, vol. 27, pp. 257-292, 1995.

66. S. Matsumoto, S. Maruyama, and H. Saruwatari, A Molecular Dynamics Simulation of a Liquid Droplet on a Solid Surface, *Proceedings of the ASME-JSME Thermal Engineering Joint Conference*, Maui, U.S.A., March 19-24, 1995, Vol. 2, pp. 557-562, 1995.

67. S. Maruyama, T. Kurashige, S. Matsumoto, Y. Yamaguchi and T. Kimura, Liquid Droplet in Contact with a Solid Surface, *Micro. Thermophys. Eng.*, vol. 2, no.1, pp. 49-62, 1998.

68. J. N. Israelachvili, *Intermolecular and Surface Forces*, Academic Press, London, 1985.

69. S. Maruyama and T. Kimura, A Molecular Dynamics Simulation of a Bubble Nucleation on Solid Surface, *Proceeding of the Eurotherm Seminar n° 57 on Microscale Heat Transfer*, Poitiers, France, July 8-10, 1998, in print, 1999.

70. G. Saville, Computer Simulation of the Liquid-Solid-Vapour Contact Angle,

J. Chem. Soc. Faraday Trans. 2, vol. 73, pp. 1122-1132, 1977.

71. M. J. P. Nijmeijer, C. Bruin and A. F. Bakker, Wetting and Drying of an Inert Wall by a Fluid in a Molecular-Dynamics Simulation, *Physical Rev. A*, vol. 42, no. 10, pp. 6052-6059, 1990.

72. J. H. Sikkenk, J. O. Indekeu, J. M. J. van Leeuwen, E. O. Vossnack and A. F. Bakker, Simulation of Wetting and Drying at Solid-Fluid Interfaces on the Delft Molecular Dynamics Processor, *J. Statistical Physics*, vol. 52, nos. 1/2, pp. 23-44, 1988.

73. P. A. Thompson, W. B. Brickerhoff and M. O. Robbins, Microscopic Studies of Static and Dynamic Contact Angle, *J. Adhesion Sci Technol.*, vol. 7, no. 6, pp. 535-554, 1993.

74. Fan, C. F. and Cagin, T., 1995, Wetting of Crystalline Polymer Surfaces: A Molecular Dynamics Simulation, *J. Chem. Phys.*, vol. 103, no 20, pp. 9053-9061.

75. S. Kotake and S. Wakuri, Molecular Dynamics Study of Heat Conduction in Solid Materials, *JSME Int. J.*, Series B, vol. 37, no.1, pp. 103-108, 1994.

76. B. Hafskjold and T. Ikeshoji, On the Molecular Dynamics Mechanism of Thermal Diffusion in Liquids, *Molecular Physics*, vol. 80, no. 6, pp. 1389-1412, 1993.

77. T. Ikeshoji and B. Hafskjold, Non-Equilibrium Molecular Dynamics Calculations of Heat Conduction in Liquid and through Liquid-Gas Interface, *Molecular Physics*, vol. 81, no. 2, pp. 251-261, 1994.

78. K. Yasuoka and M. Matsumoto, Molecular Dynamics of Homogeneous Nucleation in the Vapor Phase. I. Lennard-Jones Fluid, *J. Chem. Phys.*, vol. 109, no. 19, pp. 8451-8462, 1998.

79. K. Yasuoka and M. Matsumoto, Molecular Dynamics of Homogeneous Nucleation in the Vapor Phase. II. Water, *J. Chem. Phys.*, vol. 109, no. 19, pp. 8463-8470, 1998.

80. K. Yasuoka and M. Matsumoto, Molecular Dynamics Simulation of Homogeneous Nucleation in Supersaturated Water Vapor, *Fluid Phase Equilibria*, vol. 144, no. 1-2, pp. 369-376, 1998.

81. Y. Viisanen, R. Stray and H. Reiss, Homogeneous Nucleation Rates for Water, *J. Chem. Phys.*, vol. 99, no. 6, pp. 4680-4692, 1993.

82. T. Ikeshoji, B. Hafskjold, Y. Hashi and Y. Kawazoe, Molecular Dynamics Simulation for the Cluster Formation Process of Lennard-Jones Particles: Magic Numbers and Characteristic Features, *J. Chem. Phys.*, vol. 105, no. 12, pp. 5126-5137, 1996.

83. T. Ikeshoji, B. Hafskjold, Y. Hashi and Y. Kawazoe, Molecular Dynamics Simulation for the Formation of Magic-Number Clusters with the Lennard-Jones Potential, *Phys. Rev. Lett.*, vol. 76, no. 11, pp. 1792-1795, 1996.

84. T. Kinjo and M. Matsumoto, Cavitation Processes and Negative Pressure, *Fluid Phase Equilibria*, 144, pp. 343-350, 1998.

85. J. Tamba, T. Ohara and T. Aihara, MD Study on Interfacelike Phenomena in Supercritical Fluid, *Micro. Thermophys. Eng.*, vol. 1, no.1, pp. 19-30, 1997.

86. T. Ohara and T. Aihara, MD Study on Dynamic Structure of Water, *International Symposium on Molecular and Microscale Heat Transfer in*

Materials Processing and Other Applications (ed. I. Tanasawa and S. Nishio), Begell House, New York, pp. 75-84, 1997.

87. Y. Yamaguchi, S. Maruyama and S. Hori, A Molecular Dynamics Simulation of Metal-Containing Fullerene Formation, *Proceedings of 5th ASME-JSME Thermal Engineering Joint Conference*, San Diego, U.S.A., March 15-19, 1999, AJTE99-6508, 1999.

88. S. Hirai, K. Okazaki, S. Kuraoka and K. Kawamura, Molecular Dynamics Simulation for the Formation of Argon Clathrate-Hydrate Structure, *Micro. Thermophys. Eng.*, vol. 1, no. 4, pp. 293-301, 1997.

89. R. Car and M. Parrinello, Unified Approach for Molecular Dynamics and Density-Functional Theory, *Phys. Rev. Lett*, vol. 55, no. 22, pp. 2471-2474, 1985.

90. U. Saalmann and R. Schmidt, Non-Adiabatic Quantum Molecular Dynamics: Basic Formalism and Case Study, *Z. Phys. D*, vol. 38, pp. 153-163, 1996.

91. M. Shibahara and S. Kotake, Quantum Molecular Dynamics Study on Light-to-Heat Absorption Mechanism: Two Metallic Atom System, *Int. J. Heat Mass Transfer*, vol. 40, no. 13, pp. 3209-3222, 1997.

92. M. Shibahara and S. Kotake, Quantum molecular dynamics study of light-to-heat absorption mechanism in atomic systems, *Int. J. Heat Mass Transfer*, vol. 41, no. 6-7, pp. 839-849, 1998.

93. H. Daiguji and E. Hihara, An EXAFS (Extended X-ray Absorption Fine Structure) Study of Lithium Bromide Aqueous Solutions Using Molecular Dynamics Simulation, *Heat Transfer Japanese Research*, in print, 1999.

94. H. Daiguji and E. Hihara, Molecular Dynamics Study of the Water Vapor Absorption into Aqueous Electrolyte Solution, *Micro. Thermophys. Eng.*, vol. 3, no. 2, pp. 151-165, 1999.

95. S. Maruyama, T. Yoshida, M. Kohno and M. Inoue, FT-ICR Studies of Laser Desorbed Carbon Clusters, *Proceedings of 5th ASME-JSME Thermal Engineering Joint Conference*, San Diego, U.S.A., March 15-19, 1999, JTE99-6513, 1999.

96. S. Maruyama, Y. Yamaguchi, M. Kohno and T. Yoshida, Formation Process of Empty and Metal-Containing Fullerene. —Molecular Dynamics and FT-ICR Studies, *Fullerene Science and Technology*, vol. 7, no. 4, pp. 621-636, 1999.

NUMERICAL METHODS IN MICROSCALE HEAT TRANSFER: MODELING OF PHASE-CHANGE AND LASER INTERACTIONS WITH MATERIALS

C.P. Grigoropoulos
M. Ye

1 INTRODUCTION

Rapid advancements in electronics technology have attracted increasing research interest in microscale phenomena. This work presents a summary of recent developments in numerical modeling of microscale heat transfer processes, with an emphasis on interactions between short-pulsed lasers and materials. The chapter will begin with a brief account of macroscopic (i.e. continuum) heat conduction models. For very short time and length scales, continuum approaches may break down, requiring description of the fundamental microscopic transfer. Simplified formulations stemming from the Boltzmann transport equation provide insight into the electron-phonon system coupling. Molecular dynamics and Direct Monte Carlo simulations are being used to analyze phase-change and rapid ablation plume expansion phenomena.

2   MODELS

2.1   Macroscopic Transport

In choosing a particular solution approach, a compromise is often needed between accuracy and computational cost. Because of their relative simplicity, classical continuum transport models, possibly modified to account for non-equilibrium effects, are often used effectively. When the diffusion time scale of the system is much larger than both the mean free path time scale and the relaxation time scale of relevant energy carriers, and when the characteristic length scale of the system is much larger than the corresponding length scales, local thermodynamic equilibrium (LTE) can be applied over space and time. In this case, macroscopic transport laws are operative [1]. The characteristic relaxation times for energy transfer typically are in the picosecond regime. The Fourier heat conduction is therefore sufficiently accurate for modeling nanosecond pulsed-laser heating of materials.

2.1.1 Fourier heat transfer. In nanosecond and longer time scales, the electrons and the lattice are at thermal equilibrium, characterized by a common temperature, T. The transient temperature field can then be calculated by solving the heat conduction equation:

$$(\rho c_P)\frac{\partial T}{\partial t} = \nabla \cdot (k(T)\nabla T) + Q'''(\vec{r},t) \tag{1}$$

where ρ, c_p, k and T represent density, specific heat for constant pressure, thermal conductivity, and temperature, respectively. The term $Q'''(\vec{r},t)$ quantifies the volumetrically distributed source term.

2.1.2 Departures from equilibrium at the melt interface. In pure element materials, the transition to the melting phase normally occurs at a specified temperature. The propagation of the solid/liquid interface is prescribed by the energy balance, which may be thought of as a kinematic boundary condition. The moving interface is assumed isothermal at the equilibrium melting temperature, T_m, if there is no overheating or undercooling. Equilibrium melting is described by:

$$T_s(\vec{r} \in S_{int}) = T_l(\vec{r} \in S_{int}) = T_m \tag{2}$$

$$k_s\frac{\partial T_s}{\partial n}\bigg|_{\vec{r} \in S_{int}} - k_l\frac{\partial T_l}{\partial n}\bigg|_{\vec{r} \in S_{int}} = \rho h_{sl}V_{int;n} \tag{3}$$

where h_{sl} is the latent heat for melting, $\partial/\partial n$ indicates the derivative of the interface along the normal direction vector $\vec{n}$ which points into the liquid at any

location of the interface, $\vec{r} \in S_{int}$. $V_{int;n}$ is the velocity of the interface along $\vec{n}$.

The Stefan statement of the phase change problem assumes that the interface dynamics is governed by the heat flow rather than the phase transition kinetics.

This assumption is true only for low melting speeds. For a flat interface, according to the quasi-chemical formulation of crystal growth from the melt [2,3] $T_m \neq T_{int}$ and the velocity of recrystallization $(V_{int}(T_{int}) > 0)$, or melting $(V_{int}(T_{int}) < 0)$ is:

$$V_{int}(T_{int}) = C \exp\left[-\frac{Q}{RT_{int}}\right]\left\{1 - \exp\left[-\frac{h_{sl}\Delta T}{RT_{int}T_m}\right]\right\} \tag{4}$$

where

$$C = R_M^o \exp\left[-\frac{h_{sl}}{RT_m}\right] \quad ; \quad \Delta T = T_m - T_{int}$$

In the above, R is the gas constant, T_{int} is the interface temperature, Q is the activation energy for viscous or diffusive motion in the liquid and R_M^o is the rate of melting at equilibrium. For small ΔT, equation (4) is linearized:

$$V_{int}(T_{int}) = \beta \Delta T \tag{5}$$

In the above, β is the slope of the interface-velocity response function near T_m. On the basis of the above arguments, the advancing melt front temperature is higher than T_m, while undercooling is observed in resolidification. To calculate the motion of the phase boundary, it is necessary to solve the heat conduction equation in the solid and liquid phases and apply Equation (5) as boundary condition at the interface. The classical theory implies symmetry for β in the melting and recrystallization processes with respect to $|\Delta T|$. Evaluation of the x-ray diffraction studies [4,5] has challenged this argument [6] by showing asymmetry in the interface response function yielding significant undercooling in the recrystallization process. It is noted, however, that departures from equilibrium are important for determining the recrystallization process but usually do not affect severely the overall energy balance.

2.1.3 Hyperbolic form for the heat conduction equation. It has been known that the Fourier law fails to predict correct temperatures for very short time, extreme temperature gradients, and temperatures near absolute zero. The hyperbolic heat equation, which predicts a wavelike solution, has been proposed for these situations:

$$\tau\frac{\partial^2 T}{\partial t^2} + \frac{\partial T}{\partial t} = \alpha\frac{\partial^2 T}{\partial x^2} \tag{6}$$

where τ is the characteristic relaxation time and α the thermal diffusivity. While Fourier's law predicts infinite speed of heat transfer, Equation (6) admits a speed $c^2 = \alpha/\tau$, thereby attempting to describe the short-time response without considering the fine microstructure of the medium. For example in [7], the hyperbolic heat equation was used to examine the very short-time temperature response of a semi-infinite region to an axisymmetric laser surface source. The performance of Fourier's law, the hyperbolic heat equation, and the equation of phonon radiative transfer for heat transport in thin films were compared in [8]. A survey of research in the wave theory of heat conduction was given in [9,10].

2.2   The Boltzmann Transport Equation and its Variations

When the time and/or length scales of the system are so small that local thermodynamic equilibrium cannot be defined within the system, a more rigorous and fundamental theory is based in the Boltzmann transport equation (BTE), which is the equation of motion for the one-particle distribution function and is appropriate to a rare gas. A general form of BTE can be expressed [11] as:

$$\frac{\partial f}{\partial t} + \vec{v} \cdot \nabla f + \vec{F} \cdot \frac{\partial f}{\partial \vec{p}} = \left(\frac{\partial f}{\partial t}\right)_{scat} \tag{7}$$

where $f(\vec{r}, \vec{p}, t)$ is the statistical distribution function of an ensemble of particles, which varies with time t, particle position vector $\vec{r}$, and momentum vector $\vec{p}$. $\vec{F}$ is the externally applied and internally supported force field. The right-hand side term represents the difference between the number of particles scattered into and out of $(\vec{r}, \vec{p})$:

$$\left(\frac{\partial f}{\partial t}\right)_{sca} = \int\!\!\int \left[f^{\ast} f^{\ast\ast} - f f^{\ast}\right] \tilde{R} d^3 p' d^3 p'' d^3 p''' \tag{8}$$

where $\tilde{R} \propto \delta(\vec{p} + \vec{p}' - \vec{p}'' - \vec{p}''') \delta(\varepsilon + \varepsilon' - \varepsilon'' - \varepsilon''')$ is the rate of the conserving momentum and energy two-particle collisional process $\vec{p} + \vec{p}' = \vec{p}'' + \vec{p}'''$. The functions $f \equiv f(\vec{r}, \vec{p}, t)$, $f' \equiv f(\vec{r}, \vec{p}', t)$, $f'' \equiv f(\vec{r}, \vec{p}'', t)$, $f''' \equiv f(\vec{r}, \vec{p}''', t)$. Under equilibrium, the collisional rate $\tilde{R}$ is zero. Key assumptions in the derivation of BTE are: 1) the range of interaction/mean free path $\ll 1$, 2) the particle trajectories are rectilinear prior to and after the collision events, 3) the distribution function f is homogeneous over the range of interaction, 4) the particles in the rare gas are not correlated (*molecular chaos*) [12].

Solution of the Boltzmann transport equation and its derivatives is one of the primary methods of computing non-equilibrium and/or non-continuum transport problems [13]. Direct solution of the Boltzmann equation or BGK-type model equations [14] has been applied to model the microscale transport during evaporation and condensation [15-17]. An alternative approach is the Maxwell moment method which seeks to satisfy the first few moments of the Boltzmann equation using an assumed test form of the velocity distribution function [18-21].

The linearized Boltzmann equation for the pure phonon field in terms of the eigenvectors of the normal-process, phonon-collision operator was solved in [22]. Major emphasis was placed on heat transport by phonon collision/scattering, and the contribution from the electron gas in conducting heat was neglected. The simplification was represented by two equations relating the temperature rise and the heat flux.

$$c_p \frac{\partial T}{\partial t} + \nabla \cdot \vec{q} = 0 \tag{9a}$$

$$\frac{\partial \vec{q}}{\partial t} + \frac{c^2 c_p}{3} \nabla T + \frac{1}{\tau_R} \vec{q} = \frac{\tau_N c^2}{5} [\nabla^2 \vec{q} + 2\nabla(\nabla \cdot \vec{q})] \tag{9b}$$

where c is the average speed of phonons (speed of sound), τ_R stands for the relaxation time for the "umklapp" process [23], which conserves phonon momentum, τ_N is the relaxation time for normal processes conserving momentum and $\vec{q}$ is the heat flux vector.

A single-energy equation, exhibiting the lagging behavior for fast transient heat transport at small scales, was derived in [9,10,24].

$$\nabla^2 T + \frac{9\tau_N}{5} \frac{\partial}{\partial t}(\nabla^2 T) = \frac{3}{\tau_R c^2} \frac{\partial T}{\partial t} + \frac{3}{c^2} \frac{\partial^2 T}{\partial t^2} \tag{10}$$

An equation of phonon radiative transfer was derived from Boltzmann transport equation by using relaxation time approximation and shown to be applicable under both short time and spatial scales [8,25].

2.3 Microscale Multi-step Heat Transfer Models

The previous models all concentrate on transport by a single type of carriers, paying relatively little attention to energy transfer processes among the energy carriers. In sub-nanosecond laser heating systems and high-field electronic devices, non-equilibrium microscale energy transfer processes between different types of energy carriers need be considered. The electron-phonon scattering time scale is on the order of $100 fs$ in most metals and semiconductors, while the phonon-phonon scattering time is on the order of $10 ps$. For pico- and femto-second laser heating of solids, the electrons initially are not in equilibrium with the lattice, necessitating the use of multiple-step energy transfer models. The early version of the phonon-electron interaction model (parabolic two-step model) was proposed [26,27], albeit without a rigorous proof.

$$c_e \frac{\partial T_e}{\partial t} = \frac{\partial}{\partial x}\left(k \frac{\partial T_e}{\partial x}\right) - G(T_e - T_l) + Q_{abs}'' \tag{11a}$$

$$c_l \frac{\partial T_l}{\partial t} = G(T_e - T_l) \tag{11b}$$

where c stands for the volumetric heat capacity, k the thermal conductivity of the electron gas, subscripts e and l stand for electron and metal lattice, and G the phonon-electron coupling factor. Q_{abs}''' is the volumetric radiation absorption term.

A hyperbolic two-step model accounting for the ballistic heat transport through the electron gas was derived [28] from solutions of the linearized Boltzmann transport equation.

$$c_e \frac{\partial T_e}{\partial t} = -\frac{\partial q}{\partial x} - G(T_e - T_l) + Q_{abs}''' \tag{12a}$$

$$c_l \frac{\partial T_l}{\partial t} = G(T_e - T_l) \tag{12b}$$

$$\tau_F \frac{\partial q}{\partial t} + k \frac{\partial T_e}{\partial x} + q = 0 \tag{12c}$$

where τ_F is the electronic relaxation time evaluated at the Fermi surface.

The nonequilibrium heat transfer processes between optical and acoustic phonons, as well as between electrons and phonons, have also been studied numerically [29].

2.4 Molecular Simulation

Molecular simulation methods have the advantage of addressing complex systems in a direct and fundamental way, and have been gaining increasing interest in recent years due to availability of fast-growing computer power. In general, there exist two kinds of simulation methodologies: deterministic molecular dynamics (MD) simulation, and stochastic Monte Carlo simulation methods. Most deterministic molecular dynamics simulations compute the evolution of a system of molecules by solving Newton's equations of motion for the molecules subjected to intermolecular and overall system constraints. Stochastic Monte Carlo simulation methods generally induce a random evolution of the system of atoms, in accordance with appropriate probability distributions allowing rearrangements of the system between the simulation steps.

In the molecular dynamics method [30], an isolated system of a fixed number of molecules N possessing a fixed energy E is studied. The molecules are subjected to the influence of intermolecular forces f_i, and their positions r_i are obtained by solving Newton's equations of motion:

$$m_i \frac{d^2 r_i}{dt^2} = f_i(t) = -\sum_{j \neq i} \frac{\partial}{\partial r_i} \phi(r_{ij}) \tag{13}$$

$$v_i(t) = \frac{dr_i(t)}{dt}$$

where $i=1,2,\ldots,N$, m_i is the molecular mass, r_{ij} is the distance separating two molecules, and ϕ is the intermolecular potential function. The position coordinates, velocity components, and internal state of each molecule are tracked in the simulation process. Static and dynamic macroscopic properties are obtained by averaging the appropriate functions over time. Deterministic molecular dynamics have also been applied to nonequilibrium systems [31-34]. An external force is applied to the system to trigger nonequilibrium, and the system's response to the force is then calculated. The main difference in modeling nonequilibrium phenomena is that the gathering of statistics on the basis of the simulation results is more complicated, as considerable computational effort is required to spatially or temporally resolve the property correlations. The MD method has been proven valuable for the simulation of dense gases and liquids, but inappropriate for dilute gases. The predictive power of the MD method is limited by the accuracy of the force law used to describe interatomic interactions. Recently, increasing emphasis has been placed on *ab-initio* MD simulations, in which the forces originating from the quantum mechanical ground-state of the electrons are used to compute classical nuclei trajectories [35,36]. The *ab-initio* molecular dynamics method combines accurate quantum mechanical calculations of the potential energy surface with efficient statistical mechanical schemes sampling atomic coordinates. However, due to the extremely expensive cost of *ab-initio* methods, the size of the systems that can be modeled is currently of the order of a few hundred atoms at most, while the time span of the simulations is about a few tens of *ps*.

The first probabilistic simulation method is the test particle Monte Carlo method developed [37]. A major disadvantage of this method is that an initial estimate must be made of the distribution function over the whole flow field. The direct simulation Monte Carlo (DSMC) method [38] is a hybrid approach in that it treats the evolution of the system based on fundamental principles, but the molecular interaction and boundary conditions stochastically. An essential approximation of DSMC is the uncoupling of the molecular motion and intermolecular collisions over a short time interval. The procedures of a typical DSMC simulation can be summarized into the following steps:

1)   The flow field is divided into small cells loaded with particles, each of which represents a fixed number of molecules of a given type.
2)   Particles are initialized in space according to some desired initial condition and given translational, rotational and vibrational energies by randomly sampling values from a desired distribution function.
3)   Particles are then transported at constant velocity for one time step, which should be sufficiently small so that a typical displacement is small compared to the local mean free path.

4)   After each time step, each particle is checked to determine whether it has crossed cell boundaries. Particles crossing boundaries are subjected to appropriate actions depending on the nature of the boundary.

5)   Particles are then binned into cells according to their current spatial locations.

6)   Candidate collision pairs are chosen at random in each cell. A selection rule is used to determine the pairs and the total number of pairs per cell that are selected for collisions.

7)   Collisions are computed for each cell, conserving both energy and momentum.

A sufficiently large number of cells must be used, ensuring that the cell size is smaller than the local mean free path. On the other hand, the number of particles used in the simulation should also be adequate. Molecular collisions can be modeled via a simple hard sphere interaction model or as a collision between molecules possessing specific force interaction potentials. For detailed information on molecular dynamics simulation methods and DSMC methods, interested readers are referred to [34,35,38,39,40,41].

3 APPLICATIONS

The applications reviewed in this section focus on the laser ablation of materials, solid-liquid and liquid-vapor phase changes, and ultra-short pulsed laser heating.

3.1 Laser Ablation of Materials

Pulsed laser irradiation of nanosecond duration is used in a variety of applications, including laser deposition of thin films and micromachining. The laser beam can drive and control rapid phase-change transformations and structural modifications at the microscopic level.   During the first stage of interaction between the laser pulse and solid material, part of the laser energy is reflected back into the ambient and part of the energy is absorbed within a short penetration depth in the material. The energy absorbed is subsequently transferred by heat conduction further into the interior of the target. A liquid pool is formed on the surface when the laser pulse energy is high enough to reach the melting temperature. At higher intensities, the liquid metal vaporizes. The vapor generated can be ionized, creating a high density plasma which further absorbs the incident laser light. The physical picture of laser interaction with evaporating materials at high fluence (F>1 GW/cm^2) is quite complicated. A number of questions are still open to study, such as the desorption mechanisms, the flow characteristics of the desorbed material, and the influence of ionic species (plasma formation) on the desorption and flow processes. Of fundamental interest is the prediction of the evaporative material removal rates, as well as the velocity, density and temperature distributions of the ejected particles as functions of the laser beam pulse energy, temporal distribution, and irradiance density on the target material surface.

A scaling model was used to predict the ablation pressure and the exerted impulse on laser irradiated targets for laser intensities exceeding the plasma threshold [42]. The model was shown to follow experimental trends for the mass loss rate and the ablation depth. A theoretical model was proposed for simulating laser-plasma-solid interactions [43], assuming that the formed plasma initially undergoes a three-dimensional isothermal expansion, followed by an adiabatic expansion. This model yielded athermal, non-Maxwellian velocity distributions of the atomic and molecular species, as well as thickness and compositional variations of the deposited material as functions of the target-substrate distance and the irradiated spot size. Several studies have focused on the role of the Knudsen layer formation in laser vaporization, sputtering and deposition (e.g. [43]). Another approach [45] dealt with the laser-induced expansion of metal vapor against a background pressure. A one-component, one-dimensional hydrodynamic model describing the expansion of laser-generated plasmas was developed in [46].

Dynamic source and partial ionization effects dramatically accelerate the expansion of laser-ablated material in the perpendicular direction to the target, leading to highly forward peaking plumes [47]. Computations of the plume propagation in vacuum and in an ambient gas were conducted in [48] using particle-in-cell hydrodynamics, continuum gas dynamics and scattering models. The first two approaches were shown to reproduce the gross hydrodynamic features of plume expansion in vacuum and in ambient gas. On the other hand, the scattering model captured trends of plume splitting that has been observed by emission and absorption spectroscopy diagnostics once a critical background gas pressure is reached. This result indicated that models adopting collisional scattering schemes may be more suitable for the long mean-free path flows encountered in typical pulsed laser deposition conditions.

A new computational approach for the thorough treatment of the heat transfer and fluid flow phenomena in pulsed laser processing of metals was established in [49]. The heat conduction in the solid substrate and the liquid melt was solved by a transient heat transfer model based on the enthalpy formulation for the treatment of phase-change problems. Since the laser beam width is $O(mm)$, compared with the $O(\mu m)$ thermal penetration depth and the $O(10nm)$ absorption depth, the heat conduction and melting of the solid were treated in 1-D:

$$\rho \frac{\partial h}{\partial t} = \frac{\partial}{\partial x}\left(k\frac{\partial T}{\partial x}\right) + (1-R)\alpha I(t)\exp(-\alpha x)$$

$$(14)$$

where ρ is density, h enthalpy, T temperature, t time, x distance from the irradiated surface, k thermal conductivity, R surface reflectivity, a the absorption coefficient, I the laser intensity.

The motions of the melting and vaporizing boundaries were then calculated via a kinetic relation [50-52]:

$$j_{ev} = n_l \left(\frac{k_B T_l}{2\pi m_a} \right)^{\frac{1}{2}} \exp\left(-\frac{h_{lv}}{k_B T_l} \right) - \theta_s n_v \left(\frac{k_B T_v}{2\pi m_a} \right)^{\frac{1}{2}} \tag{15}$$

where n_l and n_v are the number of atoms per unit volume for liquid and vapor, h_{lv} and T_v the latent heat of vaporization and temperature of vapor, m_a is the atomic mass and k_B the Boltzmann constant. The first term on the right represents the evaporation rate from the liquid surface at temperature T_l. The second term represents a damping of this evaporation rate due to the return of liquid molecules to the liquid surface. The parameter ϑ_s, called the sticking coefficient, represents the probability that a vapor atom returning to the liquid surface is finally adsorbed on the liquid surface. The value of this coefficient is taken at 0.20 according to [53].

Just above the liquid surface, a Knudsen discontinuity layer [54] was assumed. The conservation of mass, momentum and energy were used for monitoring the property changes (density, pressure and enthalpy) across the liquid/vapor interface:

$$\rho_v(u_v - u_{int}) = \rho_l(u_l - u_{int})$$
$$P_v - P_l = \rho_v(-u_v - u_{int})(u_l + u_v) \tag{16}$$
$$h_l + \frac{1}{2}u_l^2 = h_v + \frac{1}{2}u_v^2$$

In the above equations, the subscripts l and v indicate the liquid and vapor phases, correspondingly. It was assumed that the onset of liquid evaporation was characterized by the liquidus line, while the vapor phase was modeled as an ideal gas.

The ejected high pressure vapor generated shock waves against the ambient background pressure. The compressible gas dynamics was computed numerically by solving the system of Euler equations for mass, momentum and energy, supplemented by an isentropic gas equation of state. The computations were performed in a cylindrical system of coordinates, assuming axial symmetry. A two-dimensional extension of Godunov's method in Eulerian coordinates, the MUSCL scheme [55], was used in order to resolve possible discontinuities and shock wave phenomena in the flow of ejected metal vapor:

$$\frac{\partial \rho}{\partial t} + \nabla \cdot \left(\rho \vec{V} \right) = 0$$

$$\frac{\partial \vec{V}}{\partial t} + \left(\vec{V} \cdot \nabla \right) \vec{V} = 0 \qquad (17)$$

$$\frac{\partial e}{\partial t} + \left(\vec{V} \cdot \nabla \right) e = -\frac{P}{\rho} \left(\nabla \cdot \vec{V} \right)$$

where ρ, $\vec{V}$, P, and e are density, velocity vector, pressure, and specific internal energy, respectively. Aluminum, copper and gold targets considered were subjected to pulsed ultraviolet excimer laser irradiation of nanosecond duration. Computational results were obtained for the temperature distribution, evaporation rate and melting depth in the target, as well as the pressure, velocity and temperature distribution in the vapor phase. The propagation of shock waves through the vapor was captured by the numerical calculations. The effects of laser fluence and background pressure on the variables of interest were examined.

At high irradiances, the laser-material interaction is dominated by the formation of a plasma above the surface. In a subsequent paper [56], the excimer laser-beam absorption and radiation transport in the vapor phase was modeled by solving the radiation transport equation with the discrete-ordinates method. The rates for ionization were computed using the Saha-Eggert equation:

$$\frac{n_e n_i}{n_n} = \left(\frac{Q_e Q_i}{Q_n} \right)^{3/2} \left(\frac{m_e m_i}{m_e + m_i} \right)^{3/2} \exp\left(-\frac{\Psi_i}{k_B T} \right) \qquad (18)$$

where n_e, n_i, and n_n stand for the number density of the electrons, ions and neutrals, respectively. Q_e, Q_i, and Q_n represent the corresponding internal partition functions. Ψ_i is the ionization potential of the neutral atom. The inverse bremsstrahlung mechanism was considered as the main mechanism of plasma absorption, and the Planck mean absorption coefficient was used to average the spectral absorption. An aluminum target was subjected to pulsed, ultraviolet excimer laser irradiation of 26 ns duration at the wavelength of 248 nm with fluence of 25 J/cm^2 and an ambient pressure of 10^{-3} atm. The modeling results show that about 10 percent of the laser irradiation energy was absorbed in the plume, and about 35 percent of this energy transferred away by thermal radiation. The highest temperature in the hot vapor layer can reach O(10) eV at the end of the laser pulse, and the plume can be regarded as an opaque medium because the optical penetration depth at the laser light frequency, as well as the Planck mean averaged penetration depth, is short, O(10)μm.

The models [49,56] were modified to analyze the high-intensity laser ablation of liquids by accounting for the explosive-boiling process [57]. Laser ablation of liquids is quite different from typical metal ablation. Because of the low thermal diffusivity of the liquids, the heated state persists on a longer time scale. The lower boiling and critical temperatures make superheating more significant. Upon careful examination of several mechanisms of vaporization (normal vaporization, normal boiling, explosive boiling, and subsurface heating), it was concluded that explosive boiling is the only physically sound mechanism to explain laser sputtering at high fluences and short pulse lengths [58]. It was assumed that the maximum attainable temperature of liquids was determined by the spinodal limit. The vaporization rate was described by kinetic theory for liquid surface below the spinodal limit; otherwise the liquid in the metastable region was assumed to vaporize instantaneously. Figure 1 below shows a schematic of the system studied [57].

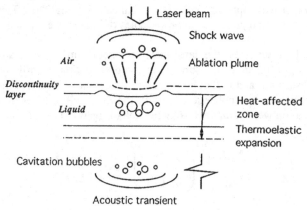

Figure 1 Diagram indicating various phenomena involved in the pulsed laser-induced ablation of absorbing liquids [56]. With Permission from Springer-Verlag.

The normalized pressure contours in the air are plotted in Figures 2a-c for a K_2CrO_4 aqueous solution with mass concentration coefficient, $m=0.01$, irradiated by an excimer laser beam of $26ns$ pulse width, fluence $2.3J/cm^2$. High-pressure vapor emerging from the irradiated spot with pressure $P_v(\approx 6MPa)$ expands into the ambient air, inducing shock-wave propagation with an average speed of about $900m/s$ and decaying maximum pressure. Figure 2d-f displays the corresponding velocity vectors in the air. Figure 3 compares the measured propagation distance of acoustic pulses in the air with computational results.

Figure 4 displays effective limits of three major approximations [38]. The continuum approach holds if the Knudsen number K_n is small. Taking the radius of the laser spot (0.5 mm) as the characteristic length L, it may be expected that

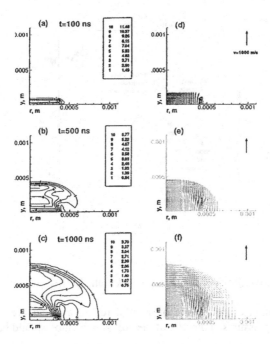

Figure 2 Normalized pressure contours (a-c) and velocity vectors (d-f) for an aqueous K_2CrO_4 solution (mass fraction $m=0.01$) and excimer laser fluence $F = 2.3 J/cm^2$ at $t = 0.1 \mu s$ for (a) and (d); $t = 0.5 \mu s$ for (b) and (e); $t = 1.0 \mu s$ for (c) and (f). The ambient pressure $P_\infty = 1 atm$ and the laser-spot diameter is 0.8mm [56]. With permission from Springer-Verlag.

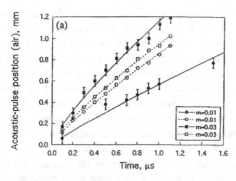

Figure 3 Comparison of propagation distance of acoustic pulses in the air. The solid lines represent power curve fitting of experimental data. The dashed lines and open markers are results from numerical calculation [56]. With permission from Springer-Verlag.

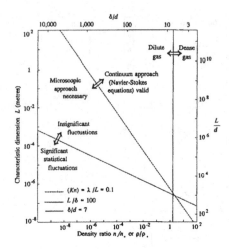

Figure 4 Effective limits of major approximations [37].

the hydrodynamic equations are only valid for an ambient pressure above 10^{-4} atm. At lower pressures, a direct particle collision method should be used.

Molecular dynamics simulations for laser ablation of copper, aluminum and silicon were preformed [59]. A thermal shock wave traveling to the interior of the target was observed, with propagation speed equal to the elastic wave velocity. This type of compression wave was also shown, in molecular dynamics simulation of laser ablation of silicon [60]. It was found that many isolated small voids are initially generated in the liquid phase, and that they later grow, coalescing with neighboring voids [59].

3-D Monte Carlo simulations [61,62] were used to study the flow of laser-desorbed particles from a binary target into vacuum. A thermal desorption mechanism was considered. Their simulation results showed that an evaporation jet is formed, and that the desorbed particles were accelerated and focused toward the jet axis. Heavy particles fly predominantly along the direction of the jet, while light particles have a higher chance of being scattered. A transfer of kinetic energy from the light to the heavy species was observed. The mean number of collisions and the angular and kinetic energy distributions of desorbed particles were found strong functions of the number of monolayers desorbed per laser pulse and of the laser spot width. The Direct Simulation Monte Carlo (DSMC) method was used to model the ablation of a target material (Si) and the expansion of the plume into background Ar gas in the presence of a substrate [63]. The 2-D images obtained from simulation are qualitatively similar to images of pulsed laser deposition plumes obtained using emission spectroscopy and laser-induced fluorescence (LIF). The laser-induced plasma plume expansion in vacuum was also studied via the DSMC method [64]. The simulation domain was divided into

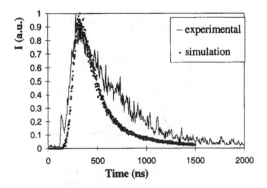

Figure 5 Simulated (symbol) and experimental (full line) time-of-flight curves of copper particles at a distance of 2mm above the target surface obtained after a KrF excimer laser irradiation at a fluence of $17J/cm^2$ [63]. Courtesy of Florence Garrelie. With permission from AIP.

45000 cells. A total number of 60000 particles was used at the beginning of the simulation and was progressively increased up to 960000 particles at longer times when the density dropped significantly. The laser energy absorption by vapor phase was calculated by the Beer-Lambert law. The simulated time-of-flight at a distance of $2mm$ from the surface plane agreed well with the experimental values, under the assumption that 35% of the total number of particles contributed to the kinetic energy gain into the expansion process through collisional recombination, as shown in Figure 5. This value also agrees well with the modeling results in [56]. Monte Carlo simulations of plume evolution in laser ablation of graphite and barium titanate yielded reasonably good agreement between calculated and experimental results for species arrival times and intensities at a hypothetical detection plane [65]. A series of 3-D Monte Carlo simulations of pulsed laser deposition processes were performed in [66,67]. The recombination and dissociation processes were shown to significantly influence the composition of the desorption jet and spread the angular and mean energy distributions of the particles arriving at the distant detector.

The velocity distribution of laser ablated particles is usually described by Maxwell-Boltzmann distribution function. However, correlating the temperature and stream velocity by fitting measured energy distributions may result in overestimated translational temperatures. MD simulations of laser ablation of a molecular solid were carried out using a 2-D breathing sphere model [68]. The molecules were represented by spherical particles interacting with each other via a pair potential. The vibrational modes of a given molecule were approximated by one internal breathing degree of freedom. The laser irradiation was simulated by the vibrational excitation of randomly chosen molecules, assuming that the

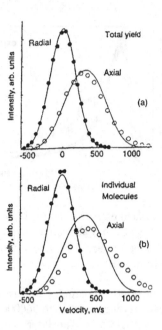

Figure 6 The ablation of a molecular solid with laser pulses of $15ps$ duration and $337nm$ wavelength is studied at a laser fluence 50% higher than the ablation threshold. Distributions of the velocities in the axial and radial directions are shown. The points are data from the simulation and the curves are obtained using Equation (19) with $T = 400K$ and $u_{max} = 650 m/s$. Data points for the total yield (a) and individual molecules only (b) are shown [67]. Courtesy of Barbara J. Garrison. With permission from AIP.

absorbed laser radiation was internally converted to vibrational energy. The simulation results showed that ejection of molecules depends on initial depth in the substrate, and that post-desorption collisions have an effect on the velocity distribution. The Maxwell-Boltzmann distribution was modified to account for a range of stream velocities in the ejected plume:

$$dN(v,T,u_{max}) = \frac{m}{4\pi k_B T u_{max}} \exp\left\{-\frac{m(v_x^2 + v_y^2)}{2k_B T}\right\}$$

$$\times \left\{erf\left[\sqrt{\frac{m}{2k_B T}}\, v_z\right] - erf\left[\sqrt{\frac{m}{2k_B T}}(v_z - u_{max})\right]\right\} d\,v_x d\,v_y d\,v_z \tag{19}$$

where m denotes the particle mass, v_x, v_y, v_z are the velocity components, T the equilibrium temperature of the plume, and u the stream velocity. The modified distribution function has two parameters that are independent of the desorption angle and have the clear physical meaning of the temperature and the maximum stream velocity of the plume, respectively. The proposed distribution function provided a consistent description of the axial and radial velocity distributions, as shown in Figure 6.

3.2 Solid-Liquid and Liquid-Vapor Phase-Change

Molecular dynamics simulation study of melting of silicon by a 23 ps laser pulse was performed in [69]. The Stillinger-Weber potential was employed for the atomic interaction. The simulation results showed that the crystal reaches the superheating limit at 14 ps and melts over the next $4ps$. A melting temperature of $1750K$ was obtained, which is comparable to the experimental value of $1683K$. However, the latent heat of melting was found to be $932J/g$, while the experimental value is about two times larger.

A molecular dynamics simulation of a 2-D atomic layer system heated by thermal radiation was carried out in [70]. The initial configuration of the model system is shown in Figure 7. The unit cell for computation is a rectangular region of length Lx and Lz, and an infinite number of unit cells are aligned periodically in the x-direction. Initially, there are 1000 atoms in the solid portion of the unit cell. Vacuum condition is assumed above the solid for the initial state. The substrate of the atomic layer is thermally insulated. The interatomic force is characterized by the Lennard-Jones-type, two-body potential function:

$$\phi(r) = 4\varepsilon \left[(\sigma/r)^{12} - (\sigma/r)^{6} \right] \qquad (18)$$

where r is the distance separating adjacent atoms, σ and ε are interatomic force constants. The initial x- and z- components of the velocity of the atoms are assumed to follow the Maxwell-Boltzmann distribution for thermal equilibrium and at $40K$. At time $t=0$, atomic movement commences, based on Newton's equation of motion. Thermal radiation is treated as an electromagnetic wave characterized by wavelength, phase and energy. Figure 8 shows the temporal evolution of the positions of solid-liquid and liquid-vapor interfaces. A solid-liquid interface forms at the surface of the layer and propagates towards the target interior. The fluctuation of the liquid-vapor interface was attributed to the fact that the liquid and the vapor were in a critical state, and that the number of atoms used in the simulation was too small in the x-direction.

Excitation of vibration and rotation between lattice particles was assumed the primary energy absorption mechanism in molecular dynamics simulation of melting and vaporization of a crystalline solid by laser radiation [71]. The excitation of atomic energy state due to laser irradiation was treated as a change in the potential energy acting between the lattice atoms. The force interaction was otherwise characterized by the Lennard-Jones 6-12 potential. After allowing the crystal to reach thermal equilibrium, laser irradiation was initiated, and the

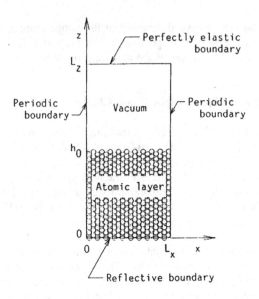

Figure 7 Schematic of the atomic layer system for numerical simulation [69]. Courtesy of Toshiro Makino.

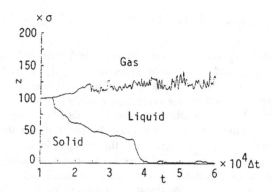

Figure 8 Temporal evolution of solid-liquid and liquid-gas interfaces [69]. Courtesy of Toshiro Makino.

Hamiltonian equations of motion were solved via a finite difference method to determine the temporal evolution of the system. Lattice atoms gained sufficient energy to break away from the lattice and escape to the ambient, leaving behind a cylindrical hole in the crystal.

In several molecular dynamics simulations (e.g. [72,73]), the transport of laser energy was effected via "energy carriers." A Monte Carlo/molecular

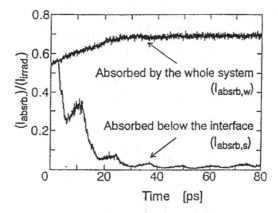

Figure 9 Vapor shieding effect in the irradiation of molybdenum with a laser pulse of $75ps$ duration and intensity $I_o = 10^{13} W/m^2$. Comparison of the laser energy absorbed by the whole atomic system and by the solid phase atoms [70].

dynamics simulation method was developed for laser melting and evaporation of materials [72]. The Morse potential function was used for the two-body interatomic potential for simulation of atomic motion. The incident laser energy was transferred by massless energy carriers, the trajectories of which were tracked by the test particle Monte Carlo method. Figure 9 shows the so-called "vapor shielding effect," which reduces the laser energy absorbed by the solid due to energy absorption by vapor.

Applications of numerical modeling have been developed for laser melting of materials. In the hard-disk-drive industry, increasing storage densities of hard disk drives require ultra-low head fly heights over disk surfaces. To prevent the stiction failure, a laser textured zone consisting of numerous surface features is created on the disk surface. The underlying mechanisms responsible for the formation of the surface topography are still not precisely quantifiable, as the detailed disk surface chemistry prior to and after the laser irradiation is largely unknown. The finite element method [74] was used to model pulsed laser texturing [75]. The continuity, momentum and energy equations were solved with finite difference method for a 2-D pulsed-laser texturing system [76] using the highly simplified marker and cell (HS-MAC) method [77] to track the free surface deformation. A schematic representation of the pulsed laser melting system is shown in Figure 10. Figure 11 shows the transient surface deformation for laser energy $2.9\mu J$, laser spot diameter $33\mu m$, and $46ns$ laser duration.

Laser annealing of semiconductors is of considerable interest for both applied and fundamental research. Computational models treating nonequilibrium melting and solidification phenomena were developed in [78,79]. The enthalpy equation was solved for heat transfer, the propagation velocity of the liquid-solid interface was determined from the classical crystal growth theory. The change of phase or state in each finite difference cell was allowed in accordance with a set

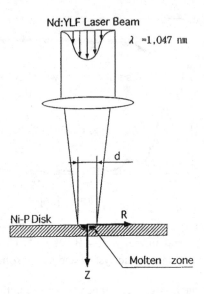

Figure 10 Schematic representation of the pulsed laser melting system [75].

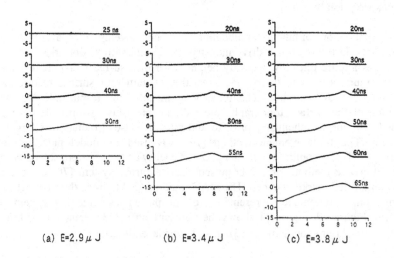

Figure 11 Transient surface deformation for pulse energies, $E = 2.9, 3.4, and 3.8 \mu J$. The laser pulse diameter, $d = 33 \mu m$ and pulse length, $t_p = 23 ns$ [75].

of prescribed conditions, subject only to the requirement of energy conservation. This was done by introducing the state array, which can be symbolically represented by:

$$A_{jk} = A_{jk}(H_j, T_i, \{K_l\}, \{N\}, \{C\})$$

where A_{jk} specifies the conditions under which a transition in each finite difference cell from the jth state to the kth state can be made. H_j is the enthalpy of the jth state, T_i the interfacial temperature, $\{K_l\}$ a set of interfacial rate parameters, $\{N\}$ the nucleation parameters, and $\{C\}$ the conditions in adjacent cells.

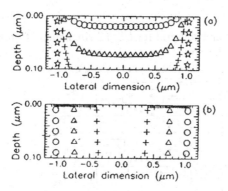

Figure 12 The location of the solid-liquid interface under laser irradiation of $800mJ/cm^2$. Melting sequence is shown in (a) at $10ns$ (O), $15ns$ (Δ), $20ns$ (+), $30ns$ (*) ans solidification sequence is shown in (b) at $35ns$ (O), $65ns$ (Δ), and $95ns$ (+). Note that the x and y axes have different scales [79]. Courtesy of James S. Im. With permission from AIP.

A 2-D model for excimer laser melting and solidification of thin S_i films on S_iO_2 was applied in [80]. Figure 12 shows calculated results for evolution of the solid-liquid interface during melting and solidification when the film is irradiated at a high energy density of $800mJ/cm^2$. The evolution of the interface is mainly parallel to the surface (i.e. vertical) during the melt-in period, but clearly lateral during solidification.

Another category of problems involving phase-change relates to droplet evaporation and condensation. The transport near a liquid nitrogen droplet evaporating in superheated nitrogen vapor at a pressure of about $0.2atm$ was studied with the DSMC method [81]. The shrinkage of the droplet was assumed to be slow enough so that the vaporization process could be treated as quasi-static and heat-transfer-controlled. The system considered in their model is shown schematically in Figure 13. The simulation in the non-continuum region near the

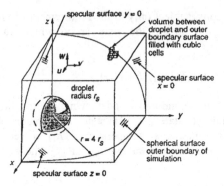

Figure 13 Boundaries and cell grid for DSMC simulation of droplet evaporation [80]. Courtesy of Van P. Carey.

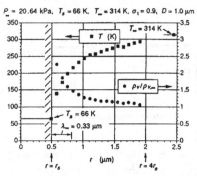

Figure 14 Calculated temperature and density profiles near an evaporating nitrogen droplet as predicted by the simulation [81]. Courtesy of Van P. Carey.

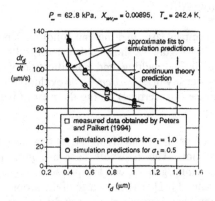

Figure 15 Comparison of simulation predictions [82] with continuum theory and droplet growth rates determined from the data in [83] for a water-argon mixture. Courtesy of Van P. Carey.

droplet was matched to the continuum transport at the outer boundary of the simulation domain. An isothermal, zero-energy flux boundary condition was used at the interface of the evaporating droplet. Computations were carried out for droplet diameters from $0.2 \mu m$ to $1.5 \mu m$ at ambient temperatures ranging from $120K$ to $400K$, corresponding to Knudsen number and Jakob number ranges of $0.09 < K_{nD} < 1.2$ and $0.2 < J_a < 1.7$. Figure 14 shows the calculated temperature and density fields adjacent to $1 \mu m$-diameter droplet evaporating in vapor at an ambient temperature of $313K$ [82]. It was found that non-continuum effects and interface vapor generation at the droplet surface are important mechanisms in the transport controlling the evaporation rate. Both effects reduce the rate of heat transfer to the droplet, with the latter effect also diminishing the temperature slip at the droplet surface. In a subsequent study [83], calculations were carried out at the conditions of the experiment [84]. Figure 15 compares the droplet growth rate obtained from Monte Carlo simulations with the experimental data, as well as the prediction of standard continuum theory using Fourier's law and Fick's law. The Knudsen numbers fall in the range of 0.114 to 0.045 for the conditions shown in Figure 15. The simulation results for an accommodation coefficient close to unity provide a better fit to the experimentally determined values than the predictions for $\sigma_f = 0.5$, and the difference between these two increases as the droplet size decreases and non-continuum effects become larger. It should be noted that the continuum theory substantially overpredicts the growth rate.

3.3 Ultra-Short Laser Pulse Heating

The energy exchange between electrons and the lattice observed in a number of experimental studies [85-87] was first treated theoretically (Equations 11). This phenomenological two-step model describes the electron and the lattice temperatures during short-pulse laser heating of materials but predicts infinite speed of energy transport. A more general hyperbolic two-step model (Equations 12) was derived from the solution of linearized Boltzmann transport equation. The hyperbolic two-step radiation-heating model, the parabolic two-step model, the hyperbolic one-step model, and the classical parabolic one-step model were then used to analyze the one-dimensional problem of short-pulse laser heating of metal films. Figure 16 shows comparison of the predicted electron temperature profile with experimental data at the rear surface of the thin film. The electron temperature rises initially due to the energy transport by the hot electrons and then drops due to the electron-phonon interactions. The computational results of two-step models agreed well with the experimental data, but the one-step models failed to predict the correct trend. Figure 17 shows a regime map of energy transport mechanisms during laser heating of gold with different heating times and at different temperatures [88]. The POS model applies for slow heating processes, the PTS model applies for fast heating processes at high temperatures, and the HOS model applies for fast heating processes at low temperatures. The HTS model provides a general, unified picture of the energy transport. Recently, the electron-phonon interaction model was also applied to model laser-induced electron emission from semiconductors [89].

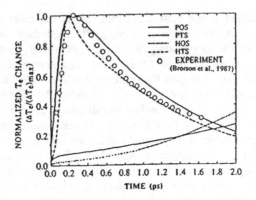

Figure 16 Comparison of electron-temperature changes with experimental data at the rear surface of a $0.1\,\mu m$ thick gold film subjected to a laser pulse of length $t_p = 96\,fs$ and fluence $F = 10\,J/m^2$ [27]. Courtesy of Chang-Lin Tien.

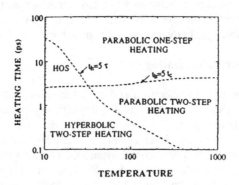

Figure 17 Regime map of energy transport mechanisms in the laser heating of gold [87]. Courtesy of Chang-Lin Tien.

4 CONCLUSIONS

Classical continuum theory could model certain microscale heat transfer problems by incorporating necessary modifications to account for non-equilibrium effects. This is usually a preferred method in engineering, with many successful applications in the modeling of nanosecond pulsed laser ablation, melting, and annealing processes.

The Boltzmann transport equation can be used to study nonlocal and nonequilibrium heat transfer phenomena and energy transfer processes of the order of $0.1ev$. BGK-type models and the Maxwell moment method have been employed to analyze such microscale transport processes as evaporation and condensation. Several models analyzing the energy transport mechanisms by relevant energy carriers have been developed by applying various approximations to the BTE. The hyperbolic two-step model is an example of this approach. The microscale multi-step models constitute useful tools for computing nonequilibrium heat transfer processes.

Upon further refinement of length scales, the BTE breaks down and molecular simulation is needed. Stochastic and deterministic molecular simulation methods have tremendous potential for advancing the study of fundamental aspects in microscale heat transfer processes, such as energy transfer between molecules, molecular kinetics at the interface of phase changes, energy transfer in condensed phases, and light interaction with materials. With the increasing available computational power, molecular simulation methods will eventually find engineering applications for modeling and optimizing realistic systems and processes. The biggest problems with DSMC methods lie in the treatment of forces and/or collision processes among particles and the treatment of particle interactions with interfaces and boundaries. Accurate specifications of collisional probability and the scattering characteristics will allow good prediction of the transport and thermodynamics properties. However, obtaining accurate probability distributions for multi-species systems is a challenging task. Molecular interactions with the liquid or solid interface have been usually modeled in the DSMC method via an accommodation coefficient, thereby drastically idealizing the interactions. Further efforts are required to predict accurate probability distribution and accommodation coefficients for complicated systems. Molecular simulations may provide such information.

Molecular dynamics simulations are the most fundamental predictive tools available. But much more detailed information about the molecular structure must be known, and MD methods are computationally expensive. Specification of accurate intermolecular potentials is critical for accurate prediction in MD simulations. Much more work needs to be done to obtain realistic intermolecular potentials for complex systems. Solving the Schrodinger equation may help provide this information. Detailed experimental validation of MD simulation results is difficult but much needed.

The accuracy of MD methods is also limited by the force law that is employed. Newton's law has been commonly used in MD simulations. The ab-initio MD method, based on quantum mechanics laws, has been developed to account for the quantum nature of materials. It is desirable to know the applicable regimes of the traditional MD methods since *ab-initio* MD simulations may be far more complicated for prediction of realistic systems.

5 ACKNOWLEDGMENT

Support for this work by the National Science Foundation and the Department of Energy is gratefully acknowledged.

6 REFERENCES

1.   C.-L. Tien, A. Majumdar, F.M. Gerner, *Microscale Energy Transport*, Taylor & Francis, Bristol, 1997.
2.   K.A. Jackson and B. Chalmers, Kinetics of solidification. *Can. J. Phys.* Vol. 34, 473-490, 1956.
3.   K.A. Jackson, Theory of melt growth. In *"Crystal Growth and Characterization"* (R. Ueda, and J.B. Mullin eds.), North-Holland, Amsterdam, 1975.
4.   B.C. Larson, C.W. White, T.S. Noggle and D. Mills, Synchrotron X-ray Diffraction Study of Silicon during Pulsed Laser-Annealing. *Phys. Rev. Lett.* Vol. 48, p. 337, 1982.
5.   B.C. Larson, J.Z. Tischler and D.M. Mills, Nanosecond Resolution Time-resolved X-ray Study of Silicon during Pulsed Laser Irradiation. *J. Mater. Res.,* Vol. 1, 144-154, 1986.
6.   P.S. Peercy, M.O. Thompson and J.Y. Tsao, Dynamics of Rapid Solidification in Silicon. *Proceedings, Materials Research Society,* (M.O. Thompson, S.T. Picraux, and J.S. Williams eds.) MRS, Pittsburgh, Vol. 75, 15-30, 1987.
7.   W.S. Kim, L.G. Hector, Jr., M.N. Ozisik, Hyperbolic Heat Conduction Due to Axisymmetric Continuous or Pulsed Surface Heat Sources, *J. Appl. Phys.,* Vol. 68, pp. 5478-5485, 1990.
8.   A.A. Joshi and A. Majumdar, Transient Ballistic and Diffusive Phonon Heat Transport in Thin Films, *J. Appl. Phys.,* Vol. 74, pp. 31-39, 1993.
9.   D.D. Joseph and L. Preziosi, Heat Waves, *Reviews of Modern Physics*, Vol. 61, pp. 41-73, 1989.
10.   D.D. Joseph and ·L. Preziosi, Addendum to the Paper on Heat Waves, *Reviews of Modern Physics*, Vol. 62, pp. 375-391, 1990.
11.   M. Ziman, *Electrons and Phonons*, Oxford Univ. Press, London, 1960.
12.   R.L. Liboff, *Kinetic Theory*, Prentice Hall, Englewood Cliffs, p. 130, 1988.
13.   C. Cercignani, *The Boltzmann Equation and its Applications*, Springer-Verlag, 1988.
14.   P.L. Bhatnagar, E.P. Gross and M. Krook, A Model for Collision Processes in Gases, I, Small Amplitude Processes in Charged and Neutral One-Component Systems, *Physical Review*, Vol. 94, pp. 511-525, 1954.
15.   V.G. Chernyak and A.Y. Margilevskiy, The Kinetic Theory of Heat and Mass Transfer from a Spherical Particle in a Rarefied Gas, *Int. J. Heat Mass Transfer*, Vol. 32, pp. 2127-2134, 1989.

16. P. Gajewski, A. Kulicki, A. Wisniewski and M. Zgorzelki, Kinetic Theory Approach to the Vapor-Phase Phenomena in a Nonequilibrium Condensation Process, *Phys. Fluids*, Vol. 17, pp. 321-327, 1974.

17. K. Aoki, Y. Sone and T. Yamada, Numerical Analysis of Gas Flows Condensing on Its Plane Condensed Phase on the Basis of Kinetic Theory, *Phys. Fluids A*, Vol. 2, pp. 1867-1878, 1990.

18. R.E. Sampson and G.S. Springer, Condensation on and Evaporation from Droplets by a Moment Method, *J. Fluid Mech.*, Vol. 36, pp. 577-584, 1969.

19. P.N. Shankar, A Kinetic Theory of Steady Condensation, J. Fluid Mech., Vol. 40, pp. 395-400, 1970.

20. J.B. Young, The Condensation and Evaporation of Liquid Droplets in a Pure Vapor at Arbitrary Knudsen Number, *Int. J. Heat Mass Transfer*, Vol. 34, pp. 1649-1661, 1991.

21. J.B. Young, The Condensation and Evaporation of Liquid Droplets at Arbitrary Knudsen Number in the Presence of an Inert Gas, *Int. J. Heat Mass Transfer*, Vol. 36, pp. 2941-2956, 1993.

22. R.A. Guyer and J.A. Krumhansl, Solution of the Linearized Boltzmann Equation, *Physical Review*, Vol. 148, pp. 766-778, 1966.

23. C. Kittel., *Introduction to Solid State Physics*, 7th ed. Wiley, New York, 1996.

24. D.Y. Tzou, *Macro- to Microscale Heat Transfer: the Lagging Behavior*, Taylor and Francis, Washington, 1996.

25. A. Majumdar, Microscale Heat Conduction in Dielectric Thin Films, *J. Heat Transfer*, Vol. 115, pp. 7-16, 1993.

26. M.I. Kaganov, I.M. Lifshitz and M.V. Tanatarov, Relaxation between Electrons and Crystalline Lattices, *Soviet Physics JETP*, Vol. 4, pp. 173-178, 1957.

27. S.I. Anisimov, B.L. Kapeliovich and T.L. Perel'man, Electron Emission from Metal Surfaces Exposed to Ultrashort Laser Pulses, *Sov. Phys. JETP*, Vol. 39, pp. 375-377, 1974.

28. T.Q. Qiu and C.-L. Tien, Heat Transfer Mechanisms during Short-Pulse Laser Heating of Metals, *J. Heat Transfer*, Vol. 115, pp. 835-841, 1993.

29. K. Fushinobu and A. Majumdar, Heat Generation and Transport in Sub-Micron Semiconductor Devices, *Heat Transfer on the Microscale*, HTD-Vol. 253, ASME, 1993.

30. B.J. Alder and T.E. Wainwright, Studies in Molecular Dynamics I. General Methods, *J. Chem. Phys.*, Vol. 31, pp. 459-466, 1959.

31. A.W. Lees and S.F. Edwards, The Computer Study of Transport Processes under Extreme Conditions, *J. Phys. C*, Vol. 5, pp. 1921-1929, 1972.

32. E.M. Gosling, I.R. McDonald and K. Singer, On the Calculation by Molecular Dynamics of the Shear Viscosity of a Simple Fluid, *Mol. Phys.*, Vol. 26, pp. 1475-1484, 1973.

33. W.T. Ashurst and W.G. Hoover, Argon Shear Viscosity via a Lennard-Jones Potential with Equilibrium and Nonequilibrium Molecular Dynamics, *Phys. Rev. Lett.*, Vol. 31, pp. 206-208, 1973.

34. W.G. Hoover, Nonequilibrium Molecular Dynamics, *Ann. Rev. Phys. Chem.*, Vol. 34, pp. 103-127, 1983.

35. R. Car, Modeling Materials by Ab Initio Molecular Dynamics, in Quantum Theory of Real Materials, (ed. J.R. Chelikowsky, S.G. Louie), pp. 23-37, Kluwer Academic, Norwell, 1996.

36. P.L. Silvestrelli and M. Parrinello, Ab Initio Molecular Dynamics Simulation of Laser Melting of Graphite, *J. Appl. Phys.*, Vol. 83, pp. 2478-2483, 1998.

37. J.K. Haviland and M.L. Lavin, Applications of the Monte Carlo Method to Heat Transfer in a Rarefied Gas, *Phys. Fluids*, Vol. 5, pp. 1399-1405, 1962.

38. G.A. Bird, *Molecular Gas Dynamics and the Direct Simulation of Gas Flows*, 2nd ed., Oxford University Press, Oxford, UK, 1994.

39. W.G. Hoover, *Molecular Dynamics*, Springer Verlag, Berlin, 1986.

40. J.M. Haile, *Molecular Dynamics Simulation: Elementary Methods*, John Wiley & Sons, Inc., New York, 1992.

41. D. Frenkel and B. Smit, *Understanding Molecular Simulation: From Algorithms to Applications*, Academic Press, San Diego, 1996.

42. C.R Phipps and R.W. Dreyfus, The High Irradiance Regime, in *Laser Ionization Mass Analysis*, edited by A. Vertes, R. Gijbels, and F. Adams, John Wiley, New York, 1993.

43. R.K. Singh and J. Narayan, Pulsed-Laser Evaporation Technique for Deposition of Thin Films: Physics and Theoretical Model, *Physical Review B (Condensed Matter)*, Vol. 41, pp. 8843-8859, 1990.

44. R. Kelly and A. Miotello, Comments on Explosive Mechanisms of Laser Sputtering, *Applied Surface Science*, Vol. 96-98, pp. 205-215, 1996.

45. M. Aden, E. Beyer, G. Herziger and H. Kunze, Laser-Induced Vaporization of a Metal Surface, *Journal of Physics D (Applied Physics)*, Vol. 25, pp. 57-65, 1992.

46. A. Vertes, R. Gijbels and F. Adams, *Laser Ionization Mass Analysis*, Wiley, New York, 1993.

47. K.R. Chen, J.N. Leboeuf, R.F. Wood, D.B. Geohegan, J.M. Donato, C.L. Liu and A.A. Puretzky, Accelerated Expansion of Laser-Ablated Materials near a Solid Surface, *Physical Review Letters*, Vol. 75, pp. 4706-4709, 1995.

48. J.N. Leboeuf, K.R. Chen, J.M. Donato, D.B. Geohegan, C.L. Liu, A.A. Puretzky and R.F. Wood, Modeling of Plume Dynamics in Laser Ablation Processes for Thin Film Deposition of Materials, *Phys. Plasmas*, Vol. 3, pp. 2203-2209, 1996.

49. J.R. Ho, C.P. Grigoropoulos and J.A.C. Humphrey, Computational Study of Heat Transfer and Gas Dynamics in the Pulsed Laser Evaporation of Metals, *J. Appl. Phys.*, Vol. 78, pp. 4696-4709, 1995.

50. V.P. Carey, *Liquid-Vapor Phase-Change Phenomena*, Hemisphere, Washington, DC, 1992.

51. L.D. Landau and E.M. Lifshitz, *Statistical Physics*, 3rd Ed., Pergamon, London, 1980.

52. D. Tabor, *Liquids and Solids*, 3rd Ed., Cambridge Univ. Press, 1991.

53. M. von Allmen, *Laser-Beam Interactions with Materials*, Springer-Verlag, Heidelberg, 1987.
54. B.R. Finke, M. Finke, P.D. Kapadia, J.M. Dowden and G. Simon, Numerical Investigation of the Knudsen-Layer, Appearing in the Laser-Induced Evaporation of Metals, SPIE: Laser-Assisted Processing II, Vol. 1279, pp. 127-134, 1990.
55. P. Colella, A direct Eulerian MUSCL scheme for gas dynamics, *SIAM J. Sci. Stat. Comput.*, Vol. 6, pp. 104-117, 1985.
56. J.R. Ho, C.P. Grigoropoulos and J.A.C. Humphrey, Gas Dynamics and Radiation Heat Transfer in the Vapor Plume Produced by Pulsed Laser Irradiation of Aluminum, *J. Appl. Phys.*, Vol. 79, pp. 7205-7215, 1996.
57. D. Kim, M. Ye, and C.P. Grigoropoulos, Pulsed Laser-Induced Ablation of Absorbing Liquids and Acoustic-Transient Generation, *Applied Physics A (Materials Science Processing)*, Vol. A67, pp. 169-181, 1998.
58. R. Kelly and A. Miotello, Comments on Explosive Mechanisms of Laser Sputtering, *Appl. Surf. Sci.*, Vol. 96-98, pp. 205-215.
59. E. Ohmura, I. Fukumoto and I. Miyamoto, Molecular Dynamics Simulation on Laser Ablation of Metal and Silicon, *J. Japan Society Precision Eng.*, Vol. 64, pp. 886-891, 1998.
60. R.F.W. Herrmann, J. Gerlach and E.E.B. Campbell, Molecular Dynamics Simulation of Silicon, *Nuclear Instruments & Methods in Physics Research, Section B*, Vol. 122, pp. 401-404, 1997.
61. H.M. Urbassek and D. Sibold, Gas-Phase Segregation Effects in Pulsed Laser Desorption from Binary Targets, *Physical Review Letters*, Vol. 70, pp. 1886-1889, 1993.
62. D. Sibold and H.M. Urbassek, Effect of Gas-Phase Collisions in Pulsed-Laser Desorption: A Three-Dimensional Monte Carlo Simulation Study, *J. Appl. Phys.*, Vol. 73, pp. 8544-8551, 1993.
63. D.L. Capewell and D.G. Goodwin, Monte Carlo Simulations of Reactive Pulsed Laser Deposition, *SPIE*, Vol. 2403, pp. 49-59, 1995.
64. F. Garrelie, J. Aubreton, and A. Catherinot, Monte Carlo Simulation of the Laser-Induced Plasma Plume Expansion under Vacuum: Comparison with Experiments, *J. Appl. Phys.*, Vol. 83, pp. 5075-5082, 1998.
65. A. Ranjan, S. Sinha, P.K. Ghosh, J.W. Hastie, D.W. Bonnell, A.J. Paul and P.K. Schenck, Monte Carlo Simulations of Plume Evolution from Laser Ablation of Graphite and Barium Titanate, *Chem. Phys. Lett.*, Vol. 277, pp. 545-550, 1997.
66. T.E. Itina, W. Marine and M. Autric, Monte Carlo Simulation of Pulsed Laser Ablation from Two-Component Target into Diluted Ambient Gas, *J. Appl. Phys.*, Vol. 82, pp. 3536-3542, 1997.
67. T.E. Itina, W. Marine and M. Autric, Monte Carlo Simulation of the Effects of Elastic Collisions and Chemical Reactions on the Angular Distributions of the Laser Ablated Particles, *Appl. Surf. Sci.*, Vol. 127-129, pp. 171-176, 1998.
68. L.V. Zhigilei and B.J. Garrison, Velocity Distribution of Molecules Ejected in Laser Ablation, *Appl. Phys. Lett.*, Vol. 71, pp. 551-553, 1997.

69. M.D. Kluge, J.R. Ray and A. Rahman, Pulsed Laser Melting of Silicon: A Molecular Dynamics Study, *J. Chem. Phys.*, Vol. 87, pp. 2336-2339, 1987.

70. T. Makino and H. Wakabayashi, Numerical Simulation of Melting Behavior of an Atomic Layer Irradiated by Thermal Radiation, *Thermal Science and Engineering*, Vol. 2, pp. 158-165, 1994.

71. S. Kotake and M. Kuroki, Molecular Dynamics Study of Solid Melting and Vaporization by Laser Irradiation, *Int. J. Heat Mass Transfer*, Vol. 36, pp. 2061-2067, 1993.

72. D.K. Chokappa, S.J. Cook and P. Clancy, Nonequilibrium Simulation Method for the Study of Directed Thermal Processing, *Phys. Rev. B*, Vol. 39, pp. 10075-10087, 1989.

73. K. Ezato, T. Kunugi and A. Shimizu, Monte Carlo/Molecular Dynamics Simulation on Melting and Evaporation Processes of Material due to Laser Irradiation, in ASME National Heat Transfer Conference, Vol. HTD, pp. 171-178, 1996.

74. P. Bach and O. Hassager, A Lagrangian Finite Element Method for the Simulation of Flow of Newtonian Liquids, *AIChE Journal*, Vol. 30, pp. 507-509, 1984.

75. T.D. Bennett, D.J. Krajnovich, C.P. Grigoropoulos, P. Baumgart and A.C. Tam, Marangoni Mechanisms in Pulsed Laser Texturing of Magnetic Disk Substrates, *J. Heat Transfer*, Vol. 119, pp. 589, 1997.

76. M. Iwamoto, M. Ye, C.P. Grigoropoulos and R. Greif, Numerical Analysis of Pulsed Laser Heating for the Deformation of Metals, *Num. Heat Transfer A*, Vol. 34, pp.791-804, 1998.

77. C.W. Hirt, B.D. Nichols and N.C. Romero, Sola-A Numerical Solution Algorithm for Transient Fluid Flows, Technical Report, Los Alamos Scientific Laboratory, 1975.

78. R.F. Wood and G.A. Geist, Modeling of Nonequilibrium Melting and Solidification in Laser-Irradiated Materials, *Physical Review B*, Vol. 34, pp. 2606-2620, 1986.

79. R.F. Wood, G.A. Geist and C.L. Liu, Two-Dimensional Modeling of Pulsed-Laser Irradiated a-Si and Other Materials, *Physical Review B*, Vol. 53, pp. 15863-15870, 1996.

80. V.V. Gupta, H.J. Song and J.S. Im, Numerical Analysis of Excimer-Laser-Induced Melting and Solidification of Thin Si Films, *Appl. Phys. Lett.*, Vol. 71, pp. 99-101, 1997.

81. V.P. Carey and N.E. Hawks, Stochastic Modeling of Molecular Transport to an Evaporating Microdroplet in a Superheated Gas, *J. Heat Transfer*, Vol. 117, pp. 432-439, 1995.

82. V.P. Carey, Modeling of Microscale Transport in Multiphase Systems, *Heat Transfer 1998, Proceedings of the 11th International Heat Transfer Conference*, Vol. 1, pp. 23-40, 1998.

83. V.P. Carey, S.M. Oyumi and S. Ahmed, Post-Nucleation Growth of Water Microdroplets in Supersaturated Gas Mixtures: A Molecular Simulation Study, *Int. J. Heat Mass Transfer*, Vol. 40, pp. 2393-2406, 1997.

84. F. Peters and B. Paikert, Measurement and Interpretation of Growth and Evaporation of Monodispersed Droplets in a Shock Tube, *Int. J. Heat Mass Transfer*, Vol. 37, pp. 293-302, 1994.

85. G.L. Eesley, Observation of Non-Equilibrium Electron Heating in Copper, *Phys. Rev. Lett.*, Vol. 51, pp. 2140-2143, 1983.

86. S.D. Brorson, J.G. Fujimoto and E.P. Ippen, Femtosencond Electronic Heat-Transfer Dynamics in Thin Gold Film, *Phys. Rev. Lett.*, Vol. 59, pp. 1962-1965, 1987.

87. P.B. Corkum, F. Brunel, N.K. Sherman and T. Srinivasan-Rao, Thermal Response of Metals to Ultrashort-Pulse Laser Excitation, *Physical Review Letters*, Vol. 61, pp. 2886-2889, 1988.

88. C.-L. Tien, T.Q. Qiu and P.M. Norris, Microscale Thermal Phenomena in Contemporary Technology, in *Molecular and Microscale Heat Transfer* (ed. S. Kotake and C.-L. Tien), p. 1, Begell House, New York, 1994.

89. S.S. Mao, X. Mao, R. Greif and R.E. Russo, Breakdown of Equilibrium Approximation for Nanosecond Laser-Induced Electron Emission from Silicon, *Appl. Phys. Lett.*, Vol. 74, pp. 1331-1333, 1998.

EIGHT

CURRENT STATUS OF THE USE OF PARALLEL COMPUTING IN TURBULENT REACTING FLOWS: COMPUTATIONS INVOLVING SPRAYS, SCALAR MONTE CARLO PROBABILITY DENSITY FUNCTION AND UNSTRUCTURED GRIDS

M.S. Raju

1 INTRODUCTION

The state of the art in multi-dimensional combustor modeling, as evidenced by the level of sophistication employed in terms of modeling and numerical accuracy considerations, is also dictated by the available computer memory and turnaround times afforded by present-day computers. With the aim of advancing the current multi-dimensional computational tools used in the design of advanced technology combustors, a solution procedure is developed that combines the novelty of the coupled CFD/spray/scalar Monte Carlo PDF (Probability Density Function) computations on unstructured grids with the ability to run on parallel architectures. In this approach, the mean gas-phase velocity and turbulence fields are determined from the solution of a conventional CFD method, the scalar fields of species and enthalpy from a modeled PDF transport equation using a Monte Carlo method, and a Lagrangian-based dilute spray model is used for the liquid-phase representation.

The gas-turbine combustor flows are often characterized by a complex interaction between various rate-controlling processes associated with turbulent transport, mixing, chemical kinetics, evaporation and spreading rates of spray, convective and radiative heat transfer, among others [1]. The phenomena to be modeled as controlled by these processes often interact with each other at various disparate time and length scales. In particular, turbulence plays an important role in determining the rates of mass and heat transfer, chemical reactions, and liquid-phase evaporation in many practical combustion devices. The influence of turbulence in a diffusion flame manifests itself in several forms, ranging from the so-called wrinkled or stretched flamelets regime to the distributed combustion regime, depending upon how turbulence interacts with various flame scales [2-3].

Most of the turbulence closure models for reactive flows have difficulty in treating nonlinear reaction rates [2-3]. The use of assumed shape PDF methods was found to provide reasonable predictions of pattern factors and NO_X emissions at the combustor exit [4]. However, their extension to multi-scalar chemistry becomes quite intractable. The solution procedure based on the modeled joint-composition PDF transport equation has an advantage in that it treats the nonlinear reaction rates without any approximation. This approach holds the promise of modeling various important combustion phenomena relevant to practical combustion devices such as flame extinction and blow-off limits, and unburnt hydrocarbons (UHC), CO, and NO_X predictions [4].

With the aim of demonstrating the viability of the PDF approach to the modeling of practical combustion flows, we have undertaken the task of extending this technique to the modeling of sprays, unstructured grids, and parallel computing as a part of the NCC (National Combustion Code) development program [5-7]. NCC is being developed in the form of a collaborative effort between NASA LeRC, aircraft engine manufacturers, and several other government agencies [8].

The use of parallel computing offers enormous computational power and memory as it can make use of hundreds of processors in concert to solve a complex problem. The trend towards parallel computing is driven by two major developments: the widespread use of distributed computing and the recent advancements in MPPs (Massively Parallel Processors). The solver is designed to be massively parallel and automatically scales with the number of available processors. Also, the ability to perform the computations on unstructured meshes allows representation of complex geometries with relative ease. The grid generation time associated with gridding up practical combustor geometries, which tend to be very complex in shape and configuration, could be reduced considerably by making use of existing automated unstructured grid generators. The solver accommodates the use of an unstructured mesh with mixed elements: triangular and/or quadrilateral for 2D (two-dimensional) geometries and tetrahedral for 3D. A solution procedure based on an unstructured grid formulation with parallel computing is becoming an accepted practice for the numerical solution of complex multidimensional reacting flows (e.g., gas-turbine combustor flows) [8-10].

A complete overview of the overall solution method with a particular emphasis on the PDF and spray algorithms, parallelization, and several other numerical issues related to the coupling between the CFD, spray, and PDF solvers is presented in this chapter. Some of the underlying differences between distributed computing and MPPs are discussed along with the underlying approaches to parallel programming involving Cray MPP Fortran and message-passing libraries such as PVM (Parallel Virtual Machine) and MPI (Message Passing Interface). The parallel performance of the three PDF, spray, and CFD modules is discussed for the case of a swirl-stabilized spray flame in both distributed and MPP computing environments. However, for a detailed presentation of the results and discussion involving the application of this method to several flows, the interested reader is referred to the published papers [1,7,25]. The chapter is concluded with some remarks on our ongoing research work and some suggestions for future research.

2 GOVERNING EQUATIONS FOR THE GAS PHASE

Here, we summarize the conservation equations for the gas-phase in Eulerian coordinates derived for the multicontinua approach [11]. This is done for the purpose of identifying the interphase source terms arising from the exchanges of mass, momentum, and energy with the liquid-phase.

The conservation of the mass leads to:

$$[\bar{\rho}V_c]_{,t} + [\bar{\rho}V_c u_i]_{,x_i} = s_{mlc} = \sum_k n_k\, m_k \qquad (1)$$

For the conservation of the jth species, we have:

$$[\bar{\rho}V_c y_j]_{,t} + [\bar{\rho}V_c u_i y_j]_{,x_i} - [\bar{\rho}V_c D y_{j,x_i}]_{,x_i} - \bar{\rho}V_c \dot{w}_j = s_{mls} = \sum_k \epsilon_j\, n_k\, m_k \qquad (2)$$

where

$$\sum_j \dot{w}_j = 0 \ and \ \sum_j \epsilon_j = 1$$

For the momentum conservation, we have:

$$[\bar{\rho}V_c u_i]_{,t} + [\bar{\rho}V_c u_i u_j]_{,x_j} + [pV_c]_{,x_i} - [\theta V_c \tau_{ij}]_{,x_j} - [(1-\theta)V_c \tau_{lij}]_{,x_j} = s_{mlm} =$$

$$\sum_k n_k\, m_k\, u_{ki} - \sum_k \frac{4\pi}{3}\, \rho_k\, r_k^3\, n_k\, u_{ki,t} \qquad (3)$$

where $\theta = $ the void fraction of the gas which is defined as the ratio of the equivalent volume of gas to a given volume of a gas and liquid mixture. For dilute sprays, the void fraction is assumed to be equal to one. The shear stress τ_{ij} in Eq. (3) is given by:

$$\tau_{ij} = \mu[u_{i,x_j} + u_{j,x_i}] - \frac{2}{3}\delta_{ij}u_{i,x_j}$$

For the energy conservation, we have:

$$[\bar{\rho}V_ch]_{,t} + [\bar{\rho}V_cu_ih]_{,x_i} - [\theta V_c\lambda T_{,x_i}]_{,x_i} - [(1-\theta)V_c\lambda_l T_{,x_i}]_{,x_i}$$
$$- [\theta V_cp]_{,t} = s_{mle} = \sum_k n_k\, m_k\Big(h_s - l_{k,eff}\Big) \tag{4}$$

3 SCALAR JOINT PDF EQUATION

The transport equation for the density-weighted joint PDF of the compositions, $\tilde{p}$, is:

$$[\bar{\rho}\tilde{p}]_{,t} \quad + \quad [\bar{\rho}\tilde{u}_i\tilde{p}]_{,x_i} \quad + \quad [\bar{\rho}w_\alpha(\underline{\psi})\tilde{p}]_{,\psi_\alpha} =$$
$$\{Transient\}\ \{Mean\ convection\}\quad \{Chemical\ reactions\}$$

$$-[\bar{\rho}<u_i''\mid\underline{\psi}>\tilde{p}]_{,x_i} - [\bar{\rho}<\frac{1}{\rho}J^\alpha_{i,x_i}\mid\underline{\psi}>\tilde{p}]_{,\psi_\alpha}$$

$$\{Turbulent\ convection\}\quad \{Molecular\ mixing\}$$

$$- [\bar{\rho}<\frac{1}{\rho}s_\alpha\mid\underline{\psi}>\tilde{p}]_{,\psi_\alpha} \tag{5}$$

$$\{Liquid - phase\ contribution\}$$

where

w_α	=	chemical source term for the α-th composition variable,
$<u_i''\mid\underline{\psi}>$	=	conditional average of Favre velocity fluctuations,
$<\frac{1}{\rho}J^\alpha_{i,x_i}\mid\underline{\psi}>$	=	conditional average of scalar dissipation, and
$<\frac{1}{\rho}s_\alpha\mid\underline{\psi}>$	=	conditional average of spray source terms.

The terms on the left-hand side of the above equation could be evaluated without any approximation, but the terms on the right-hand side of the equation require modeling. The first term on the right represents transport in physical space due to turbulent convection [3]. Since the joint PDF, $\tilde{p}$, contains no information on velocity, the conditional expectation of $<u_i''\mid\underline{\psi}>$ needs to be modeled. It is modeled based on a gradient-diffusion model with information supplied on the turbulent flow field from the flow solver [3].

$$- <u_i''\mid\underline{\psi}>\tilde{p} = \Gamma_\phi\tilde{p}_{,x_i} \tag{6}$$

The fact that the turbulent convection is modeled as a gradient-diffusion makes the turbulent model no better than the $k - \epsilon$ model. The uncertainties associated the use of a standard $k - \epsilon$ turbulence model to swirling flows are well known [12]. Some of the modeling uncertainties associated with the use of the standard $k - \epsilon$ model would be addressed in our future studies with the implementation of a non-linear $k - \epsilon$ developed for the modeling of swirling flows [12].

The second term on the right-hand side represents transport in the scalar space due to molecular mixing. A mathematical description of the mixing process is rather complicated, and the interested reader is referred to Ref. [3]. Molecular mixing is accounted for by making use of the relaxation to the ensemble mean submodel [2].

$$< \frac{1}{\rho}J^{\alpha}_{i,x_i} \mid \underline{\psi} >= -C_{\phi}\omega(\phi_{\alpha} - \bar{\phi}_{\alpha}) \tag{7}$$

where $\omega = \epsilon/k$, and C_{ϕ} is a constant. For a conserved scalar in a homogeneous turbulence, this model preserves the PDF shape during its decay, but there is no relaxation to a Gaussian distribution [3]. However, the results of Ref. [4] indicate that the choice between the different widely-used mixing models is not critical in the distributed reaction regime of premixed combustion as long as the turbulent mixing frequencies are above 1000 Hz. Most of the practical combustors seem to operate at in-flame mixing frequencies of 1000 Hz and above. The application of this mixing model seemed to provide some satisfactory results when applied to flows representative of those encountered in the gas-turbine combustion [4].

The third term on the right-hand side represents the contribution from the spray source terms:

$$< \frac{1}{\rho}s_{\alpha} \mid \underline{\psi} >= \frac{1}{\bar{\rho}\Delta V} \sum n_k m_k(\epsilon_{\alpha s} - \phi_{\alpha}) \tag{8}$$

where $\phi_{\alpha} = y_{\alpha}, \alpha = 1, 2, ..., s = \sigma - 1$

$$< \frac{1}{\rho}s_{\alpha} \mid \underline{\psi} >= \frac{1}{\bar{\rho}\Delta V} \sum n_k m_k(-l_{k,eff} + h_{ks} - \phi_{\alpha}) \tag{9}$$

where $\phi_{\sigma} = h$ and is defined by:

$$h = \sum_{i=1}^{\sigma-1} y_i h_i \tag{10}$$

where

$$h_i = h^o_{fi} + \int_{T_{ref}}^{T} C_{pi}y_i dT,$$

$$C_{pi} = \frac{R_u}{W_i}(A_{1i} + A_{2i}T + A_{3i}T^2 + A_{4i}T^3 + A_{5i}T^4),$$

$h_{f_i}^o$ is the heat of formation of ith species, R_u is the universal gas constant, $\epsilon_{\alpha s}$ is a mass fraction of the evaporating species at the droplet surface, and $l_{k,eff}$ is the effective latent heat of vaporization as modified by the heat loss to the droplet interior:

$$l_{k,eff} = l_k + 4\pi \frac{\lambda_l r_k^2}{m_k} \left(\frac{\partial T_k}{\partial r} \right)_s \tag{11}$$

Here we assumed that the spray source terms could be evaluated independent of the fluctuations in the gas-phase compositions of species and enthalpy. Eqs. (8)-(10) represent the modeled representation for the conditional averages of the spray contribution to the PDF transport equation.

4 LIQUID-PHASE EQUATIONS

The spray model is based on the multicontinua approach which allows for resolution on a scale greater than the average spacing between two neighboring droplets [11]. A Lagrangian scheme is used for the liquid-phase equations as it eliminates errors associated with numerical diffusion. The vaporization model of a polydisperse spray takes into account the transient effects associated with the droplet internal heating, the forced convection effects associated with droplet internal circulation and the phenomena associated with boundary layers and wakes formed in the intermediate droplet Reynolds number range [13]. The present formulation is based on a deterministic particle-tracking method and on a dilute spray approximation which is applicable for flows where the droplet loading is low. Not considered in the present formulation are the effects associated with the droplet breakup, the droplet/shock interaction, the multicomponent nature of liquid spray and the phenomena associated with dense spray effects and super-critical conditions. The spray method provided some favorable results when applied to both unsteady and steady-state calculations [1,13-15].

For the particle position of the kth drop group, we have:

$$\frac{dx_{ik}}{dt} = u_{ik} \tag{12}$$

For the droplet velocity:

$$\frac{du_{ik}}{dt} = \frac{3}{16} \frac{C_D \mu_{gs} Re_k}{\rho_k r_k^2} [u_{ig} - u_{ik}] \tag{13}$$

where

$$Re_k = 2 \frac{r_k \rho_g}{\mu_{gs}} [(u_g - u_k) \cdot (u_g - u_k)]^{1/2} \tag{14}$$

$$C_D = \frac{24}{Re_k} \left(1 + \frac{Re_k^{2/3}}{6} \right) \tag{15}$$

For droplet size, the droplet regression rate is determined from one of three different correlations depending upon the droplet-Reynolds-number range. When $Re_k > 20$, the regression rate is determined based on a gas-phase boundary-layer analysis [16] valid for Reynolds numbers in the intermediate range. The other two correlations, valid when $Re_k \leq 20$, are taken from Clift et al [17].

$$\frac{ds_k}{dt} = -2\frac{\mu_l}{\rho_k} \left[\frac{2}{\pi}Re_k\right]^{1/2} f(B_k) \quad if \ Re_k > 20$$

$$\frac{ds_k}{dt} = -\frac{\mu_l}{\rho_k} \left[1 + (1 + Re_k)^{1/3}\right] Re_k^{0.077} ln(1 + B_k)$$

$$if \ 1 < Re_k \leq 20 \tag{16}$$

$$\frac{ds_k}{dt} = -\frac{\mu_l}{\rho_k} \left[1 + (1 + Re_k)^{1/3}\right] ln(1 + B_k) \quad if \ Re_k < 1$$

where B_k is the Spalding transfer number defined in Eq. (22). The function $f(B_k)$ is obtained from the solution of Emmon's problem [18]. The range of validity of this function was extended in Raju and Sirignano [13] to consider the effects of droplet condensation.

The internal droplet temperature is determined based on a vortex model [16]. The governing equation for the internal droplet temperature is given by:

$$\frac{\partial T_k}{\partial t} = 17\frac{\lambda_l}{C_{pl}\rho_l r_k^2} \left[\alpha\frac{\partial^2 T_k}{\partial \alpha^2} + (1 + C(t)\alpha)\frac{\partial T_k}{\partial \alpha}\right] \tag{17}$$

where

$$C(t) = \frac{3}{17} \left[\frac{C_{pl}\rho_l}{\lambda_l}\right] r_k \frac{dr_k}{dt} \tag{18}$$

where α represents the coordinate normal to the streamsurface of a Hill's Vortex in the circulating fluid, and $C(t)$ represents a nondimensional form of the droplet regression rate. The initial and boundary conditions for Eq. (17) are given by:

$$t = t_{injection}, \quad T_k = T_{k,o} \tag{19}$$

$$\alpha = 0, \quad \frac{\partial T_k}{\partial \alpha} = \frac{1}{17}\left[\frac{C_{pl}\rho_l}{\lambda_l}\right] r_k^2 \frac{\partial T_k}{\partial t} \tag{20}$$

$$\alpha = 1, \quad \frac{\partial T_k}{\partial \alpha} = -\frac{3}{32}\frac{\rho_k}{\lambda_l} \left[\frac{C_p(T_g - T_{ks})}{B_k} - l_k\right] \frac{ds_k}{dt} \tag{21}$$

where $\alpha = 0$ refers to the vortex center, and $\alpha = 1$ refers to the droplet surface.

The Spalding transfer number is given by:

$$B_k = \frac{C_p(T_g - T_{ks})}{l_{k,eff}} = \frac{(y_{fs} - y_f)}{(1 - y_{fs})} \tag{22}$$

$$y_{fs}^{-1} = 1 + \frac{M_a}{M_f}\left(\chi_{fs}^{-1} - 1\right) \tag{23}$$

where M_a is the molecular weight of the gas excluding fuel vapor.

Based on the assumption that phase equilibrium exists at the droplet surface, the Clausius-Clapeyron relationship yields

$$\chi_{fs} = \frac{P_n}{P}exp\left[\frac{l_k}{R_u}\left(\frac{1}{T_b} - \frac{1}{T_{ks}}\right)\right] \tag{24}$$

In Eq. (14), the molecular viscosity is evaluated at a reference temperature using Sutherland's equation

$$\mu(T_{ref}) = 1.4637 \; 10^{-6}\frac{T_{ref}^{3/2}}{T_{ref} + 120} \tag{25}$$

where

$$T_{ref} = \frac{1}{3}T_g + \frac{2}{3}T_{ks} \tag{26}$$

The droplets may evaporate, move along the wall surfaces, and/or reflect with reduced momentum upon droplet impingement with the combustor walls. In our present computations, subsequent to the droplet impingement with the walls, the droplets are assumed to flow along the wall surfaces with a velocity equal to that of the surrounding gas.

5 DETAILS OF DROPLET FUEL INJECTION

The success of any spray model depends a great deal on the specification of the appropriate injector exit conditions. However, a discussion involving the physics of liquid atomization is beyond the scope of this subject matter. In our present computations, the liquid fuel injection is simulated by introducing a discretized parcel of liquid mass in the form of spherical droplets at the beginning of every fuel-injection time step.

For certain cases, the fuel-injection time step, Δt_{il}, needs to be determined based on the resolution permitted by the length and time scales associated with several governing parameters such as average grid spacing and average droplet spacing and velocity. However, our experience showed that for the case of a steady-state solution, a time step based on the average droplet lifetime yields better convergence [13-15]. Its value typically ranges between 1 and 2 milli-seconds for the case of reacting flows.

The spray computations facilitate fuel injection through the use of a single fuel injector comprised of different holes [14-15]. However, multiple fuel injection in a steady-state calculation could be simulated by simply assigning different initial conditions for the spatial locations of the droplet groups associated with each one of the different holes. For a polydisperse spray, the spray computations require inputs for the number of droplet groups in a given stream and for the initial droplet locations and velocities. However, the number of droplets in a given group and their sizes could be either input directly or computed from a properly chosen function for the droplet size distribution. The specified initial inputs should be representative of the integrated averages of the experimental conditions [1,7,14-15].

One correlation typical of those used for the droplet size distribution is taken from Ref. [19]:

$$\frac{dn}{n} = 4.21 \ 10^6 \left[\frac{d}{d_{32}}\right]^{3.5} e^{-16.98\left(\frac{d}{d_{32}}\right)^{0.4}} \frac{dd}{d_{32}} \tag{27}$$

where n is the total number of droplets and dn is the number of droplets in the size range between d and $d + dd$. The Sauter mean diameter, d_{32}, could be either specified or estimated from the following correlation [20]:

$$d_{32} = B_d \frac{2\pi\sigma_l}{\rho_g V_T^2} \lambda_m^* \tag{28}$$

where B_d is a constant, V_T is the average relative velocity between the liquid interface and the ambient gas, and λ_m^* is a function of the Taylor number, $(\rho_l\sigma_l^2)/(\rho_g\mu_l^2 V_T^2)$.

A typical droplet size distribution obtained from the above correlation in terms of the cumulative percentage of droplet number and mass as a function of the droplet diameter is shown in Fig. 1 [1].

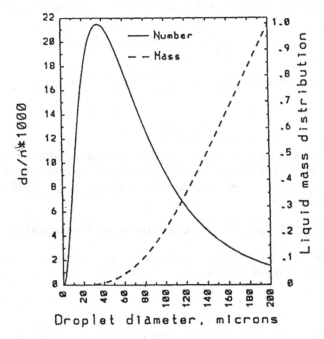

Figure 1 Droplet-size distribution.

6 CFD SOLUTION ALGORITHM

The gas-phase mass and momentum conservation equations together with the standard $k - \epsilon$ turbulence equations with wall functions are solved by making of a modified version of the Pratt and Whitney's CORSAIR - an unstructured CFD solver. It is a finite-volume solver with an explicit fourth-stage Runge-Kutta scheme. Further details of the code can be found in Refs. [9-10].

7 PDF SOLUTION ALGORITHM

In order to facilitate the integration of the Monte Carlo PDF method in a finite-volume context, the volume integrals of convection and diffusion in Eq. (5) were first recast into surface integrals by means of a Gauss's theorem [21]. Partial integration of the PDF transport equation would yield:

$$\tilde{p}_p(\underline{\psi}, t + \Delta t) = (1 - \frac{c_p \Delta t}{\bar{\rho} \Delta V}) \tilde{p}_p(\underline{\psi}, t) + \sum_n \frac{c_n \Delta t}{\bar{\rho} \Delta V} \tilde{p}_n(\underline{\psi}, t)$$

$$- \Delta t [w_\alpha(\underline{\psi}) \tilde{p}]_{,\psi_\alpha} - \Delta t [< \frac{1}{\rho} J_{i,x_i}^\alpha \mid \underline{\psi} > \tilde{p}]_{,\psi_\alpha} - \Delta t [< \frac{1}{\rho} s_\alpha \mid \underline{\psi} > \tilde{p}]_{,\psi_\alpha} \quad (29)$$

where subscript n refers to the nth-face of the computational cell. The co-efficient c_n represents the transport by convection and diffusion through the nth-face of the computational cell, p. The convection/diffusion coefficients in the above equation are determined by one of the following two expressions:

$$c_n = \Gamma_\phi(\frac{2\underline{a}_n \cdot \underline{a}_n}{\Delta V_p + \Delta V_s}) + max[0, -\bar{\rho} \underline{a}_n \cdot \underline{u}_n]$$

$$c_n = max[|0.5\bar{\rho} \underline{a}_n \cdot \underline{u}_n|, \Gamma_\phi(\frac{2\underline{a}_n \cdot \underline{a}_n}{\Delta V_p + \Delta V_s})] - 0.5\bar{\rho} \underline{a}_n \cdot \underline{u}_n$$

and

$$c_p = \sum_n c_n$$

In both the above expressions for c_n, a cell-centered finite-volume derivative is used to describe the viscous fluxes; but an upwind differencing scheme is used for the convective fluxes in the first expression and a hybrid differencing scheme in the second.

7.1 Numerical Method Based on Approximate Factorization

The transport equation is solved by making use of an approximate factorization scheme [3]. Eq. (29) can be recast as:

$$\tilde{p}_p(\underline{\psi}, t + \Delta t) =$$
$$(I + \Delta t R)((I + \Delta t S)(I + \Delta t M)(I + \Delta t T) \tilde{p}_p(\underline{\psi}, t) + O(\Delta t^2) \quad (30)$$

where I represents the unity operator and T, M, S, and R denote the operators associated with spatial transport, molecular mixing, spray, and chemical reactions, respectively. The operator is further split into a sequence of intermediate steps:

$$\tilde{p}_p^*(\underline{\psi}, t) = (I + \Delta t T)\tilde{p}_p(\underline{\psi}, t) \tag{31}$$

$$\tilde{p}_p^{**}(\underline{\psi}, t) = (I + \Delta t M)\tilde{p}_p^*(\underline{\psi}, t) \tag{32}$$

$$\tilde{p}_p^{***}(\underline{\psi}, t) = (I + \Delta t S)\tilde{p}_p^{**}(\underline{\psi}, t) \tag{33}$$

$$\tilde{p}_p(\underline{\psi}, t + \Delta t) = (I + \Delta t R)\tilde{p}_p^{***}(\underline{\psi}, t) \tag{34}$$

The operator-splitting method provides the solution for the transport of $\tilde{p}$ by making use of a Monte Carlo technique. In the Monte Carlo simulation, the density-weighted PDF at each grid cell is represented by an ensemble of N_m stochastic elements where the ensemble-averaged PDF over N_m delta functions replaces the average based on a continuous PDF [3].

$$\tilde{p}_{pm}(\psi) = <\tilde{p}_p(\psi)> = \frac{1}{N_m}\sum_{n=1}^{N_m}\delta(\psi - \phi^n) \tag{35}$$

The discrete PDF $\tilde{p}_{pm}(\psi)$ is defined in terms of N_m sample values of ϕ^n, $n = 1, 2, 3...N_m$. The statistical error in this approximation is proportional to $N_m^{-1/2}$.

Using the operator-splitting method, the solution for the PDF transport equation is obtained sequentially according to the intermediate steps given by Eqs. (31)-(34).

7.2 Convection/Diffusion Step

The first step associated with convection/diffusion is given by:

$$\tilde{p}_p^*(\underline{\psi}, t) = (I + \Delta t T)\tilde{p}_p(\underline{\psi}, t) =$$
$$(1 - \frac{c_p \Delta t}{\bar{\rho}\Delta V})\tilde{p}_p(\underline{\psi}, t) + \sum_n \frac{c_n \Delta t}{\bar{\rho}\Delta V}\tilde{p}_n(\underline{\psi}, t) \tag{36}$$

This step is simulated by replacing a number of particles (= the nearest integer of $\frac{c_n \Delta t N_m}{\bar{\rho}\Delta V}$) at $\phi_p(t)$ by randomly selected particles at $\phi_n(t)$.

7.3 Numerical Issues Associated With Fixed Versus Variable Time Step

It is obvious from the above equation that a necessary criterion for stability requires satisfaction of $\frac{c_p \Delta t}{\bar{\rho} \Delta V} < 1$. When the computations are performed with a fixed time step, this criterion tends to be too restrictive for most applications. Depending on the flow configuration, the allowable maximum time increment Δt is likely to be limited by a region of the flow field where convective fluxes dominate (such as close to injection holes). But in the main stream, the flow is usually characterized by much lower velocities. Resolution considerations require a higher concentration of the grid in certain regions of the flowfield than the others. For example, more grid lines are clustered in regions where boundary layers are formed. In such regions the allowable maximum time increment might be limited in a direction dominated by the largest of the diffusive fluxes as determined by $\Gamma_\phi / \Delta x$. This problem gets magnified if the cells also happen to be highly skewed.

Such restrictions on the allowable maximum time step could lead to a frozen condition when the Monte Carlo simulation is performed with a limited number of stochastic particles per cell. For clarity, let us consider the following criterion:

$$N_m > \frac{\bar{\rho} \Delta V}{c_n \Delta t} \qquad (37)$$

which has to be satisfied at all grid nodes. It is estimated that about 10^3 stochastic particles per cell are needed in order to avoid the so-called frozen condition for performing a typical 3-D gas-turbine combustor calculation. The frozen condition is referred to a state in which no transfer of stochastic particles takes place between the neighboring cells when N_m falls below a minimum required. Scheurlen et al [21] were the first ones to recognize the limitations associated with the use of a fixed time step in the Monte Carlo PDF computations.

However, our experience has shown that this problem can be overcome by introducing the concept of local time-stepping which is a convergence improvement technique widely used in many of the steady-state CFD computations. In this approach, the solution is advanced at a variable time step for different grid nodes. In our present computations, it is determined based on

$$\Delta t = min(C_{tf} \Delta t_f, \frac{\rho \Delta V}{C_t(c_n + s_{mlc})}) \qquad (38)$$

where C_{tf} and C_t are calibrated constants and were assigned the values of 4 and 2.5, respectively, Δt_f is the local time step obtained from the flow (CORSAIR) module, and $s_{mlc} = \sum n_k m_k$. The time step is chosen such that it permits transfer of enough particles across the boundaries of the neighboring cells while ensuring that the time step used in the PDF computations does not deviate very much from the time step used in the flow solver.

7.4 Molecular Mixing Step

The second step associated with molecular mixing is given by:

$$\frac{d\phi_\alpha}{dt} = -C_\phi \omega (\phi_\alpha - \bar{\phi}_\alpha) \tag{39}$$

The solution for this equation is updated by:

$$\phi_\alpha^{**} = \phi_\alpha^* + (\phi_\alpha^* - \bar{\phi}_\alpha^*)e^{-C_\phi \omega \Delta t} \tag{40}$$

where C_ϕ was assigned a value of 1.

7.5 Spray Step

The third step associated with the spray contribution is given by:

$$\frac{d\phi_\alpha}{dt} = \frac{1}{\bar{\rho}\Delta V}\sum n_k m_k (\epsilon_\alpha - \phi_\alpha) \tag{41}$$

where $\phi_\alpha = y_\alpha, \alpha = 1, 2, ..., s = \sigma - 1$

$$\frac{d\phi_\alpha}{dt} = \frac{1}{\bar{\rho}\Delta V}\sum n_k m_k (-l_{k,eff} + h_{ks} - \phi_\alpha) \tag{42}$$

where $\phi_\sigma = h$. The solution for the above equations is upgraded by a simple explicit scheme:

$$\phi_\alpha^{***} = \epsilon_\alpha \frac{\Delta t \sum n_k m_k}{\bar{\rho}\Delta V} + \phi_\alpha^{**}(1 - \frac{\Delta t \sum n_k m_k}{\bar{\rho}\Delta V}) \tag{43}$$

where $\alpha \le \sigma - 1$

$$\phi_\alpha^{***} = \frac{\Delta t \sum n_k m_k}{\bar{\rho}\Delta V}(-l_{k,eff} + h_{ks}) + \phi_\alpha^{**}(1 - \frac{\Delta t \sum n_k m_k}{\bar{\rho}\Delta V}) \tag{44}$$

where $\alpha = \sigma$. After a new value for enthalpy is updated, the temperature is determined iteratively from the solution of Eq. (10).

7.6 Reaction Step

Finally, the fourth step associated with chemical reactions is given by:

$$\frac{d\phi_\alpha}{dt} = -\nu_f \frac{W_f}{\rho} A(\frac{\rho\phi_f}{W_f})^a (\frac{\rho\phi_o}{W_o})^b e^{-(\frac{E_a}{T})} \tag{45}$$

where $\phi_\alpha = y_f$.

$$\frac{d\phi_\alpha}{dt} = -\nu_o \frac{W_o}{\rho} A \left(\frac{\rho\phi_f}{W_f}\right)^a \left(\frac{\rho\phi_o}{W_o}\right)^b e^{-\left(\frac{E_a}{T}\right)} \tag{46}$$

where $\phi_\alpha = y_o$.

$$\frac{d\rho\phi_\alpha}{dt} = 0 \tag{47}$$

where $\phi_\alpha = h$.

The numerical solution for Eqs. (45)-(47) is integrated by an implicit Euler scheme [22]. The resulting non-linear algebraic equations are solved by the method of quasi-linearization [23].

7.7 Details of Combustion Chemistry

In this section, we present an example of how combustion chemistry is handled for the case of n-heptane when it is modeled by a single-step global mechanism of Westbrook and Dryer [24]. The corresponding rate constants in Eqs. (45)-(46) are given by $A = 0.286 \ 10^{+10}$, $a = 0.25$, $b = 1.25$, and $E_a = 0.151 \ 10^{+05}$. This global combustion model is reported to provide adequate representation of temperature histories in flows not dominated by long ignition delay times. For example, the overall reaction representing the oxidation of the n-heptane fuel is given by:

$$C_7H_{16} + 11(O_2 + 3.76N_2) \rightarrow$$
$$7CO_2 + 8H_2O + 41.36N_2 \tag{48}$$

Because of the constant-Schmidt-number assumption made in the PDF formulation, based on atomic balance of the constituent species, the mass fractions of N_2, CO_2, and H_2O can be shown to be related to the mass fractions of O_2 and C_7H_{16} by the following expressions:

$$y_{H_2O} = K_2 - K_1 K_2 y_{O_2} - K_2 y_{C_7H_{16}}$$
$$y_{CO_2} = K_2 K_3 - K_1 K_2 K_3 y_{O_2} - K_2 K_3 y_{C_7H_{16}} \tag{49}$$
$$y_{N_2} = 1 - K_2 - K_2 K_3 - y_{O_2}(1 - K_1 K_2 - K_1 K_2 K_3) -$$
$$y_{C_7H_{16}}(1 - K_2 - K_2 K_3)$$

where $K_1 = 4.29$, $K_2 = 0.08943$, and $K_3 = 2.138$.

Using Eq. (49) results in considerable savings in computational time as it reduces the number of variables in the PDF equation from five (four species and one energy) to three (two species and one energy).

7.8 Revolving Time-Weighted Averaging

It is noteworthy that although local time-stepping seems to overcome some of the problems associated with the PDF computations, the application of the Monte Carlo method requires the use of a large number of particles because the statistical error associated with the Monte Carlo Method is proportional to the inverse square root of N_m, thereby making the use of the Monte Carlo method computationally very time consuming. However, a revolving averaging procedure used in our previous work [25] seems to alleviate the need for using a large number of stochastic particles, N_m, in any one given time step. In this averaging scheme, the solution provided to the CFD solver is based on an average of all the particles present over the last N_{av} time steps instead of an average solely based on the number of particles present in any one single time step. This approach seemed to provide smooth Monte Carlo solutions to the CFD solver, thereby improving the convergence of the coupled CFD and Monte Carlo computations. The reason for improvement could be attributed to an effective increase in the number of stochastic particles used in the computations from N_m to $N_{av}N_m$. Here, it is assumed that the solution contained within different iterations of the averaging procedure is statistically independent of each other.

8 SPRAY SOLUTION ALGORITHM

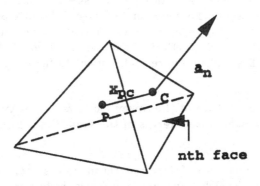

Figure 2 A vector illustration used in the particle search analysis.

In order to evaluate the initial conditions that are needed in the integration of the liquid-phase equations, we first need to know the gas-phase properties at each particle location. But in order to evaluate the gas-phase properties, it is first necessary to identify the computational cell where a particle is located. It is a trivial task to search for the computational cell of the particle location in rectangular coordinates. However, a search for the particle location becomes a complicated problem when the computational cell is no longer rectangular in the physical domain. An efficient particle-search algorithm is developed and implemented into the Lagrangian spray solver in order to facilitate particle movement in an unstructured grid of mixed elements. The search is initiated in the form of a local search from the computational cell of the previous time step as the starting point. The location of the computational cell is determined by evaluating the dot product of $\underline{x}_{pc} \cdot \underline{a}_n = |x_{pc}| \, |a_n| \, cos\,(\phi)$, where $\underline{x}_{pc}$ is the vector defined by the distance between the particle location and the center of the n-face of the computational cell, $\underline{a}_n$ is the outward area normal of the n-face as shown in Fig. 2, and ϕ is the angle between the two vectors.

A simple test for the particle location requires that the dot product be negative over each and every one of the n-faces of the computational cell. If the test fails, the particle search is carried over to the adjacent cells of those faces for which the dot product turns out to be positive. Some of those n-faces might represent the boundaries of the computational domain while the others represent the interfaces between two adjoining interior cells. The search is first carried over to the adjacent interior cells in the direction pointed out by the positive sign of the dot products. The boundary conditions are implemented only after making sure that all the possibilities lead to a search outside of the computational boundaries. This implementation ensures against any inadvertent application of the boundary conditions before locating the correct interior cell.

After the gas-phase properties at the particle location are known, the ordinary differential equations of particle position, size, and velocity are advanced by making use of a second-order accurate Runge-Kutta method. The partial differential equations governing the droplet internal temperature distribution are integrated by an Euler method. Finally, after the liquid-phase equations are solved, the liquid-phase source contributions to the gas-phase equations are evaluated.

8.1 The Flow Structure of the Spray Code & Time-Averaging of the Interphase Source Terms of the Gas-Phase Equations

The spray solver makes use of three different time steps: Δt_{ml} is the allowable time step, Δt_{gl} is the global time step, and Δt_{il} is the fuel-injection time step. Δt_{ml} needs to be evaluated based on the smallest of the different time scales which are associated with various rate-controlling phenomena of a rapidly vaporizing droplet, such as those imposed by an average droplet lifetime, the local grid spacing and a relaxation time scale associated with droplet velocity among others. This restriction usually leads to a small time step which typically has values in the neighborhood of 0.01 milliseconds. However, our experience has shown that the convergence for the steady-state computations

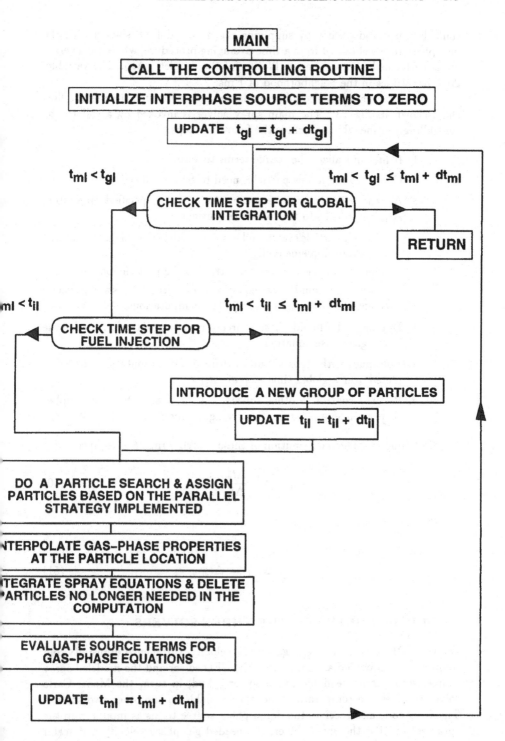

Figure 3 The flow structure of the spray code.

could be improved greatly by supplying the flow and PDF solvers with the
interphase terms obtained from a time-averaging procedure, where the averaging
is performed over an average lifetime of the droplets, Δt_{gl}. The variable,
Δt_{gl}, has values in the neighborhood of 1 ms.

The averaging scheme could be explained better through the use of a flow
chart shown in Fig. 3. The main spray solver is invoked by a call to the
controlling routine which executes the following steps:

1. It first initializes the source terms to zero.

2. Checks to see if new particles need to be introduced.

3. Advances liquid-phase equations over a pre-specified time step,
   Δt_{ml}, with calls to the following routines:

   – Does a particle search and assigns particles based on the parallel
     strategy implemented.

   – Interpolates gas-phase properties at the particle location.

   – Advances liquid-phase equations and, also, deletes any particles
     that are no longer needed in the computations.

4. Evaluates the liquid-phase source-term contributions, S_{ml}, for use
   in the gas-phase equations.

5. Continues with steps (2) and (3) until the computations are completed
   over a global time step of Δt_{gl}.

6. Returns control to other solvers, e.g., flow or PDF, and supplies
   them with source terms, S_{gl}, averaged over Δt_{gl}.

The time-averaged contribution of these source terms, S_{gl}, is given by:

$$S_{gl} = \sum_{m=1}^{M} \frac{\Delta t_{ml}}{\Delta t_{gl}} S_{ml} \tag{50}$$

where

$$\sum_{m=1}^{M} \Delta t_{ml} = \Delta t_{gl} \tag{51}$$

9 COUPLING BETWEEN THE THREE SOLVERS

For the PDF solver, the mean gas-phase velocity, turbulence diffusivity and
frequency are provided as inputs from the CFD solver and the modeled spray
source terms from the liquid-phase solver. And, in turn, the Monte-Carlo
solver supplies the temperature and species fields to the other two solvers.
The CFD code also receives the liquid-phase source terms as inputs from the
spray solver. For the spray solver, the needed gas-phase velocity and scalar

fields are supplied by the other two solvers. The liquid-phase, PDF, and CFD solvers are advanced sequentially in an iterative manner until a converged solution is obtained. It should also be noted that both the PDF and spray solvers are called once at every specified number of CFD iterations. All three PDF, CFD, SPRAY codes were coupled and parallelized in such a way in order to achieve maximum efficiency.

10 PARALLELIZATION

The trend towards the use of the parallel computing from that of serial vector machines is driven by several factors: the increased capabilities of RISC (Reduced Instruction Set Computing) processors, the limited increases in scalar/vector technology, the increased capabilities of network communication, the increased memory size with the availability of easily affordable DRAM (Dynamic Random Access Memory) chip storage capacity, and the scalability of both memory size and problem requirements with the the number of processors. This led to two major developments in parallel computing: the widespread use of distributed computing and MPPs.

The use of distributed computing is becoming widespread with the proliferation of workstation clusters which are tied into a network and the availability of computer software such as PVM and MPI. For example, PVM was developed at ORNL (Oak Ridge National Laboratory). Both PVM and MPI provide a set of different Fortran and C++ library routines which are available in the public domain and allow a network of heterogeneous computers to be used as a single large parallel computer. They provide the needed user-level message-passing interface for communicating between different PEs. Their main features include automatic data conversion, barrier synchronization, buffer-management functions, task-control functions, and data transmittal and receipt functions. Thus, large computational problems can be solved by using the aggregate power of many computers.

On the other hand, MPPs make use of hundreds of homogeneous processors in concert to solve a problem. For example, Cray T3D is a massively parallel computer with an aggregate of 64 PEs (Processor Elements). Each PE consists of a DEC Alpha chip 21064 with 8 Mwords of memory. The 64 PE Cray T3D delivers an aggregate peak performance of 19.2 Gflops on 64-bit data and supports a total of 512 Mwords of memory. The Cray T3D has 32 nodes configured in a three-dimensional (3-D) torus network topology with each node having two PEs. The topology permits fast interconnect network for communication and data movement. The unique features of MPPs also permit the support of special programming languages such as Cray MPP Fortran, which resembles Fortran 90 in many ways. Cray MPP Fortran provides several easier-to-implement programming tools such as shared memory functions and doshared directives, among several others. It is designed to support and exploit certain platform-dependent hardware-specific intrinsics. Therefore, programs written in such programming languages provide significant performance gains over those written in Fortran 77 with PVM on platforms such as Cray T3D.

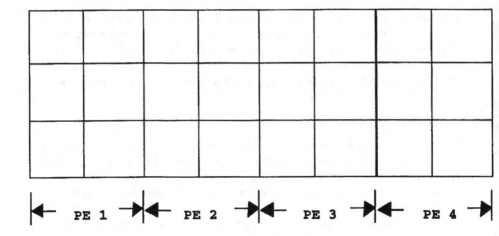

Figure 4 An illustration of the parallelization strategy employed in the gas flow computations.

10.1 Gas-Phase Domain Decomposition Methods

There are several ways to partition a grid. The most commonly used domain decomposition is referred to as 1-D partitioning, where the total domain is simply divided equally amongst the available PEs. Fig. 4 illustrates a simple example of the domain decomposition strategy adopted for the gas-phase computations. In this case, we assumed the number of available PEs to be equal to four. Any communication overhead associated with 1-D partitioning is limited to data transfer across the interfaces of the connecting subdomains.

10.2 Some of the Programming Differences Between PVM & MPI Versus Cray MPP Fortran

Here we highlight some of the basic differences and approaches to programming using PVM and Cray MPP Fortran. The concept of data sharing is to allow global variables and dummy arguments to be distributed across all PEs so that each PE can manipulate its share of data independently of other PEs. The challenge of programming is to keep data processing local to each PE and to keep PE-to-PE communication to a minimum while maintaining scalability. In distributed computing all data are private; no PE knows of any other PE's memory. But the need for communication between PEs couldn't be completely eliminated as, for example, during the numerical integration step, each PE needs to know in general some information about what is contained in one or more layers of adjacent grid nodes located on the other side of the divided interface. This process is usually accomplished by preparing what are known as receive-and-send tables. Send tables contain information about what grid data need to be sent to the other PEs while receive tables contain information about what data need to be expected from the other PEs. During the information exchange process, the information contained in the send table is processed first by transmitting the required data to the other PEs before receiving the needed information from the other PEs. After receiving the data, it is stored in appropriate arrays. So for parallel implementation, a data preparation stage is clearly needed in terms of fetching, storing, and providing information on where to look for the stored data that was received. The PVM library also provides appropriate functions for use in the global information exchange and summation purposes. Such information might be needed in many places such as global mass preservation check, residuals evaluation, and propagating information on some reference variables. While PVM provides the needed user-level message-passing interface, it is up to the user to resolve several issues arising from the domain decomposition strategy adopted.

However, in Cray MPP Fortran, the need for using explicit communication calls is significantly minimized as it supports the use of shared memory. In shared memory, all PEs have access to the shared data without any need for invoking explicit communication calls regardless of where the memory is actually located. The use of shared data reduces the programming effort by a considerable degree as it eliminates the need for certain data preparation and information exchange stages associated with the use of private data. For a grid of mesh size I x J, the shared data could be distributed such that each PE owns one block of contiguous elements, (I/N\$PES) x J, where N\$PES is the total number of available PEs for a given application. In Fig. 4, N\$PES = 4, I = 8, and J = 3. Similarly, arrays of different dimensions containing the ith dimension could all be blocked using 1-D partitioning. Instructions in a shared loop are divided up among processors, so each PE has a subset of the entire instruction space. It should be kept in mind that only those arrays and variables that are needed to be accessed by other PEs should be defined as the shared data while all others should be declared as private in order to achieve effective utilization of the available computer resources.

Even though Cray MPP Fortran provides a means for efficient implementation, its application is mainly limited to certain Cray computer platforms. Programming with PVM and MPI is gaining more ground as it offers a wider platform independence which is an important factor to consider in light of the fast pace at which work-station-cluster environment is changing.

10.3 Some Basic Guidelines to Parallel Implementation

There are several issues associated with the parallelization of the PDF and spray computations. The goal of the parallel implementation is to extract maximum parallelism so as to minimize the execution time for a given application on a specified number of processors [26]. Several types of overhead costs are associated with parallel implementation which include data dependency, communication, load imbalance, arithmetic, and memory overheads.

Arithmetic overhead refers to the extra arithmetic operations required by the parallel implementation, and memory overhead refers to the extra memory needed. Excessive memory overhead reduces the size of a problem that can run on a given system, and the other overheads result in performance degradation [26]. Any given application usually consists of several different phases that must be performed in certain sequential order. The degree of parallelism and data dependencies associated with each of the subtasks can vary widely [26]. The goal is to achieve maximum efficiency with a reasonable programming effort.

10.4 Parallel Implementation of the PDF Solver

Both the spray and kinetics steps of the PDF method lend themselves perfectly to parallel computing as no particle interaction of any kind occurs during their integration. During the mixing step, the interaction between the particles is limited to only those particles present in a given cell. This step also lends itself to parallel computing without any associated overhead.

During the spatial transport step of the PDF solution method, particles are moved across the neighboring cells based on the computed values of the convection and turbulent-diffusion coefficients as determined from the numerical scheme used. The communication overhead in this step is limited to data transfer across the interfaces of the neighboring subdomains without any data dependency.

The combined overheads associated with parallel implementation are minimal since the time spent in the subroutines that deal with random number generation, convection and turbulent-diffusion, and boundary conditions is only a small fraction of the total time that it takes for the entire PDF solution method. Therefore, a very high degree of parallelism could easily be achieved if enough care is exercised in distributing the spatial grid points uniformly amongst all the available PEs. Thus, the Monte Carlo simulation is ideally suited for parallel computing, and the run time could be considerably minimized by performing the computations on a massively parallel computer.

10.5 Parallel Implementation of the Spray Solver

In an approach where an Eulerian scheme is employed for the gas-phase computations and a Lagrangian scheme for the liquid-phase computations, the spray computations are difficult to parallelize as the spray distribution tends to be both spatially non-uniform and temporally dynamic in nature for the reasons cited below:

- Most of the spray is usually confined to a small region near the atomizer location.

- The Lagrangian particles tend to move in and out of different parts of the computational domain processed by different PEs.

- Some new particles might be added to the computation at the time of fuel injection while some others might be taken out of computation. A particle is removed when it exits out of the computational boundaries or when it becomes small enough to the point where it is considered to be no longer needed in the computation.

Two different domain decomposition strategies were developed in order to explore their effectiveness on the parallel performance of the spray computations:

1. Strategy I: The Lagrangian particles are assigned uniformly amongst the available processors. However, the particle search and the computations involving the gas-phase property evaluation at the particle location as well as the spray-source terms, which are used in the CFD and PDF equations, were evaluated on the processor of the computational grid where the particle is located. This strategy leads to uniform load balancing during the integration of the liquid-phase equations but may result in excessive message passing during the other operations.

   This strategy yielded reasonable parallel performance when the computations were performed on a massively parallel computer like Cray T3D [1]. But its performance turned out to be rather poor when the computations were performed on the NASA LeRC LACE cluster [7]. The poor performance was identified to have resulted from the poor inter-processor communication capabilities of the workstation cluster. For that reason, a second strategy was explored in order to improve the parallel performance of the spray code.

2. Strategy II: The Lagrangian particles are assigned to the processor of the computational grid where the particle is located. This strategy may lead to non-uniform load balancing during the integration of the liquid-phase equations but is likely to result in less message passing since the inter-processor communications are limited to a single operation associated with the particle search.

10.6 Results and Discussion on the Parallel Performance

The applicability of the PDF approach for several test cases was documented in Refs. [1, 7, 25]. In a separate study, Chen et al [10] documented the performance of the CFD solver for several benchmark test cases. In this section, we summarize only the results of the parallel performance of the CFD, PDF, and spray solvers for two test cases.

The first case refers to the Cray MPP Fortran calculation of an open swirl-stabilized spray flame which was performed on a structured grid of 60x60x3

(=10,800) nodes with a total of 2.7 million particles (=250 particles/cell) requiring about 27 Mwords of computer memory for the major shared array allocation [1]. The computations were performed on the Cray T3D at NASA LeRC with the number of processors ranging between 8 to 32.

Table 1   Cpu time (sec) per cycle versus number of PEs on Cray T3D

Solver	Characteristic	Number of processors		
		8	16	32
PDF solver	1 step/cycle	12.37	6.26	3.19
CFD solver	4 iterations/cycle	6.16	3.33	1.88
Spray solver (Strategy I)	50 steps/cycle	1.31	1.15	1.07

Table 1 summarizes the cpu times per cycle taken by the PDF, SPRAY, and CFD solvers. The cpu time for the PDF solver scales linearly with the number of processors, thereby indicating the realization of a very high degree of parallelization. Even better scaling would be obtained if the i-th dimension of the grid is changed from 60 to 64, a power of 2. What is noteworthy is that the cpu times for the CFD solver also tend to scale linearly with the number of PEs used. The speed-up in the spray computations is considerably smaller than the other two solvers. Much of the slowdown in the spray computations is caused by the use of Cray MPP Fortran cdir$ critical function, which was used to prevent any racing conditions from occurring during the summation of source-term contributions from different particles located in a given cell. By rewriting this part of the code, the spray computations could be speeded up by a factor of 5 to 10.

Table 2   Cpu time (sec) per cycle versus number of PEs on LACE cluster

Solver	Characteristic	Number of processors			
		2	4	8	16
PDF solver	1 step/cycle	2.08	1.16	0.67	0.42
CFD solver	5 steps/cycle	4.25	2.2	1.6	1.4
Spray solver (Strategy I)	100 steps/cycle	1.33	2.67	5.37	12.56
Spray solver (Strategy II)	100 steps/cycle	0.60	0.58	1.1	2.50

The computations take about two hours of cpu time on 32 PEs to reach a converged solution. Based on our previous 2-D computations, it is expected for the cpu times on a CRAY T3D/32-PEs to be comparable with the performance of a CRAY Y-mP/1-PE.

The second case refers to the Fortran 77 with PVM calculation of a confined swirl-stabilized spray flame. The axisymmetric computations were performed on an unstructured grid of 3600 nodes and a total of 0.36 million Monte

Carlo particles ($=100$ particles/cell). The computations were performed on NASA LeRC LACE cluster with the number of processors ranging between 2 and 16.

Table 2 summarizes the cpu times per cycle taken by the PDF, SPRAY, and CFD solvers versus the number of processors. The PDF solver shows good parallel performance with an increase in the number of processors. The CFD solver also shows reasonable parallel performance but the gains seem to taper off more progressively with the increase in the number of PEs from 4 to 8. For the spray computations, strategy II seems to result in a considerable improvement in parallel performance over strategy I. Initially, with strategy II, there is a slight performance gain with the increase in PEs from 2 to 4. However, there is a progressive deterioration in the performance when the number of PEs further increased from 4 to 16.

These results clearly demonstrate the need for a fast interconnect network for communication and data movement and the need for keeping the inter-processor communications to a minimum, especially when the computations are performed over a work-station-cluster environment. It is evident from the results that the maximum achievable parallel efficiency for the combined PDF/spray/CFD computations may be constrained by the spray performance, especially when a Lagrangian representation is used for the spray and an Eulerian for the gas-phase.

11 CONCLUDING REMARKS

A solution procedure has been outlined for the computation of turbulent spray flames on unstructured grids with parallel computing. The numerical method outlines several techniques designed to overcome some of the high computer time and storage limitations associated with the combined PDF/spray/CFD computations of practical combustor flows. Because of the ease with which grids could be generated for complex combustion geometries, the present method is well suited for its application to the modeling of practical combustion devices. The commercially available grid-generation software like CFD-GEOM [27] have the ability to generate the required interior grids from the data taken directly from a typical CAD/CAM package.

There are several important aspects of spray combustion research that need to be addressed in order to provide better prediction tools needed in the design of advanced combustors. We conclude this chapter by making a few comments on our ongoing work and the planned research for the near future.

We are planning to extend our spray calculations to multi-component sprays under super-critical conditions. Since most of the aviation fuels are mostly multi-component, accurate representation of vaporization models becomes important in the prediction of some important combustion phenomena such as combustion instabilities, flame ignition, etc. [13]. Under super-critical conditions, flow-field evolution is governed by both compressibility effects as well as variable inertial effects. Increase in pressure introduces both thermodynamic non-idealities and transport anomalies. Near the critical point, fluid properties exhibit liquid-like densities, gas-like diffusivities, and strongly pressure-dependent solubilities. Surface tension and heat of vaporization approach zero, and isothermal compressibility and constant pressure specific heat increase significantly. These phenomena have a significant impact on the vaporization and overall dynamics associated with a given system. It is important to include the high pressure effects as most of the advanced technology combustors are planned to operate under elevated (near or above critical) pressure conditions.

There are several advantages to a Lagrangian representation for sprays:

- It is efficient on serial computers,

- It is the most widely used in dilute spray modeling, and

- Its solution is free of numerical diffusion.

As is evident from our earlier discussion, it is difficult to parallelize the overall calculation procedure, especially when it is used in conjunction with an Eulerian formulation for the gas-phase and a Lagrangian formulation for the spray. The overall parallel performance of the combined solution could be vastly improved by developing a spray code based on an Eulerian formulation. The Eulerian approach offers several advantages over a Lagrangian formulation:

- It results in a highly efficient parallel algorithm for the combined PDF/spray/CFD solution,

- It is more convenient for including some of the dense spray effects, and

- It offers faster convergence to steady-state solutions.

Not even the recent advances in parallel computing can provide the tremendous cpu time needed for a multi-species computation of a practical combustion flow. Therefore, it is important to develop reduced chemistry mechanisms that could be of interest in several important combustion phenomena such as emissions (NOX and CO), unburnt hydrocarbons (UHC), flame ignition and extinction limits. One approach that is developed as a part of the NCC effort is based on the ILDM (Intrinsic Low Dimensional Manifolds) method [10]. If the ILDM approach proved to be useful, significant cpu time savings could be realized by integrating the ILDM tables with the Monte Carlo PDF method.

12 ACKNOWLEDGEMENT

The research funding for this work was provided by NASA Lewis Research Center with Dr. N.-S. Liu acting as the technical monitor.

13 NOMENCLATURE

A	pre-exponent of an Arrhenius reaction rate term
a	non-unity exponent of an Arrhenius reaction rate term
$\underline{a}_n$	outward area normal vector of the nth surface, m^2
B_k	Spalding transfer number
b	non-unity exponent of an Arrhenius reaction rate term
C_D	drag coefficient
C_p	specific heat, J/(kg K)
C_ϕ	a constant in Eq. (39)
c_n	convection/diffusion coefficient of the nth face, kg/s
D	turbulent diffusion coefficient, m^2/s
d	drop diameter, m
E_a	activation energy of an Arrhenius reaction rate term
h	specific enthalpy, J/kg
J_i^α	diffusive mass flux vector, kg/ms
k	turbulence kinetic energy, m^2/s^2
l_k	latent heat of evaporation, J/kg
$l_{k,eff}$	effective latent heat of evaporation, J/kg (defined in Eq. (11))
M_i	molecular weight of ith species, kg/kg-mole
m_k	droplet vaporization rate, kg/s
m_{ko}	initial mass flow rate associated with kth droplet group
N_{av}	number of time steps employed in the PDF time-averaging scheme
N_f	number of surfaces contained in a given computational cell
N_m	total number of Monte Carlo particles per grid cell
N_p	total number of computational cells
n_k	number of droplets in kth group
P	pressure, N/m^2
P_r	Prandtl number
p	joint scalar PDF
R_u	gas constant, J/(kg K)
Re	Reynolds number
r_k	droplet radius, m
r_{ko}	initial drop radial location, m
s_k	droplet radius squared, r_k^2, m^2
s_{mlc}	liquid source contribution of the gas-phase continuity equation
s_{mle}	liquid source contribution of the gas-phase energy equation
s_{mlm}	liquid source contribution of the gas-phase momentum equations
s_{mls}	liquid source contribution of the gas-phase species equations
s_α	liquid source contribution of the α variable
T	temperature, K
t	time, s
u_i	ith velocity component, m/s
u_{ik}	ith velocity component of kth drop group, m/s
V_c	volume of the computational cell, m^3

w_α chemical reaction rate, 1/s

$\dot{w}_j$ gas-phase chemical reaction rate, 1/s

x_i Cartesian coordinate in the ith direction, m

y_j mass fraction of jth species

$\underline{x}$ spatial vector

χ mole fraction

Δt local time step used in the PDF computations, s

Δt_f local time step in the flow solver, s

Δt_{gl} global time step in the spray solver, s

Δt_{il} fuel injection time step, s

Δt_{ml} allowable time step in the spray solver, s

ΔV computational cell volume, m^3

δ Dirac-delta function

ϵ rate of turbulence dissipation, m^2/s^3

ϵ_j species mass fraction at the droplet surface

$\epsilon_{\alpha s}$ species mass fraction at the droplet surface

Γ_ϕ turbulent diffusion coefficient, kg/ms

λ thermal conductivity, J/(ms K)

μ dynamic viscosity, kg/ms

ω turbulence frequency, 1/s

ϕ represents a set of scalars of the joint PDF

$\underline{\psi}$ independent composition space

ρ density, kg/m^3

σ dimensionality of $\underline{\psi}$-space

τ stress tensor term, kg/ms^2

θ void fraction

Subscripts

f represents conditions associated with fuel

g global or gas-phase

i index for the coordinate or species components

j index for the species component

k droplet group or liquid-phase

l liquid-phase

m conditions associated with N_m

n nth-face of the computational cell

o initial conditions or oxidizer

p conditions associated with the properties of a grid cell

s represents conditions at the droplet
surface or adjacent computational cell

t conditions associated with time

α index for the scalar component of the joint PDF equation

$,$ partial differentiation with respect
to the variable followed by it

Superscripts

~ Favre averaging

⁻ time averaging or average based on the Monte Carlo
   particles present in a given cell

// fluctuations

14 REFERENCES

1. M.S. Raju, Application of Scalar Monte Carlo Probability Density Function Method For Turbulent Spray Flames, Numerical Heat Transfer, Part A, vol. 30, pp. 753-777, 1996.

2. R. Borghi, Turbulent Combustion Modeling, Prog. Energy Combust. Sci., vol. 14, pp. 245-292, 1988.

3. S.B. Pope, PDF Methods for Turbulent Reactive Flows, Prg. Energy Combust. Sci., vol. 11, pp. 119-192, 1985.

4. S.M. Correa, Development and Assessment of Turbulence-Chemistry Models in Highly Strained Non-Premixed Flames, AFOSR/NA Contractor Report, 110 Duncan Avenue, Bolling AFB, DC 20332-0001, 31 October 1994.

5. M.S. Raju, LSPRAY - a Lagrangian Spray Solver - User's Manual, NASA/CR-97-206240, NASA Lewis Research Center, Cleveland, Ohio, November 1997.

6. M.S. Raju, EUPDF - an Eulerian-Based Monte Carlo Probability Density Function (PDF) Solver - User's Manual, NASA/CR-1998-20401, NASA Lewis Research Center, Cleveland, Ohio, April, 1998.

7. M.S. Raju, Combined Scalar Monte Carlo PDF/CFD Computations of Spray Flames on Unstructured Grids With Parallel Computing, AIAA/ASME/SAE/ASEE 33rd Joint Propulsion Conference, Seattle, Wash., July 6-9, 1997.

8. N.S. Liu and R.M. Stubbs, Preview of National Combustion Code, AIAA 97-3114, AIAA/ASME/SAE/ASEE 33rd Joint Propulsion Conference, Seattle, Wash., July 6-9, 1997.

9. R. Ryder, CORSAIR User's Manual: Version 1.0, SID: Y965, Pratt and Whitney Engineering, United Technologies Corporation, 25 January 1993.

10. K.-H. Chen, A.T. Norris, A. Quealy, and N.-S. Liu, Benchmark Test Cases For the National Combustion Code, AIAA 98-3855, AIAA/ASME/SAE/ASEE 34th Joint Propulsion Conference, Cleveland, Ohio, July 13-15, 1998.

11. W.A. Sirignano, Fluid Dynamics of Sprays, Journal of Fluids Engineering, vol. 115, no. 3, pp. 345-378, September 1993.

12. T.-H. Shih, K.-H. Chen, and N.-S. Liu, A Non-Linear $k - \epsilon$ Model for Turbulent Shear Flows, AIAA/ASME/SAE/ASEE 34th Joint Propulsion Conference, Cleveland, Ohio, July 13-15, 1998.

13. M.S. Raju and W.A. Sirignano, Multi-Component Spray Computations in a Modified Centerbody Combustor, Journal of Propulsion and Power, vol. 6, no. 2, pp. 97-105, 1990.

14. M.S. Raju, AGNI-3D: A Computer Code for the Three-Dimensional Modeling of a Wankel Engine, Computers in Engine Technology: Proceedings IMechE, London, United Kingdom, pp. 27-37, 1991.

15. M.S. Raju, Heat Transfer and Performance Characteristics of a Dual-Ignition Wankel Engine, Journal of Engines - Section 3, SAE Trans., vol. 101, pp. 466-509, 1992.

16. A.Y. Tong and W.A. Sirignano, Multi-component Transient Droplet Vaporization With Internal Circulation: Integral Formulation and Approximate Solution, Numerical Heat Transfer, vol. 10, pp. 253-278, 1986.

17. R. Clift, J.R. Grace, and M.E. Weber, Bubbles, Drops, and Particles, Academic, New York, 1978.

18. H. Schlichting, Boundary-Layer Theory, McGraw-Hill Series in Mechanical Engineering: McGraw-Hill, Inc., New York, 1968.

19. Y. El Banhawy and J.H. Whitelaw, Calculation of the Flow Properties of a Confined Kerosene-Spray Flame, AIAA J., vol. 18, no. 12, PP. 1503-1510, 1980.

20. F.V. Bracco, Modelling of Engine Sprays, SAE paper 850394, 1985.

21. M. Scheurlen, B. Noll, and S. Wittig, Application of Monte Carlo Simulation For Three-Dimensional Flows, AGARD-CP-510: CFD Techniques For Propulsion Applications, February 1992.

22. D.A. Anderson, J.C. Tannehill, and R.H. Fletcher, Computational Fluid Mechanics and Heat Transfer: Series in Computational Methods in Mechanics and Thermal Sciences, Hemisphere Publishing Corporation, Washington, D.C., 1984.

23. M.S. Raju, W.Q. Liu, and C.K. Law, A Formulation of Combined Forced and Free Convection Past Horizontal and Vertical Surfaces, Int. J. Heat and Mass Transfer, vol. 27, pp. 2215-2224, 1984.

24. C.K. Westbrook and F.L. Dryer, Chemical Kinetic Modelling of Hydrocarbon Combustion, Progress in Energy and Combustion Science, vol. 10, no. 1, pp. 1-57, 1984.

25. A.T. Hsu, Y.-L.P. Tsai, and M.S. Raju, A Probability Density Function Approach for Compressible Turbulent Reacting Flows, AIAA Journal, vol. 32, no. 7, pp. 1407-1415, 1994.

26. J.S. Ryan and S.K. Weeratunga, Parallel Computation of 3-D Navier-Stokes Flowfields for Supersonic Vehicles, AIAA 93-0064, AIAA 31st Aerospace Sciences Meeting, Reno, Nevada, 1993.

27. CFD-GEOM, Version 2.0, CFD Research Corporation, 3325 Triana Blvd., Huntsville, Alabama 35805, July 1996.

NINE

OVERVIEW OF CURRENT COMPUTATIONAL STUDIES OF HEAT TRANSFER IN POROUS MEDIA AND THEIR APPLICATIONS - FORCED CONVECTION AND MULTIPHASE HEAT TRANSFER

H. Hadim
K. Vafai

1 INTRODUCTION

The topic of thermal convection and multiphase transport in porous media has gained increasing research interest during the past two decades. This is due to the presence of porous media in a wide range of geophysical and engineering applications of current interest. These applications include, but are not limited to, geothermal energy extraction, drying processes (wood and food products), oil reservoir engineering, groundwater contamination, thermal energy storage, storage of radioactive nuclear waste materials, heat pipes, building insulation, metal casting (alloy solidification), separation processes in chemical industries, regenerative heat exchangers, simulation of complex structures, and heat transfer enhancement especially in high heat flux applications such as cooling of electronic equipment.

A concise review covering most of the aspects of interest to convection and multiphase transport in porous media has been reported earlier by Tien and Vafai [1]. Since then, reviews of several topics within this broad research area can be found in earlier books presented by Kakac et al. [2], Nield and Bejan [3], Kaviany [4] and more recently Ingham and Pop [5].

In general, rigorous studies of transport in porous media including all the important physical phenomena must be performed using either sophisticated experimental techniques or robust numerical modeling. For a large class of problems, experimental techniques are limited in the scope that can be covered, and they are not practical for many problems, especially those requiring extensive parametric studies. Consequently, a large number of recent investigations have been focused on developing robust numerical models that are able to simulate these various physical aspects.

In this chapter, an up-to-date review of current computational work involving most of the important aspects of forced convection and multiphase transport in porous media and their applications mainly over the past decade is presented. This chapter is divided into two major sections related to forced convection and multiphase transport, while a follow-up chapter in this volume is related to natural and mixed convection in porous media. Generalized models which include all the important physical phenomena governing convective flow and energy transport in porous media are presented. Due to space limitations, detailed numerical simulations of these models have been omitted, and the reader is referred to related studies which have been reviewed in this chapter. The literature review in each section is divided into individual sub-topics related to geometrical configurations of fundamental and practical interest, important physical phenomena, major computational models, and practical applications.

2 FORCED CONVECTION IN POROUS MEDIA

2.1 Mathematical Formulation and Numerical Solution

A thorough understanding of the fluid mechanics and heat transfer characteristics in porous media is quite complicated. In this respect, the complex microscopic transport phenomena at the pore level are important to better represent existing physical phenomena. However, the complexity of the porous structure usually precludes a detailed microscopic investigation of the transport phenomena at the pore level. Therefore, the general transport equations are commonly integrated over a representative elementary volume, which accommodates the fluid and the solid phases within a porous structure [6]. Though the loss of information with respect to the microscopic transport phenomena is inevitable with this approach, the integrated quantities, coupled with a set of proper constitutive equations which represent the effects of microscopic interactions on the integrated quantities, do provide a rigorous and effective basis for analyzing the transport phenomena in porous media [7].

Earlier studies of transport in porous media utilized the Darcy flow model, which is based on the assumption of creeping flow through an infinitely extended homogeneous and uniform porous medium. However, it is now well established that in many applications, non-Darcian effects including inertial resistance within the porous matrix and viscous effects at the boundary become significant under various conditions. Additional important physical phenomena, including variable porosity, thermal dispersion, and non-local thermal equilibrium, have also been shown to have a significant effect on fluid flow and heat transfer. The literature review on heat transfer in porous media indicates that the state of local thermal equilibrium condition was heavily employed in most of the investigations without a sound justification. However, in a number of applications, the modeling of energy transport in a porous medium demands accounting for the temperature of the individual phases. This requires representation of the individual phases by separate energy equations with an additional term in each equation to assimilate the energy exchange between the two phases. A general model which takes into account these important physical phenomena is presented in the following section.

2.1.1 The General Model. It is customary to use the local volume-averaging technique in order to develop a more rigorous set of governing equations for the transport processes in porous media. It is usually necessary to make use of the two different averages of a quantity in the governing equations. These are the local volume average and the intrinsic phase average [6]. While the local volume average of a quantity Φ associated with phase Ψ is defined as

$$< \Phi > = \frac{1}{V} \int_{V_\Psi} \Phi \ dV \tag{1}$$

The intrinsic phase average of a quantity Φ associated with phase Ψ is defined as

$$< \Phi >^\Psi = \frac{1}{V_\Psi} \int_{V_\Psi} \Phi \ dV \tag{2}$$

where V is an averaging volume which is bounded by a closed surface in the porous medium, and V_Ψ represents the volume associated with phase Ψ.

In using the volume averaging technique in the governing equations, it is important to distinguish between the properties that are associated with a single phase, such as the solid phase temperature, in which case one should use the intrinsic phase average, and the properties which have a characteristic average value over the averaging volume, such as the so-called "superficial fluid velocity," in which case one can use the spatial average or the local volume average value. These will help in presenting an accurate form of the governing equations.

The volume-averaging technique as outlined by Whitaker [6] and later reformulated by Vafai and Tien [7] and Vafai and Sözen [8] is adopted in developing the governing equations for mass, momentum and energy equations for the full general model of incompressible flow and heat transfer in porous media including non-Darcy effects, variable porosity, thermal dispersion, and non-local thermal equilibrium. The vectorial forms of these equations are given as

conservation of mass:

$$\nabla \cdot < V >= 0 \tag{3}$$

conservation of momentum:

$$\frac{1}{\varepsilon} \frac{DV}{Dt} = -\frac{1}{\rho_f} \nabla < P >^f + v_f \nabla^2 < V > -\frac{\mu_f < V >}{\rho_f K} - \frac{F\varepsilon}{\sqrt{K}} |< V >| < V > \tag{4}$$

conservation of energy:

(i) fluid phase

$$\varepsilon(\rho c)_f \frac{D<T_f>^f}{Dt} = -\nabla \cdot (k_{feff} \cdot \nabla <T_f>^f) + h_{sf} a_{sf} (<T_s>^s - <T_f>^f) \qquad (5)$$

(ii) Solid phase

$$(1-\varepsilon)(\rho c)_s \frac{D<T_s>^s}{Dt} = -\nabla \cdot (k_{seff} \cdot \nabla <T_s>^s) - h_{sf} a_{sf} (<T_s>^s - <T_f>^f) \qquad (6)$$

In the above equations, $\mathbf{V}$ represents the velocity vector, T the temperature, $<P>^f$ the average pressure read off a pressure gage, t is the time, ρ_f and ρ_s the fluid and solid densities respectively, μ_f the viscosity of the fluid, v_f the fluid kinematic viscosity, c_f and c_s are the fluid and solid specific heats at constant pressure respectively, ε the porosity. The intrinsic phase average of temperature, $<T_f>^f$, corresponds more closely with measurable quantities while the fluid velocity is frequently represented by the volume phase average $<V>$. The geometric function F and the permeability of a porous medium K are based on Ergun's model [9] and are expressed as [10]

$$F = \frac{1.75}{\sqrt{150\varepsilon^3}} \qquad (7)$$

$$K = \frac{\varepsilon^3 d_p^2}{150(1-\varepsilon)^2} \qquad (8)$$

where d_p is the bead diameter.

The fluid's effective thermal conductivity consists of the stagnant and the dispersion conductivities and it is formulated based on the experimental findings of Wakao and Kaguei [11] as follows (the volume-average symbol is omitted for convenience):

$$k_{feff,x} = \varepsilon k_f + 0.5 \left(\frac{|V| d_p}{v_f} \right) Pr\, k_f \qquad (9)$$

$$k_{feff,y} = \varepsilon k_f + 0.1 \left(\frac{|V| d_p}{v_f} \right) Pr\, k_f \qquad (10)$$

where $|V| = \sqrt{u^2 + v^2}$, and the Prandtl number $Pr = v_f / \alpha_f$. The solid's effective thermal conductivity consists of the stagnant component only

$$k_{seff} = (1-\varepsilon)k_s \qquad (11)$$

The specific surface area of the porous medium, a_{sf}, is based on geometrical consideration [12]:

$$a_{sf} = \frac{6(1-\varepsilon)}{d_p} \qquad (12)$$

The solid-to-fluid heat transfer coefficient is based on the empirical correlation reported by Wakao et al. [13] and can be expressed as follows:

$$h_{sf} = k_f \left[2 + 1.1 \, \text{Pr}^{1/3} \left(\frac{|V| d_p}{\nu_f} \right)^{0.6} \right] / d_p \qquad (13)$$

It is common practice to consider an exponential decaying function to approximately simulate porosity variation near the wall. A typical form of such an exponential decay is given by [14, 10]

$$\varepsilon = \varepsilon_{\infty} [1 + a \, e^{(by/d_p)}] \qquad (14)$$

The free-stream porosity value ε_{∞} is a function of the bed-to-particle D_h/d_p ratio and varies between 0.259 and 0.43, depending on the packing arrangement. In addition, for uniform solid-sphere particles, the values of a $=1.7$ and b$=6$ were found to yield close approximation to the experimental data [14].

2.1.2 Numerical Solution. The governing equations in the generalized model described above for incompressible flow and heat transfer through a porous medium are a mixed elliptic-parabolic system of equations which are solved simultaneously. Like the majority of partial differential equations in fluid mechanics and heat transfer, the governing equations involving a general model are nonlinear. Various numerical methods have been developed for solving the generalized governing equations for fluid flow and heat transfer processes in porous media. These well-established methods belong to three main categories: the finite difference techniques, the finite volume methods, and the finite element formulations. Within each category, there are several possible discretization algorithms which are by no means unique, and different types are expected to give the same solutions. Also, the computational problems such as stability and convergence encountered in solving the governing equations need to be dealt with here as well. Different techniques have been developed to overcome these problems. Due to space limitations, a comprehensive discussion of these techniques could not be presented here, and the reader is referred to the related studies reviewed in this chapter as well as a large number of references which discuss these techniques in more detail (e.g. [15]), including the comprehensive handbook edited by Minkowycz et al., [16]. However, a standard approach based on the finite difference technique is presented as follows.

For the general model presented above, Eqs. (3)-(6) and the associated boundary conditions are solved using the finite difference method. For forced convection, the momentum equations and the energy equation are not coupled; therefore, they are solved separately. Central differencing is used for the spatial derivatives except for the convective terms which are discretized using upwind differencing. The nonlinear term in the momentum equation is linearized as discussed in [15]. The resulting set of algebraic equations for the momentum equation is solved using the tridiagonal matrix algorithm. Next, the energy equations are discretized using similar schemes after specifying the exit boundary condition. At grid points on the right boundary (i.e. the exit), a three-point backward differencing is employed for the spatial x derivatives. The strong parabolic nature of forced convective flows validates this procedure. Further validation can be achieved by extending the computational domain in the axial direction until the numerical results within the physical domain are no longer affected by further increase in the computational length. Due to their temperature dependence, the source terms appearing in Eqs. (5) and (6) are updated after each iteration. The energy equations are solved for the fluid and the solid phase temperatures using the successive over-relaxation scheme (SOR).

Some fundamental issues related to computational grid generation for thermal convection problems in porous media are discussed as follows. For forced convection in porous media, it has been shown that the solutions of the governing equations exhibit viscous sub-layers which are adjacent to the impermeable walls and whose thickness varies with the governing parameters describing the properties of the fluid and porous matrix including Reynolds number, Darcy number (or particle diameter), and Prandtl number. Hence, as a result of imposing no-slip conditions at a solid surface, steep velocity gradients exist near the walls, especially in the presence of channeling phenomena due to variable porosity. Meanwhile, primary variables such as velocity and temperature usually exhibit significant development in the entrance region, and the upstream variations are significantly larger than the downstream ones. Consequently, a skewed grid distribution is used along all boundaries of the computational domain such that the node density is higher in the vicinity of the walls in the transverse direction and at the inlet in the longitudinal direction. The generated grid should be smoothly varying, close to orthogonal, with local grid aspect ratio close to unity. When the physical models considered involve rectangular configurations, simple stretching functions can be used to generate such non-uniform grid layouts. Appropriate values of the stretching factors are usually obtained by trial and error.

As discussed by many investigators, the complexity of the local fluid dynamics and heat transfer interactions between the fluid and the solid matrix leads to empiricism to some extent which must be employed when the volume-averaging technique is used for developing the macroscopic governing equations. As a result, such local volume-averaged and bulk semi-empirical features may seem to cause a conflict between macroscopic physical phenomena predicted and volume-averaging concepts used. For example, the fully developed momentum boundary layer thickness in a porous medium is found to be of the order of $K^{1/2}$ (i.e. $10^{-2} d_p$ - $10^{-1} d_p$) [7]. This thickness is less than the particle (or pore) size which may seem contradictory to the volume-averaging concept [4]. Thus, for accurate capturing of the boundary layer phenomenon, a grid size of the order of $0.1 K^{1/2}$ (i.e. $10^{-3} d_p$– $10^{-3} d_p$) is needed near the solid surface. For such small grid size, non-uniform grid size distribution (e.g. skewed grid) must be used. However, abrupt changes in grid size may lead to numerical instabilities and inaccuracies [15]. Consequently, a large number of grid points is usually required near the solid surface.

2.1.3 Vorticity-Stream Function Formulation. The governing equations for fluid flow may be expressed in two different classical formulations of the Navier-Stokes equations based on the dependent variables used. First is the primitive variable formulation expressed in terms of pressure and velocity, and the second form is the so-called vorticity-stream function, which is derived from the Navier-Stokes equations by incorporating the definitions of the vorticity and the stream function. For two-dimensional problems, the vorticity-stream function formulation has some attractive features. The pressure is not utilized, and, instead of dealing with the continuity equation and two momentum equations, only two equations need to be solved. The main shortcoming of this method is in solving three-dimensional problems. Though it is possible to extend this approach by the use of the so-called vector potential for the three-dimensional case, additional complications arise for problems involving porous media. Extending the generalized model to three dimensions leads to physical inconsistencies [7]. Thus, for three-dimensional applications, which are not based on the generalized equations, the extension of the equations in primitive variable formulation is recommended.

For the case when local thermal equilibrium is assumed between the fluid and the solid matrix, the governing equations can be written as [7]

$$\nabla \cdot \vec{u} = 0 \tag{15}$$

$$\frac{\rho_f}{\varepsilon} \frac{D\vec{u}}{Dt} = -\nabla P + \mu_{eff} \nabla^2 \vec{u} - \frac{\mu_f}{K} \vec{u} - \frac{\rho_f F \varepsilon}{\sqrt{K}} \left| \vec{u} \right| \vec{u} \tag{16}$$

$$\sigma \frac{\partial T}{\partial t} + \vec{u} \cdot \nabla T = \alpha_{eff} \nabla^2 T \tag{17}$$

where

$$\left| \vec{u} \right| = \sqrt{u^2 + v^2}$$

In the above equations, $\vec{u}$ is the velocity vector, α_{eff} the effective thermal diffusivity, μ_{eff} the apparent viscosity of the medium, σ the thermal capacitance of the porous matrix. While the effective viscosity of the fluid-saturated porous medium μ_{eff} associated with the Brinkman's term in the momentum equation may have a different value than the fluid viscosity μ_f, as a first approximation μ_{eff} is usually taken as equal to μ_f due to a lack of robust data. This approximation provides a good agreement with the experimental data reported by Lundgren [17], as well as Neale and Nader [18]. It is worth noting that here, the thermal dispersion effect is accounted for by lumping it with the thermal conductivity. That is, we have taken one of the customary approaches in which the thermal dispersion contribution is embedded into the effective thermal conductivity. Hence, the effective thermal diffusivity can be decomposed into two parts: one stands for the stagnant thermal diffusivity of the fluid-saturated porous medium and the other incorporates the additional thermal transport due to the transverse mixing.

In the two-dimensional Cartesian coordinates, upon cross-differentiation and eliminating pressure in the momentum equations and introducing the stream function and vorticity, the vorticity transport and stream function equations in their general forms can be written as

$$\frac{\rho_f}{\varepsilon}\frac{D\zeta}{Dt} = \mu_f \nabla^2 \zeta - \frac{\mu_f}{K}\zeta - \frac{\rho_f F\varepsilon}{\sqrt{K}}(\frac{\partial}{\partial y}(|\vec{u}|u) - \frac{\partial}{\partial x}(|\vec{u}|v)) \qquad (18)$$

$$\nabla^2 \Psi = \zeta \qquad (19)$$

where

$$\zeta = \frac{\partial u}{\partial y} - \frac{\partial v}{\partial x} \qquad (20)$$

and

$$u = \frac{\partial \Psi}{\partial y} \ , \ v = -\frac{\partial \Psi}{\partial x} \qquad (21)$$

Several numerical procedures based on finite difference, finite volume, or finite element techniques have been developed for solving the governing equations described above as discussed earlier.

2.1.4 Free Surface Analysis in Porous Media. The prediction of the fluid interface displacement in porous media has become increasingly important in some manufacturing processes such as resin transfer molding (RTM) and structural reaction injection molding (SRIM). Among the first investigations of free surface transport phenomena through porous media, Vafai and Chen [19] developed a phenomenological model of the problem using a finite difference scheme based on the marker-and-cell (MAC) method. Later, Chen and Vafai [20-21] used the model to examine the effects of pressure differences and permeabilities on free surface transport based on Darcy flow [21] and then extended the analysis to include the non-Darcy effects using the Brinkman-Forchheimer-extended Darcy model. They also included the effect of interfacial tension at the free surface [21]. A summary of the main features of this model is presented here while more details could be obtained in the related references.

The governing momentum and energy equations which account for the inertial and boundary effects are [7]

continuity equation:

$$\frac{\partial u}{\partial x} + \frac{\partial v}{\partial y} = 0 \qquad (22)$$

momentum equations:

$$-\frac{\varepsilon}{\rho_f}\nabla P + \frac{\mu_f}{\rho_f}\nabla^2\vec{u} - \frac{\mu_f}{K}\vec{u} - \frac{\rho_f F\varepsilon}{\sqrt{K}}|\vec{u}|\vec{u} = 0 \tag{23}$$

energy equation:

$$u\frac{\partial T}{\partial x} + v\frac{\partial T}{\partial y} = \alpha_{eff}\left(\frac{\partial^2 T}{\partial x^2} + \frac{\partial^2 T}{\partial y^2}\right) \tag{24}$$

The boundary, initial, and interface conditions required to solve the governing equations are expressed as follows:

Boundary conditions:

$$p = p_e, v = 0, \text{ and } T = T_e \text{ at } x = 0 \tag{25a}$$

$$\frac{\partial u}{\partial x} = 0, \text{ and } -k_e\frac{\partial T}{\partial n} = h(T - T_\infty) \text{ at } x = x_0 \tag{25b}$$

$$u = v = 0, \text{ and } T = T_w \text{ at } y = 0 \text{ and } y = 2H \tag{25c}$$

$$P_\xi = \sigma_{eff}\kappa_\xi + 2\mu\left[n_x^2\frac{\partial u}{\partial x} + n_x n_y\left(\frac{\partial u}{\partial y} + \frac{\partial v}{\partial x}\right) + n_y^2\frac{\partial v}{\partial y}\right] \quad \text{at } x = x_0 \tag{25d}$$

The general form of the initial condition for the free surface location is based on practical applications such as in injection molding. It is based on some fluid residing in the porous channel where a high pressure is applied at the entrance of the channel. A pertinent benchmark for the numerical solutions is the analytical solution presented by Srinivasan and Vafai [22]. This analytical solution is modified to account for the existing encroaching fluid. The derivation of the modified solution is somewhat lengthy, and here we present only the final form as

$$\frac{x_0}{L} = \frac{1 - \sqrt{1 + 2(1-\delta)\left\{\frac{-K}{\mu_1\varepsilon}\left(\frac{\Delta p}{L^2}\right)t + \frac{(1-\delta)}{2}\left(\frac{x_i}{L}\right)^2 - \frac{x_i}{L}\right\}}}{(1-\delta)} \tag{26}$$

where x_i represents the initial location of the encroaching fluid and δ is the mobility ratio. Also,

$$u = v = 0, T = T_e, \text{ and } x = x_i \text{ at } t = 0. \tag{27}$$

In the above equations, x_0 represents the free surface position, p_e the entrance pressure, P_ξ the capillary pressure at the free surface, σ the surface tension coefficient, κ_ξ the sum of curvature at the free surface, T_e the entrance temperature, T_w the wall temperature, T_∞ the ambient temperature, x_i the initial position of the free surface, h the heat transfer coefficient and k_e represents the effective thermal conductivity.

Equation (25d) represents the complete formulation for the pressure field at the free surface including the surface tension and viscous effects. The surface tension effect introduced in this work is used to account for the curvature variation when the no-slip boundary condition is imposed. Based on the work by Vafai and Chen [19], the pressure caused by the viscous force will be insignificant. Therefore, Eq. (25d) reduces to

$$P_\xi = \sigma_{eff} \kappa_\xi \tag{28}$$

The numerical methodology for this free surface problem is based on the finite difference formulation of the governing equations as described in detail by Vafai and Chen [19]. A brief description of the methodology is presented as follows :

x-momentum:

$$-\frac{\varepsilon}{\rho_f}\frac{\partial p_f}{\partial x} + \frac{\mu_f}{\rho_f}\nabla^2 u - \frac{\varepsilon\mu_f}{K}u - \frac{\rho_f F \varepsilon}{\sqrt{K}}|\vec{u}|u = 0 \tag{29}$$

y-momentum:

$$-\frac{\varepsilon}{\rho_f}\frac{\partial p_f}{\partial y} + \frac{\mu_f}{\rho_f}\nabla^2 v - \frac{\varepsilon\mu_f}{K}v - \frac{\rho_f F \varepsilon}{\sqrt{K}}|\vec{u}|v = 0 \tag{30}$$

Equations (29) and (30) can be written as

$$-\frac{\theta}{\rho_f}\frac{\partial p_f}{\partial x} + f(x,y) = 0 \tag{31}$$

$$-\frac{\theta}{\rho_f}\frac{\partial p_f}{\partial y} + g(x,y) = 0 \tag{32}$$

where f(x,y) and g(x,y) are defined as

$$f(x,y) = \frac{\mu_f}{\rho_f}\nabla^2 u - \left[\frac{\mu_f \varepsilon u}{K\rho_f} + \frac{F u \varepsilon |\vec{u}|}{\sqrt{K}}\right] \tag{33}$$

$$g(x,y) = \frac{\mu_f}{\rho_f}\nabla^2 v - \left[\frac{\mu_f \varepsilon v}{K\rho_f} + \frac{F v \varepsilon |\vec{u}|}{\sqrt{K}}\right] \tag{34}$$

Taking divergence of equations (33) and (34) will result in the following equation:

$$\overline{\nabla}^2 p_f = \frac{\rho_f}{\varepsilon} \left\{ \frac{\partial f(x,y)}{\partial x} + \frac{\partial g(x,y)}{\partial y} \right\} \tag{35}$$

The SOR method is employed in solving the discretized momentum and pressure equations, and the MAC (Marker and Cell) method is utilized to track the temporal free surface position. During the process, extrapolation of the velocity fields in the empty cells is required to carry out the numerical iteration of the momentum and pressure equations. Calculation of the marker particle velocities is accomplished using the obtained velocity field to move the free surface position. An implicit scheme is used to solve the energy equation. The implementation of the present numerical scheme will not be presented here for the sake of brevity. More details can be found in Vafai and Chen [19].

In the following sections, an updated review of computational studies related to various important aspects of forced convection in porous media is presented. A summary of this literature review is presented in Table 1.

2.2 External Flow

Forced convection over external boundaries in the presence of a porous medium has been an important research topic for the past several decades. This fundamental problem involves important physical phenomena which occur in various practical applications. For this class of problems, earlier studies were based on a Darcy flow model, and the boundary layer approximation and the similarity solution technique were commonly applied. Further studies were related to non-Darcy effects as well as other important phenomena. An earlier review was presented by Lauriat and Vafai [23].

2.2.1 Flat Plate. Forced convection over a flat plate embedded in a porous medium is a fundamental problem observed in a wide variety of engineering applications. Earlier studies of this problem were based on the Darcy flow model [3]. Since the original work of Vafai and Tien [7], it is now well established that the viscous effects at the solid boundary and the inertial drag from the solid matrix have significant effects on flow and heat transfer in porous media. These effects are enhanced for high Reynolds number, especially in porous media with high porosity and permeability. During the past decade, these effects have been studied extensively through the use of the general flow model better known as the Brinkman-Forchheimer-extended Darcy model. Using the Forchheimer-extended Darcy model, Nakayama and Ebinuma [24] investigated the transient effects on non-Darcy forced convection. Their results showed how the non-Darcy inertia effects decrease the velocity and delay covering of the plate by the steady-state thermal boundary layer. Ling and Dybbs [25] studied the effects of variable viscosity.

Table 1 Summary of Computational Studies of Forced Convection in Porous Media

Author	geometry	non-Darcy effects	variable porosity	thermal dispersion	local non-thermal equilibrium	other aspects
Vafai and Tien [7]	plate	•				
Vafai and Sozen [8]	channel	•	•	•	•	
Vafai [10]	plate	•	•	•		
Benenati and Brosilow [14]	cylinder		•			
Nakayama and Ebinuma [24]	plate	•	•			
Ling and Dybbs [25]	plate	•				
Thevenin and Sadaoui [26]	cylinder					
Thevenin [27]	cylinder					
Hossain et al. [29]	wedge	•	•			partially porous
Vafai and Kim [30]	plate	•	•			
Huang and Vafai [35]	plate	•	•			
Shenoy [39]	channel	•				power-law fluid
Hady and Ibrahim [41]	plate	•		•		:
Poulikakos and Renken [45]	channel/pipe	•	•	•		
Amiri and Vafai [46]	channel	•	•	•	•	:
Hunt and Tien [47]	pipe	•		•		
Hunt and Tien [48]	channel	•		•		
Hsu and Cheng [49]	channel/pipe	•	•	•		
Cheng and Zhu [50]	pipe		•	•		
Cheng and Vortmeyer [51]	channel	•	•	•		
Cheng et al. [52]	channel	•	•	•		

Author	geometry	non-Darcy effects	variable porosity	thermal dispersion	local non-thermal equilibrium	other aspects
Sozen and Vafai [53]	channel	•		•		partially porous
Fu and Huang [56]	slot jet	•			•	
Amiri and Vafai [58]	channel	•	•	•	•	
Lee and Vafai [59]	channel		•	•	•	
Hadim [60]	channel	•				partially porous
Sozen and Kuzay [61]	tube	•				
Lage et al. [65]	channel	•				partially porous
Huang and Vafai [67]	channel	•			•	
Chikh et al. [68]	channel	•				:
Ould-Amer et al. [69]	channel	•				:
Fu et al. [70]	channel	•	•	•		:
Tong and Sharatchandra [71]	channel	•				:
Hadim and Bethancourt [72]	channel	•				:
Guo et al. [73]	pipe					:
Al-Nimr and Alkam [74]	annulus					:
Alkam and Al-Nimr [75]	pipe				•	:
Sozen and Vafai [76]	channel			•	•	:
Al-Nimr et al. [79]	annulus		•			:
Nakayama and Shenoy [81]	channel					power-law fluid
Chen and Hadim [82]	channel		•	•		:
Alkam et al. [85]	annulus			•		:
Chou et al. [86]	duct	•		•		

2.2.2 Cylinders, Spheres, and Wedges. Several studies have been related to forced convection over a horizontal cylinder in a porous medium. Thevenin and Sadaoui [26] used the finite element method and the Darcy-Brinkman model to investigate heat transfer enhancement over a circular cylinder embedded in a fluid-saturated fibrous porous medium. Thevenin [27] extended the same study to include the transient effects. Nasr et al. [28] considered the case of a cylinder embedded in a packed bed. The non-Darcy effects and thermal dispersion were included in their model. They solved the governing equations using the general-purpose finite element program FIDAP, which employs a Galerkin-based finite element method.

Taking into account the effects of inertia from the porous matrix, viscous effects at the boundary, and convective inertia terms, Hossain et al. [29] employed three different methods to study non-Darcy forced convection, boundary layer flow over a wedge with a variable free stream. The three methods consisted of: (1) the finite difference technique combined with the Keller box method, (2) the extended series method together with Shanks' transformation, and (3) the local non-similarity method for integrating momentum and energy equations.

2.2.3 Composite Porous/nonporous Media. Problems involving an interface between a porous medium and another medium occur in several engineering applications. Vafai and Kim [30] studied the problem of convective flow and heat transfer through a composite system consisting of a fluid layer overlaying a porous substrate, which is attached to a solid surface. A general flow model that accounts for the effects of inertia and the impermeable boundary was used to describe the flow inside the porous region. Their numerical simulation focused primarily on flows that have the boundary layer characteristics even though the boundary layer approximation was not used. They studied in detail the effects of the governing parameters on the characteristics of the flow and temperature fields in the composite layer. Huang and Vafai [31] solved the same problem analytically using the Karman-Pohlhausen integral method which produced faster yet reasonably accurate results. Sahraoui and Kaviany [32, 33] studied the problem of forced convection at the interface between porous and plain media. They focused on the slip and no-slip conditions first for velocity [32] and later for temperature [33], and they presented how to model these conditions at the pore level using a configuration consisting of a two-dimensional periodic arrangement of cylinders and a finite volume-based numerical simulation.

An innovative scheme for controlling flow and heat transfer on an external surface was proposed by Huang and Vafai [34], who analyzed a composite system made of a multiple porous block structure on an external surface. They described the flow inside the porous region using the Brinkman Forchheimer-extended Darcy model. They developed a vorticity-stream function formulation of the problem which they solved using a finite-difference method. They analyzed in detail the effects of several governing dimensionless parameters on the flow pattern and heat transfer characteristics due to the existence of the multiple porous block structure. Similar studies were performed by the same authors for the configuration consisting of alternately placed porous cavity-block wafers over an external surface in a porous region [35] and for the configuration consisting of intermittent porous cavities over an external surface in a nonporous region [36]. Fu and Huang [37] presented a study of heat transfer enhancement using a laminar slot jet impinging on a porous block which is mounted on a heated surface. They used the SIMPLEC-based numerical algorithm, and they considered three different block shapes: rectangular, convex, and concave. Abu-Hijleh [38] used the finite element method to solve the problem of forced convection over a backward-facing step which includes a porous floor segment. They studied the effects of porosity and orthotropicity of the porous segment on the Nusselt number.

2.2.4 Non-Newtonian Fluids. Convective transport of non-Newtonian fluids in porous media occurs in various applications including biomechanics (e.g. blood flow through arteries), packed-bed chemical reactors, ceramic engineering, oil extraction, and fluid composite molding, among others. Shenoy [39] presented an extensive review of the literature on this topic covering both boundary layer and confined flows. Shenoy [40] used the Darcy-Forchheimer model to conduct a boundary layer analysis of forced convection of a power-law-type non-Newtonian fluid over a semi-infinite plate, and Hady and Ibrahim [41] solved the non-similarity problem including non-Darcy effects. Similar studies for non-isothermal smooth surfaces of arbitrary shape have also been reported for both Darcy and non-Darcy flows [42, 43].

2.3 Forced Convection in Internal Flows

Forced convection in confined flows through porous media has been of continuing interest to researchers due to the increasing pool of industrial applications related to this problem. Among them, the use of a porous matrix for heat transfer enhancement has received significant recent interest in both industry and academia because of its potential use in high heat flux applications including electronics cooling and cryogenics. Earlier reviews of the literature related to this topic were presented by Lauriat and Vafai [23], Nield and Bejan, [3], and Kaviany [4]. Recently an updated review of the literature related to modeling forced convective flows through confined porous media was presented by Vafai and Amiri [44].

2.3.1 Non-Darcy Effects. Among the earlier studies related to non-Darcy effects of inertia in the porous matrix and viscous shear at the boundary, Poulikakos and Renken [45] presented numerical simulations for two channel configurations: a parallel-plate channel and a circular pipe exposed to a constant wall temperature. They assumed the flow to be fully developed and neglected axial conduction in the energy equation. A comprehensive numerical study of forced convection through porous media was conducted by Amiri and Vafai [46]. They employed a full general model of the momentum and energy equations and they performed a comprehensive study of the inertia effects, boundary effects, variable porosity, thermal dispersion, and local thermal equilibrium as well as two-dimensionality effects on transport processes in the porous medium. Based on extensive computational parametric studies, the authors generated error maps for assessing the importance of various simplifying assumptions. A sample of these error maps, presented in Fig. 1, displays the percent error on the average Nusselt number for different simplified models while incorporating an exponential porosity variation and thermal dispersion effects. Other computational studies related to various important physical phenomena governing internal flow forced convection in porous media are reviewed in the following sections.

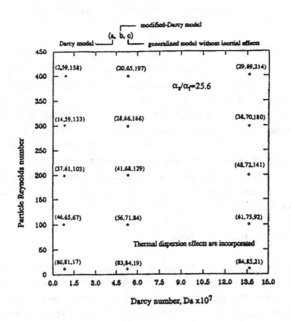

Figure 1 Percent Error on the Average Nusselt Number Using Various Models (From Amiri and Vafai [46]; Copyright 1994 Elsevier Science Ltd – Reprinted with permission).

2.3.2 Variable Porosity. Most of the existing studies of convective transport in porous media assumed constant porosity throughout the porous medium. However, in several applications involving porous media made of granular material (such as in packed-bed catalytic reactors, drying, and metal processing), this assumption is not valid near a solid surface because of the influence of the impermeable boundary on the porosity variation. For such applications, it has been established that the porosity is much higher in the vicinity of the impermeable boundary, and it decreases to an asymptotic value in the core region. Moreover, the porosity of the bed was found to exhibit sinusoidally damping decay, especially close to the wall [14]. This phenomenon introduces the channeling effect, which refers to the occurrence of a maximum velocity in a region close to an external boundary, and it has been widely discussed in the literature [10, 47]. Vafai [10] investigated analytically and numerically the effects of variable porosity and inertial forces on convective flow and heat transfer in porous media. He showed that the nature of variation in porosity significantly affects the convective flow patterns as well as the heat transfer. In their study of fully-developed non-Darcy forced convection in a cylindrical packed bed, Hunt and Tien [47] compared numerical results from three different porosity profiles with available experimental data for chemical reactors. Various aspects of the use of an actual porosity variation versus a simplified exponential variation are given in Amiri and Vafai [46]. Several other studies which considered the channeling effects included the thermal dispersion effects as well, as discussed in the following section.

2.3.3 Thermal Dispersion. Thermal dispersion is a complex phenomenon, which has been shown to be responsible for forced convection heat transfer enhancement using porous materials, especially at high Reynolds number. Thermal dispersion occurs due to mixing and recirculation of the fluid as it meanders through the tortuous paths around the solid particles. This secondary effect was neglected by most investigators in the past. During the past decade, several investigations described the physical aspects of the dispersion phenomenon and developed semi-empirical models needed to account for thermal dispersion effects. Existing models of thermal dispersion include several empirical coefficients that are usually obtained by comparing theoretical/numerical results with experimental measurements. For fibrous media, Hunt and Tien [47, 48] studied the effects of thermal dispersion on forced convection heat transfer enhancement. They presented a thermal dispersion conductivity model for porous media consisting of foam-metal through a comparison between numerical calculations and experimental data [48].

For packed beds, Cheng and co-workers [49 - 52] conducted a series of theoretical and numerical studies related to these effects. They used a wall function of exponential form to model the transverse thermal dispersion near the wall, which leads to a large temperature drop near the wall of a packed bed. They considered both a cylindrical packed bed (for the specific application to catalytic reactors) as well as a parallel plate channel, and they determined the corresponding thermal dispersion coefficients. Sözen and Vafai [53] analyzed the contribution of the longitudinal dispersion conductivity to the overall effective thermal conductivity of a porous medium. They concluded that the effects of longitudinal dispersion were less pronounced for high Reynolds number flows. Later, Amiri and Vafai [46] examined the effects of the longitudinal and transverse dispersion over a broad range of Darcy number, particle Reynolds number, and thermophysical properties. Adnani et al. [54] studied the effects of anisotropic dispersion on non-Darcy forced convection in a particle packed tube. They included both axial and radial dispersion and showed that heat transfer in the thermally developing region is affected by axial dispersion when the Peclet number is smaller than 10.

A direct numerical simulation of thermal dispersion was reported by Kuwahara et al. [55]. They used a two-dimensional periodic model consisting of square rods placed regularly in an infinite space. They obtained numerical solutions of the flow and heat transfer at the pore scale and then integrated the results over a unit structure to obtain the thermal dispersion model. Very recently, Fu and Huang

[56] analyzed the effects of a random porosity model on heat transfer performance by comparing the results from three different porosity models consisting of constant, variable and random porosities. They adopted a one-dimensional thermal model with the Van Driest's wall function for the energy equation and performed the numerical computations using the SIMPLEC algorithm.

2.3.4 Local Thermal Equilibrium. A review of the literature indicates that it is common practice to employ the local thermal equilibrium (LTE) condition which implies that the local temperature difference between the fluid and the solid phases is negligible at all locations in a fluid-saturated porous medium. However, this assumption cannot be imposed when the temperature difference between the fluid and the solid matrix is significant (e.g. fuel rods surrounded by the coolant fluid and porous heat exchanger). In such a case, the modeling of the energy transport in a porous medium demands accounting of the individual phases by separate energy equations with an additional term in each equation to assimilate the energy exchange between the two phases. In these situations, the effects of interfacial surface and the interstitial heat transfer coefficient, which are related to the internal heat exchange between the solid and the fluid phases, are major factors causing heat transfer augmentation in porous media. In such cases, the two-equation model needs to be utilized as described earlier. Several investigations of forced convective flow through porous media utilizing a rigorous formulation based on the local volume-averaged two-equation model have been reported [8, 46, 57-59].

In their numerical analysis of forced convective flow of a gas through a packed bed of spherical solid particles, Vafai and Sözen [8] showed that the local thermal equilibrium was very sensitive to the particle Reynolds number and the Darcy number, while thermophysical properties did not have a very significant effect on this condition. Based on extensive computational parametric studies, Amiri and Vafai [46] generated error maps for assessing the importance of the local thermal equilibrium assumption under various conditions. Subsequently, the two-phase energy equation was used to simulate local non-thermal equilibrium in a porous bed consisting of metallic particles of constant porosity [57] and in the transient thermal behavior of packed beds [58]. A sample of the effects of Darcy number on the instantaneous LTE condition is illustrated in Fig. 2 [58]. It is shown that the LTE condition improves with time and becomes more justifiable for small Darcy number. Recently, Lee and Vafai [59] presented a novel analytical characterization on local non-thermal equilibrium using the two-equation model for energy transport which includes the transverse conduction contribution for both the fluid and the solid temperature fields. They classified the energy transport into three mechanisms: fluid conduction, solid conduction, and internal heat exchange between the solid and the fluid phases. They derived a thermal network representation which characterized the energy transport due to these mechanisms. The authors summarized their results with an error map as illustrated in Fig. 3 which displays the error on the Nusselt number based on using the one-equation model for a wide range of Biot number and the ratio κ of fluid to solid effective thermal conductivity.

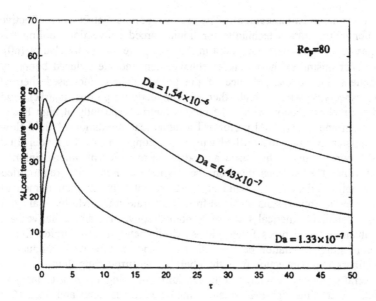

Figure 2 Effects of Darcy Number on the Instantaneous LTE Condition for $k_s/k_f = 12$ (From Amiri and Vafai [58]; Copyright 1998 Elsevier Science Ltd – Reprinted with permission).

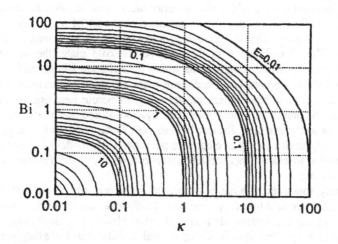

Figure 3 Error Map for the Nusselt Number Based on Using the One-Equation Model (From Lee and Vafai [59]; Copyright 1999 Elsevier Science Ltd – Reprinted with permission).

2.3.5 Heat Transfer Enhancement. The use of a porous matrix as an effective heat transfer augmentation technique for liquid forced convection cooling has been suggested by several investigators in the past as reviewed by Hadim [60]. The main mechanisms for heat transfer enhancement include reduced boundary layer thickness due to the presence of the porous matrix, increased thermal diffusion (especially with a high thermal conductivity porous matrix), and increased thermal dispersion within the porous matrix, especially at high Peclet number. Sözen and Kuzay [61] performed a numerical simulation of fluid flow and heat transfer in round tubes filled with rolled copper mesh. To model fluid flow in the porous tubes, they used a modified Ergun-Forchheimer-Brinkman equation in which the two constants from the original Ergun model were modified based on empirical data. They found good agreement between their numerical results and experimental results obtained from a separate study. Martin et al. [62] performed a detailed numerical study of combined convective and radiative heat transfer enhancement in pipes filled with small-diameter silicon carbide fibers. Their model included radiation between the fibers and the tube wall, conduction within the fibers, and convection from the fibers to the surrounding fluid.

Several studies were related to the feasibility of using a porous channel as a heat sink for cooling high-performance microelectronics. Kuo and Tien [63] analyzed heat transfer augmentation in a foam-material filled duct with heat sources, and they included the boundary, inertia and thermal dispersion effects. They showed that an increase of two to four times in heat transfer is achievable as compared to the nonporous case. Hwang et al. [64] used the general model including non-Darcy effects, thermal dispersion, and local non-thermal equilibrium to study forced convection in an asymmetrically heated sintered porous channel with a high thermal conductivity solid matrix. A design of a microporous heat sink for cooling phased-array radar systems was analyzed by Lage et al. [65]. They developed a two-dimensional model of a thin enclosure with one inlet and one outlet and containing a low-permeability aluminum alloy porous layer. They found that with the microporous cold plate, the operating temperatures of the electronic components within the radar are substantially more uniform than without a porous layer, and these operating temperatures are reduced significantly due to the enhanced heat transfer using the porous heat sink. Fedorov and Viskanta [66] presented a numerical investigation of conjugate heat transfer in a porous channel containing discrete heat sources flush-mounted into a highly conductive substrate simulating an electronic package. Their model for the energy transport included thermal dispersion and the two-equation model for local non-thermal equilibrium.

Recently, much research has been focused on the partially porous system as a potentially attractive heat transfer augmentation technique because of its increased heat transfer and reduced pressure drop as reported by Hadim [60]. Such composite porous/non-porous systems also occur in several applications including porous media heat exchangers, fault zones in geothermal systems, solidification of binary alloys, porous bearings, porous heat pipes, blood flow in lungs or in arteries and many others. The main geometric configurations considered include porous layers, porous bands and porous blocks in rectangular channels or circular cylinders.

Several numerical studies of forced convection heat transfer enhancement considered a channel containing discrete porous blocks. Huang and Vafai [67] studied this configuration using the Brinkman-Forchheimer-extended Darcy model and a stream function-vorticity formulation. They performed extensive parametric studies of the effects of important governing parameters on heat transfer enhancement in the channel. They also presented a detailed discussion of recirculation caused by the porous blocks within the channel. Chikh et al. [68] considered the configuration in which the channel is intermittently heated and the porous blocks are mounted on the heated sections. Ould-Amer et al. [69] investigated the case when heat generating nonporous blocks are mounted on one wall and porous bands are inserted between the blocks. Convective heat transfer enhancement due to the presence of a single porous block made of spherical beads and mounted on a heated wall of a two-dimensional channel was investigated by Fu et al. [70]. They used the Brinkman-Forchheimer-extended Darcy model including the effects of variable porosity and thermal dispersion, and they solved the governing equations using the SIMPLEC numerical algorithm.

Tong and Sharatchandra [71] investigated the use of a long porous insert for heat transfer enhancement in a channel. They derived an analytical solution of the fully developed velocity based on the Brinkman-extended Darcy model in the porous region and the Navier-Stokes equation in the fluid region, and they used a variable grid-size, implicit finite difference method to solve for temperature and heat transfer in the channel. They found that the heat transfer in the channel could be maximized by choosing an appropriate thickness of the porous layer combined with proper selection of the porous material properties as described by the Darcy number and the thermal conductivity ratio between the porous medium and the fluid. Hadim [60] and Hadim and Bethancourt [72] considered a channel divided by several discrete porous bands placed on top of discrete heat sources flush-mounted on one wall. They used the general Brinkman-Forchheimer-extended Darcy model and a finite volume-based numerical algorithm. Their numerical results demonstrated that a partially porous channel may be a potentially attractive heat transfer augmentation technique because of its reduced pressure drop.

Among related transient studies, Guo et al. [73] studied forced pulsating flow and heat transfer in a circular pipe partially filled with a porous medium which is attached to the pipe wall where a uniform heat flux is applied. They employed the Brinkman-Forchheimer-extended Darcy model to investigate various effects including thickness of porous layer and thermal conductivity ratio, as well as frequency and amplitude of pulsation. Their results indicated that the effective axial thermal conductivity is substantially increased by pulsation, and this augmentation is more pronounced by partial filling of the pipe with the porous material. More recently, Al-Nimr and Alkam [74] studied transient non-Darcy forced convection in the developing region of an annulus partially filled with a porous substrate which is attached either to the inner cylinder or to the outer one. They studied the transient thermal behavior caused by a step change in temperature at the cylinder wall which is in contact with the porous substrate. Alkam and Al-Nimr [75] conducted a similar study for a partially porous pipe.

2.3.6 Other Applications. Several other practical applications involving porous media have been investigated. For energy storage applications, Vafai and co-workers performed several studies simulating the hydrodynamic and thermal behavior of packed beds. Sözen and Vafai [53, 76, 77] conducted finite-difference-based numerical simulations to investigate various aspects of forced convection of an ideal gas through a packed bed including the effect of oscillating gas-phase temperature/pressure inlet boundary conditions [76], the thermal energy storage characteristics [77], and the longitudinal thermal dispersion effects [53]. Amiri et al. [57] used a two-phase energy equation model to study non-Darcy forced convection through a porous bed consisting of metallic particles of constant porosity. They considered both boundary conditions of constant wall temperature and constant wall heat flux for which they compared the results with available experimental data. Amiri and Vafai [58] simulated numerically the temporal energy transport for incompressible flow through a packed bed using a control volume method. Non-Darcy effects and thermal dispersion were accounted for, and the two-energy equation model was used for simulating the energy transport.

Al-Nimr et al. [78, 79] considered the problem of transient forced convection in the entrance region of a porous tube and a porous concentric annulus with developing thermal boundary layer. Both Darcian and non-Darcian effects on the flow were considered, and a step change in the temperature of the tube wall was assumed in their work. An innovative approach for modeling flow and heat transfer through spirally fluted tubes was presented by Srinivasan et al. [80]. They modeled the flutes within the tubes using a porous substrate with direction-dependent permeabilities, and they solved the governing equations using a Galerkin-based finite element method.

2.3.7 Non-Newtonian Flows. Most of the existing numerical studies of non-Newtonian flow and heat transfer in porous media considered external flows and are based on Darcian flow behavior in which inertia and viscous effects are neglected. A modified Brinkman-Forcheimer-extended Darcy model for power-law fluids was presented by Nakayama and Shenoy [81], who investigated non-Darcy forced convection of a power-law fluid in a porous channel. Using a Runge-Kutta-Gill method, they obtained numerical results for the case of fully developed flow and heat transfer in the channel. Recently, Chen and Hadim [82] used a modified Brinkman-Forcheimer-extended Darcy model and included variable porosity and thermal dispersion effects to perform a detailed numerical simulation of non-Darcy forced convection in a channel packed with an isotropic granular material and saturated with a power-law fluid. Their results showed the possibility of enhancing heat transfer while reducing the pressure drop when a shear thinning fluid is combined with a porous matrix with a larger particle diameter. The same authors [83-84] conducted similar studies for the case when the channel is filled with a fibrous medium and is exposed to either uniform temperature or uniform heat flux conditions at the walls. They developed a boundary layer analysis based on the integral method [83] and conducted extensive computational parametric studies [84] to examine the effects of various governing parameters. Using a similar model,

Alkam et al. [85] investigated the transient forced convection of a non-Newtonian fluid in the entrance region of a porous concentric annulus.

2.3.8 Three-dimensional Studies. Three-dimensional studies of forced convection in porous media are rather limited as they do not have a sound physical basis with the commonly used Brinkman-Forchheimer-extended Darcy model in which the choice of the volume element for the averaging process applies only to two-dimensional problems in a strict sense [7]. However, the outcomes of some of the previous work on three-dimensional configurations do indicate that the two-dimensional model can be extended to compare favorably with the three-dimensional experimental studies [57]. Chou et al. [86] presented a numerical study of fully developed non-Darcy forced convection in a square packed-sphere channel. Their results showed that when the Peclet number is small, the fully-developed Nusselt number was influenced mainly by the channeling effects while the effects of thermal dispersion became dominant when the Peclet number is high. They also found that the effects of thermal dispersion on heat transfer were weak in the entrance region and became more significant in the thermally fully developed region. Similar observations were made by Chen and Hadim [87] who conducted a more detailed study of three-dimensional forced convection in a square duct packed with an isotropic granular material. They applied a set of consistent empirical models for flow channeling and thermal dispersion in the duct, and they used a control volume numerical procedure based on the SIMPLE algorithm.

3 MULTIPHASE TRANSPORT IN POROUS MEDIA

Multiphase transport in porous media occurs in several engineering applications including drying processes (wood and food products), oil reservoir engineering, geothermal reservoirs, groundwater contamination, heat pipes, phase change in building insulation, metal casting (alloy solidification), and heat transfer enhancement. Consequently, this topic has been investigated extensively during the past two decades. Various complex physical phenomena exist in multiphase flow in porous media. Exact solutions for this class of problems are limited to highly simplified one-dimensional problems. More practical problems involve a combination of several phenomena including diffusion, capillary action, phase change, and multi-dimensional effects. These problems must be analyzed using either experimental techniques or sophisticated numerical models. Consequently, a number of recent investigations have been focused on developing robust numerical models that are able to simulate these various physical aspects.

3.1 Mathematical Formulation and Numerical Solution

3.1.1 The General Model. A general model of multiphase transport in porous media is presented in this section. Although it has been developed for analyzing the phase change and non-thermal equilibrium flow of a gas through a packed bed [88, 89], its general structure allows it to be applicable to other problems of multiphase transport in porous media. Owing to the lack of information on a number of transport coefficients, Vafai and Sozen [89] reformulated the governing equations and relaxed the local gradient of phase-averaged temperature such that simulation of practical problems could be achieved. In their work, the following volume-averaged governing balance equations and the associated coupling relations were obtained:

Vapor phase continuity equation:

$$\frac{\partial}{\partial t}\left(\varepsilon_\gamma < \rho_\gamma >^\gamma\right) + \nabla \cdot \left(< \rho_\gamma >^\gamma < \vec{v}_\gamma >\right) = - < \dot{m} > \tag{36}$$

Liquid phase continuity equation:

$$\frac{\partial \varepsilon_\beta}{\partial t} + \nabla \cdot < \vec{v}_\beta > - \frac{< \dot{m} >}{\rho_\beta} = 0 \tag{37}$$

Vapor phase equation of motion:

$$\nabla < P_\gamma >^\gamma = - \frac{< \rho_\gamma >^\gamma F \varepsilon_\gamma}{K_\gamma^{1/2}} \left[< \vec{v}_\gamma > . < \vec{v}_\gamma >\right] \frac{< \vec{v}_\gamma >}{|< \vec{v}_\gamma >|} - \frac{\mu_\gamma}{K_\gamma} < \vec{v}_\gamma > \tag{38}$$

Liquid phase equation of motion:

$$< \vec{v}_\beta > = - \frac{k_{r\beta} K}{\mu_\beta} \left\{ k_\varepsilon \nabla \varepsilon_\beta + k_{<T>} \nabla < T_f >^f + (\rho_\beta - < \rho_\gamma >^\gamma) \vec{g} \right\} \tag{39}$$

Fluid phase energy equation:

$$\left[\varepsilon_\beta \rho_\beta (c_p)_\beta + \varepsilon_\gamma < \rho_\gamma >^\gamma (c_p)_\gamma\right] \frac{\partial < T_f >^f}{\partial t} - < \dot{m} > \Delta h_{vap}$$

$$+ \left[\rho_\beta (c_p)_\beta < \vec{v}_\beta > + < \rho_\gamma >^\gamma (c_p)_\gamma < \vec{v}_\gamma >\right] . \nabla < T_f >^f = \nabla . \left[k_{feff} \nabla < T_f >^f\right]$$

$$+ h_{\sigma\beta} a_{\sigma\beta} \left[< T_\sigma >^\sigma - < T_f >^f\right] + h_{\sigma\gamma} a_{\sigma\gamma} \left[< T_\sigma >^\sigma - < T_f >^f\right] \tag{40}$$

Solid phase energy equation:

$$\varepsilon_\sigma \rho_\sigma (c_p)_\sigma \frac{\partial < T_\sigma >^\sigma}{\partial t} = \nabla . \left[k_{\sigma eff} \nabla < T_\sigma >^\sigma\right]$$

$$- h_{\sigma\beta} a_{\sigma\beta} \left[< T_\sigma >^\sigma - < T_f >^f\right] - h_{\sigma\gamma} a_{\sigma\gamma} \left[< T_\sigma >^\sigma - < T_f >^f\right] \tag{41}$$

Volume constraint relation:

$$\varepsilon_\sigma + \varepsilon_\gamma(t) + \varepsilon_\beta(t) = 1 \tag{42}$$

Equation of state for vapor phase:

$$< P_\gamma >^\gamma = < \rho_\gamma >^\gamma R_\gamma < T_f >^f \tag{43}$$

Thermodynamic relation for the saturation density of vapor:

$$\rho_{\gamma,s} = \frac{\exp\left(A - \frac{B}{T_f}\right)}{R_\gamma T_f} \qquad (44)$$

where A and B are known constants, T_f is in degrees Kelvin and $\rho_{\gamma,s}$ is in kg/m^3.

The effective thermal conductivities of the working fluid and of the solid phase are modeled as

$$k_{\sigma eff} = \varepsilon_\sigma k_\sigma$$

and $\qquad\qquad\qquad\qquad\qquad\qquad\qquad\qquad\qquad\qquad\qquad (45)$

$$k_{feff} = \varepsilon_\gamma k_\gamma + \varepsilon_\beta k_\beta$$

The permeability for the vapor phase, K_γ, and the geometric factor, F, in the vapor phase momentum equation can be expressed as functions of ε_γ and d_p in the following form [9, 10]:

$$K_\gamma = \frac{\varepsilon_\gamma^3 d_p^2}{150(1 - \varepsilon_\gamma)^2}$$

$$F = \frac{1.75}{\sqrt{150} \ \varepsilon_\gamma^{3/2}} \qquad (46)$$

The relative permeability of the liquid phase is modeled in the form suggested by Udell and Fitch [90] as

$$k_{r\beta} = S^3 \qquad (47)$$

where $\qquad\qquad S = \frac{s - s_{im}}{1 - s_{im}} \qquad$ and $\qquad s = \frac{\varepsilon_\beta}{\varepsilon} = \frac{\varepsilon_\beta}{1 - \varepsilon_\sigma} \qquad (48)$

Here S denotes the normalized saturation and s denotes the absolute saturation while s_{im} is the so-called "immobile " saturation below which the liquid phase is assumed to be immobile. Due to lack of better experimental findings, the value of 0.1 for s_{im} used by Kaviany and Mittal [91] is considered. With the value of porosity taken to be equal to 0.39 as the average asymptotic value found by Benanati and Brosilow [14] and s_{im} as 0.1, the critical value of the liquid faction, $\varepsilon_{\beta crit}$, below which the liquid phase is assumed to be immobile becomes 0.039. Based on the specific surface area of a packed bed of spheres presented by Dullien [12] as

$$a = 6(1-\varepsilon)/d_p \qquad (49)$$

which was obtained from some geometric arguments, where ε is the porosity and d_p is the particle diameter, and also due to the fact that ε_β takes very small values (less than 0.01) in the problem considered, the specific surface area between the vapor phase and the bed of particles can be accurately approximated as

$$a_{\sigma\gamma} = \frac{6(1 - \varepsilon_\gamma - \varepsilon_\beta)}{d_p} \qquad (50)$$

Also from an analysis of the representative length scales and volume scales of the liquid and vapor phases of the working fluid, one may obtain a relation between the specific surface areas $a_{\sigma\gamma}$ and $a_{\sigma\beta}$ as

$$a_{\sigma\beta} = a_{\sigma\gamma} \left(\frac{\varepsilon_\beta}{\varepsilon_\gamma}\right)^{2/3} \qquad (51)$$

and hence,

$$a_{\sigma\beta} = \frac{6(1 - \varepsilon_\gamma - \varepsilon_\beta)}{d_p} \left(\frac{\varepsilon_\beta}{\varepsilon_\gamma}\right)^{2/3} \qquad (52)$$

Empirical correlations found by Gamson et al. [92] and originally expressed in the form of Colburn-Chilton j_h-factors were found to be suitable for use. These correlations can be expressed in the following forms:

$$h_{\sigma j} = 1.064 \, (c_p)_j \, G_j \left[\frac{c_p\mu}{k}\right]_j^{-2/3} \left[\frac{d_p G}{\mu}\right]_j^{-0.41} \quad \text{for} \quad \frac{d_p G}{\mu} \geq 350 \quad \text{(turbulent)} \qquad (53)$$

$$h_{\sigma j} = 18.1 (c_p)_j \, G_j \left[\frac{c_p\mu}{k}\right]_j^{-2/3} \left[\frac{d_p G}{\mu}\right]_j^{-1} \quad \text{for} \quad \frac{d_p G}{\mu} \leq 40 \quad \text{(laminar)} \quad (54)$$

where G denotes the mass flux of phase j and j stands for β or γ for the liquid and vapor phases, respectively.

Eqs. (36) through (44) yield a system of nine highly coupled equations in nine unknowns, namely, $\varepsilon_\beta(t)$, $\varepsilon_\gamma(t)$, $<\rho_\gamma>^\gamma$, $<\vec{v}_\gamma>$, $<\vec{v}_\beta>$, $<P_\gamma>^\gamma$, $<T_f>^f$, $<T_\sigma>^\sigma$ and $<\dot{m}>$.

3.1.2 Numerical Solution. In the model described in the previous section, the system of governing equations can be solved numerically using the finite difference technique as follows. Based on the nature of the system of equations, the explicit finite difference scheme was found to be the most appropriate one to use. A forward Euler differencing scheme is applied to temporal derivative terms while a central differencing scheme is utilized for spatial derivative terms except in the convective terms for which upwind differencing is used. Another exception is the use of forward differencing in approximating the pressure gradient term in the gas-phase momentum equation. This is done to ensure stability of the numerical solution during the *early stages* of the solution time. Also, forward and backward differencing are employed for the spatial derivative terms for the left and right boundary grid points, respectively. The common procedure for ensuring stability and accuracy of the explicit schemes is by choosing a proper combination of Δx and Δt. This is done by systematically decreasing the grid size and selecting an appropriate Δt for a converging solution. Because of space constraints, more details of the numerical solution of the above model could not be presented here, and the reader is referred to related references [88, 89].

3.2 Models

A comparative analysis of the most relevant multiphase transport models in porous media was reported by Vafai and Sozen [93], while more recent reviews of this topic were reported by Dhir [94] and Wang and Cheng [95]. In their comparative study, Vafai and Sozen [93] selected a set of five models representing the most significant and classical contributions that became the fundamental basis for most of the subsequent work related to transport processes in the presence of phase change in porous media. For each model, they examined in detail its relevant terms as well as the main assumptions and corresponding simplifications made to generate the model. Based on their comprehensive comparison, Vafai and Sozen [93] concluded their study with recommendations on ways to deal with common difficulties encountered when modeling heat and mass transport in the presence of phase change in porous media.

Wang and Beckermann [96] presented an alternative approach for the theoretical analysis and numerical simulation of two-phase transport phenomena in porous media. They presented a model for two-phase transport in capillary porous media in which they viewed the two phases as constituents of a binary mixture, and they derived the conservation equations from the classical separate flow model. The formulation in this model is analogous to conventional multicomponent mixture flow theories and leads to a reduction in the number of differential equations required for the primary variables.

Wu and Pruess [97] developed a three-dimensional, fully implicit, integral finite difference simulator for single- and multi-phase flow of non-Newtonian fluids in porous/fractured media. They implemented their numerical scheme using the TOUGH2 code. They incorporated into the model several rheological models for power-law and Bingham non-Newtonian fluids. Fyhr and Rasmuson [98] developed a multiphase two-dimensional model of the coupled transport of water, vapor, air, and heat in anisotropic hygroscopic porous media. The model was used to study drying kinetics of single wood chips as a function of time and external conditions, such as temperature, pressure and velocity of the superheated steam.

Several models were developed recently to study deformable porous media. Vaziri [99] presented a finite element formulation of a multiphase fluid flow and heat transfer within a deforming porous medium. They applied the model to a one-dimensional problem involving heating of a saturated soil layer and focused on the interdependence between fluid flow, heat flow, and deformation processes. A multi-frontal parallel algorithm to solve coupled heat transfer, water and gas flow in deformable porous media was developed by Xicheng and Schrefler [100]. Their mathematical model made use of the modified effective stress concept combined with the capillary pressure relationship, and it incorporated phase change and latent heat transfer. The authors demonstrated the speed and efficiency of this method and compared it with a general domain decomposition method based on band matrix methods. Zhou et al. [101] developed a mathematical model to simulate coupled heat, moisture, and air transfer in deformable unsaturated porous media. A constitutive model that included thermo-hydro-mechanical coupling effects for a non-isothermal unsaturated medium and fully coupled heat and moisture transfer was used to establish the coupled nonlinear governing equations which were solved using a mixed-type finite element formulation. Their model included heat of wetting, heat sink due to thermal expansion of the medium, phase change between liquid and vapor, and compressibility of liquid water.

Wijeysundera et al. [102] developed a one-dimensional, transient numerical model for simulating heat and moisture transfer through a porous insulation with impermeable adiabatic vertical boundaries, and with one horizontal boundary facing a warm humid ambient and the other facing a cold impermeable surface. They identified four phases of energy and moisture transport processes which were formulated by a system of transient intercoupled equations and several thermodynamic relations using the local volume-averaging technique. Adler and Thovert [103] reviewed finite-difference-based numerical algorithms for the simulation of multiphase transport in real porous media with random geometry.

3.3 Phase Change

Transport phenomena in porous media with phase change exist in a wide variety of environmental and engineering systems including freezing and melting of soils, thermal energy storage, hypothetic nuclear reactor accidents, and alloy solidification. Vafai and Tien [104] presented a numerical study of heat and mass transfer with phase change in porous materials. Their model consisted of a system of transient inter-coupled equations governing the two-dimensional multiphase transport process in porous media. Their solution algorithm allows full simulation without any significant simplifications. They discussed in detail the inter-relationship between the temperature, vapor density, condensation rate, liquid content and the fluid velocity fields.

Several studies have been reported on the problem of solid/liquid phase change over external surfaces embedded in a porous medium. Bakier [105] studied the problem of melting from a vertical flat plate embedded in a porous medium. Both aiding and opposing flows were considered. Their numerical results showed that the melting phenomenon decreased the local Nusselt number at the solid-liquid interface. Sasaguchi et al. [106] presented a numerical model to analyze solid/liquid phase change heat transfer problems around cylinders in a rectangular porous cavity. Their model could treat solid/liquid phase change heat transfer as well as transient natural convection. Viswanath and Jaluria [107] performed a numerical investigation of the solidification phenomenon in an enclosed mold including conduction in the mold wall. They solved the conjugate, transient problem, in which the flow in the liquid was driven by thermal buoyancy using an enthalpy formulation for the energy equation and a porous medium approximation for the region undergoing phase change.

The transient heat transfer behavior during melting of ice in a porous medium within a rectangular enclosure was investigated by Chang and Yang [108] using the SIMPLEC numerical algorithm. Pak and Plumb [109] presented a one-dimensional numerical model of melting of a multi-component mixture in a packed bed containing a mixture of equal size particles of two materials. Stubos et al. [110] performed a numerical simulation of the transient response of a liquid-saturated self-heated porous bed including two-phase flow and phase change heat transfer processes. They used an enthalpy approach to develop a one-domain formulation of the problem, avoiding explicit internal boundary tracking between single- and two-phase regions. Daurelle et al. [111] developed a finite element model of two-dimensional coupled heat and mass transfer with phase change in a porous medium. They applied their model to the study of superheated steam drying.

3.4 Boiling, Evaporation and Condensation

A numerical algorithm for heat transfer problems involving boiling and natural convection in a fluid-saturated porous medium was presented by Ramesh and Torrance [112]. In their algorithm, they used a moving boundary approach to continuously track the interface between the liquid phase and the vapor phase. They used a finite difference control volume method on a deformable grid conforming to the interface. Using a test case, they discussed in detail convergence and accuracy considerations of the algorithm. Forsyth and Simpson [113] presented a numerical study of natural convection in a two-phase, two-component flow in a porous medium heated from below. Wang and Beckermann [114] applied their two-phase mixture model to investigate pressure-driven, two-phase boiling flow along a heated surface embedded in a porous medium. They used approximations analogous to the classical boundary layer theory, and they derived a set of boundary layer equations for two-phase flow which they solved using similarity transformations. They integrated numerically the resulting ordinary differential equations using a combination of the Gear stiff method and a shooting procedure. Stubos et al. [115] analyzed the effect of capillary heterogeneity induced by variation in permeability on the steady state, countercurrent, vapor-liquid flow in porous media.

Buoyancy-driven two-phase flow along a vertical plate in a capillary porous medium was investigated by Wang and Beckermann [116]. They obtained full two-phase similarity solutions for both boiling and condensing flows. Schmidt et al. [117] investigated numerically the problem of steady motion and thermal behavior of a volatile wetting liquid in an open cavity under low gravity. They used a configuration which approximated a two-phase pore of liquid on a wicking structure surface. Thermal non-equilibrium and convection were established by symmetrically superheating or subcooling the pore boundaries.

The problem of film condensation in a porous medium was considered by Majumdar and Tien [118]. In the two-phase zone, their solutions showed a boundary-layer profile for the capillary pressure. They analyzed the liquid zone using three models which assumed either slip or no slip at the wall and Darcy velocity or no shear at the interface within the two-phase zone. Chung et al. [119] presented numerical solutions of condensation in a porous medium near a cold vertical surface. In their numerical model, they assumed that a distinct two-phase zone exists between the liquid and the vapor phases, and they included the effect of the vapor flow in the two-phase zone.

A one-dimensional numerical model of steady-state evaporative heat transfer and vapor/liquid counterflow in a capillary porous medium heated volumetrically or from below was presented by Konstantinou et al. [120]. Their model was used to analyze heat transfer enhancement during boiling of liquid coolants in porous layers. Bouddour et al. [121] used the homogenization method of asymptotic expansions for periodic structures for modelling of heat and mass transfer in wet porous media in the presence of evaporation-condensation.

3.5 Environmental Problems

Several environmental applications involve various phenomena related to multiphase transport in a porous medium. As an application to radioactive waste disposal, Forsyth [122] described a numerical technique for modeling two-phase (liquid and gas), two-component (water and air) nonisothermal flow in fractured porous media. In his model, several phenomena were simulated including interphase mass transfer, latent heat, conduction, convection, gravity, and capillary effects. To circumvent numerical problems associated with phase disappearance/ appearance, he used a rigorous method based on variable substitution. Catton and Chung [123] reviewed two-phase flow processes relevant to in-core severe accident progression from nuclear reactors. They considered two-dimensional dry-out processes and concluded that one-dimensional dry-out limits were adequate for predicting the onset of core melting.

Kissling et al. [124] considered horizontal one-dimensional flow of water and steam through a block of porous material within a geothermal reservoir. Cheng and Wang [125] presented a multiphase mixture model of infiltration and transport of nonaqueous phase liquids (NAPLs) in the unsaturated subsurface. Simultaneous flow of liquid and gas phases, solutal convection, and the associated organic vapor transport in the gas phase were simulated in their numerical model. Wang and Cheng [126] used the same model for numerical analysis of the multidimensional transport phenomena occurring during steam injection in porous media including multidimensional two-phase flow, phase change heat transfer at boiling and condensation fronts, and thermal convection within the phases. Emmert et al. [127] developed a numerical model of steam and hot air injection in a vertical sand-filled column. The model was capable of solving transient multiphase fluid flow and heat transfer in saturated and unsaturated porous media.

4 CONCLUSIONS

A comprehensive review of current computational work involving most of the important aspects of forced convection and multiphase transport in porous media and their applications mainly over the past decade was presented in this chapter. Generalized models which include the important physical phenomena governing convective flow and energy transport in porous media have been presented. Due to space limitations, detailed presentation of the computational techniques used for these models has been omitted, and the reader is referred to related studies which have been reviewed in this chapter. The literature review was presented in a convenient format allowing the reader to identify the latest progress in individual areas of research in this field.

It is hoped that the present review will provide a convenient reference for the current research related to forced convection and multiphase transport processes in porous media and their applications and a better understanding of the current research directions in these areas.

5 REFERENCES

1. C.L. Tien and K. Vafai, Convective and Radiative Heat Transfer in Porous Media, *in Adv. Appl. Mech.*, vol. 27, pp. 225-281, 1990.
2. S. Kakac, S. Kilkis, F.A. Kulacki, and F. Arinc, *Convective Heat and Mass Transfer in Porous Media*, Kluwer Academic, Dordrecht, 1991.
3. D.A. Nield, and A. Bejan, A., *Convection in Porous Media*, Springer-Verlag, New York, 1992.
4. M. Kaviany, *Principles of Heat Transfer in Porous Media*, Springer-Verlag, New York, 1995.
5. D.B. Ingham and I. Pop, *Transport Phenomena in Porous Media*, Pergamon, Elsevier Science Ltd., UK, 1998.
6. S. Whitaker, Simultaneous Heat, Mass and Momentum Transfer in Porous Media: a Theory of Drying, *Adv. Heat Transfer*, vol. 13, pp. 119-203, 1977.
7. K. Vafai, and C.L. Tien, Boundary and Inertia Effects on Flow and Heat Transfer in Porous Media, *Int. J. Heat Mass Transfer*, vol. 24, pp. 195-203, 1981.
8. K. Vafai and M. Sozen, Analysis of Energy and Momentum Transport for Fluid Flow Through a Porous Bed, *J. Heat Transfer*, vol. 112, pp. 690-699, 1990.
9. S. Ergun, Fluid Flow Through Packed Columns, *Chem. Eng. Prog.*, vol. 48, pp. 89-94, 1952.
10. K. Vafai, Convective Flow and Heat Transfer in Variable Porosity Media, *J. Fluid Mechanics*, vol. 147, pp. 233-259, 1984.
11. N. Wakao and S. Kaguei, *Heat and Mass Transfer in Packed Beds*, Gordon and Breach, New York, 1982.
12. F.A.L. Dullien, *Porous Media Fluid Transport and Pore Structure*, Academic Press, New York, 1979.

13. N. Wakao, S. Kaguei, and T. Funazkri, Effect of Fluid Dispersion Coefficients on Particle-to-Fluid Heat Transfer Coefficients in Packed Beds, *Chem. Eng. Sci.,* vol. 34, pp. 325-336, 1979.

14. R.F. Benenati, and C.B. Brosilow, Void Fraction Distribution in Beds of Spheres, *AIChE J.,* vol. 8, pp. 359-361, 1962.

15. S.V. Patankar, *Numerical Heat Transfer and Fluid Flow,* Hemisphere, New York, 1980.

16. W.J. Minkowycz, E.M. Sparrow, G.E. Schneider, R.H. Pletcher, *Handbook of Numerical Heat Transfer,* John Wiley, New York, 1988.

17. T.S. Lundgren, Slow Flow Through Stationary Random Beds and Suspensions of Spheres, *J. Fluids Mech.,* vol. 51, Pt.2, pp. 273-299, 1972.

18. G. Neale and W. Nader, Practical Significance of Brinkman's Extension of Darcy's Law Coupled Parallel Flows Within a Channel and a Bounding Porous Medium, *Can. J. Chem. Eng.,* vol. 52, pp. 475-478, 1974.

19. K. Vafai and S.C. Chen, Analysis of Free Surface Transport Within a Hollow Glass Ampule, *Num. Heat Transfer A,* vol. 22, pp. 21-49, 1992.

20. S.C. Chen and K. Vafai, Analysis of Free Surface Momentum and Energy Transport in Porous Media, *Num. Heat Transfer A,* vol. 29, pp. 281-296, 1996.

21. S.C. Chen and K. Vafai, Non-Darcian Surface Tension Effects on Free Surface Transport in Porous Media, *Num. Heat Transfer A,* vol. 31, pp. 235-254, 1997.

22. V. Srinivasan and K. Vafai, Analysis of Linear Encroachment in Two-Immiscible Fluid Systems in a Porous Medium, *J. Fluids Eng.,* vol. 116, pp. 135-139, 1994.

23. G. Lauriat and K. Vafai, Forced Convective Flow and Heat Transfer Through a Porous Medium Exposed to a Flat Plate or Channel, in *Convective Heat and Mass Transfer in Porous Media,* eds. Kakac et al., Kluwer Academic, Dordrecht, pp. 299-328, 1991.

24. Nakayama, and C.D. Ebinuma, Transient Non-Darcy Forced Convective Heat Transfer from a Flat Plate Embedded in a Fluid-Saturated Porous Medium, *Int. J. Heat Fluid Flow,* vol. 11, pp. 249-253, 1990.

25. X.J. Ling and A. Dybbs, The Effect of Variable Viscosity on Forced Convection Over a Flat Plate Submersed in a Porous Medium, *J. Heat Transfer,* vol. 114, pp. 1063-1065, 1992.

26. J. Thevenin, and D. Sadaoui, About Enhancement of Heat Transfer over a Circular Cylinder Embedded in a Porous Medium, *Int. Comm. Heat Mass Transfer,* vol. 22, pp. 295-304, 1995.

27. J. Thevenin, Transient Forced Convection Heat Transfer from a Circular Cylinder Embedded in a Porous Medium, *Int. Comm. Heat Mass Transfer,* vol. 22, pp. 507-516, 1995.

28. K.J. Nasr, S. Ramadhyani, and R. Viskanta, Numerical Studies of Forced Convection Heat Transfer from a Cylinder Embedded in a Packed Bed, *Int. J. Heat Mass Transfer,* vol. 38, pp. 2353-2366, 1995.

29. M.A. Hossain, N. Banu, and A. Nakayama, Non-Darcy Forced Convection Boundary Layer Flow over a Wedge Embedded in a Saturated Porous Medium, *Num. Heat Transfer, Part A,* vol. 26, pp. 399-414, 1994.

30. K. Vafai, and S.J. Kim, Analysis of Surface Enhancement by a Porous

Substrate, *J. Heat Transfer*, vol. 112, 1990.

31. P.C. Huang, and K. Vafai, Analysis of Flow and Heat Transfer Over an External Boundary Covered with a Porous Substrate, *J. Heat Transfer*, vol. 116, pp. 768-771, 1994.

32. M. Sahraoui and M. Kaviany, Slip and No-slip Temperature Boundary Conditions at the Interface of Porous, Plain Media: Convection, *Int. J. Heat Mass Transfer*, vol. 37, pp. 1029-1044, 1994.

33. M. Sahraoui and M. Kaviany, Slip and No-slip Velocity Boundary Conditions at the Interface of Porous, Plain Media, *Int. J. Heat Mass Transfer*, vol. 35, pp. 927-943, 1992.

34. P.C. Huang, and K. Vafai, Flow and Heat Transfer Control Over an External Surface Using a Porous Block Array Arrangement, *Int. J. Heat Mass Transfer*, vol. 36, pp. 4019-4032, 1993.

35. P.C. Huang and K. Vafai, Passive Alteration and Control of Convective Heat Transfer Utilizing Alternate Porous Cavity-Block Wafers, *Int. J. Heat Fluid Flow*, vol. 15, pp. 48-61, 1994.

36. K. Vafai and P.C. Huang, Analysis of Heat Transfer Regulation and Modification Employing Intermittently Emplaced Porous Cavities, *J. Heat Transfer*, vol. 116, pp. 604-613, 1994.

37. W.S. Fu and H.C. Huang, Thermal Performance of Different Shape Porous Blocks Under an Impinging Jet, *Int. J. Heat Mass Transfer*, vol. 40, pp. 2261-2272, 1994.

38. B. Abu-Hijleh, Convection Heat Transfer from a Laminar Flow over a 2-D Backward Facing Step with Asymmetric and Orthotropic Porous Floor Segments, *Num. Heat Transfer, Part A, Applications*, vol. 31, pp. 325-335, 1997.

39. A.V. Shenoy, Non-Newtonian Fluid Heat Transfer in Porous Media, *Adv. Heat Transfer*, vol. 24, pp. 101-190, 1994.

40. A.V. Shenoy, Darcy Natural, Forced, and Mixed Convection Heat Transfer from an Isothermal Vertical Flat Plate Embedded in a Porous Medium Saturated with an Elastic Fluid of Constant Viscosity, *Int. J. Eng. Sci.*, vol. 30, pp. 455-467, 1992.

41. F.M. Hady, and F.S. Ibrahim, Forced Convection Heat Transfer on a Flat Plate Embedded in Porous Media for Power-law Fluids, *Transp. Porous Media*, vol. 28, pp. 125-134, 1997.

42. A.Nakayama and I. Pop, A Unified Similarity Transformation for Free, Forced, and Mixed Convection in Darcy and non-Darcy Porous Media, *Int. J. Heat Mass Transfer*, vol. 34, pp. 357-367, 1991.

43. Nakayama and A.V. Shenoy, A Unified Similarity Transformation for Darcy and non-Darcy Forced, Free, and Mixed Convection Heat Transfer in non-Newtonian Inelastic Fluid-Saturated Porous Media, *Chem. Eng. J.*, vol. 50, pp. 33-45, 1992.

44. K. Vafai and A. Amiri, Non-Darcian Effects in Confined Forced Convective Flows, in *Transport Phenomena in Porous Media*, eds. D.B. Ingham and I. Pop, pp. 313-329, 1998.

45. D.Poulikakos, and K.J. Renken, Forced Convection in a Channel Filled With

Porous Medium, Including the Effects of Flow Inertia, Variable Porosity, and Brinkman Friction, *J. Heat Transfer*, vol. 109, pp. 880-888, 1987.

46. Amiri, and K. Vafai, Analysis of Dispersion Effects and Non-thermal Equilibrium, Non-Darcian, Variable Porosity Incompressible Flow through Porous media, *Int. J. Heat Mass Transfer*, vol.37, pp. 939-954, 1994.

47. M.L. Hunt, and C.L. Tien, Non-Darcian Convection in Cylindrical Packed Beds, *J. Heat Transfer*, vol. 96, pp. 635-641, 1988.

48. M.L. Hunt, and C.L. Tien, Effects of Thermal Dispersion on Forced Convection in Fibrous Media, *Int. J. Heat Mass Transfer*, vol. 31, pp. 301-309, 1988.

49. C.T. Hsu, and P. Cheng, Thermal Dispersion in a Porous Medium, *Int. J. Heat Mass Transfer*, vol. 33, pp. 1587-1597, 1990.

50. P. Cheng, and H. Zhu, Effects of Radial Thermal Dispersion on Fully-developed Forced Convection in Cylindrical Packed Tubes, *Int. J. Heat Mass Transfer*, vol. 30, pp. 2373-2383, 1987.

51. P. Cheng, and D. Vortmeyer, Transverse Thermal Dispersion and Wall Channeling in a Packed Bed with Forced Convective Flow, *Chem. Eng. Sci.*, vol. 43, pp. 2523-2532, 1988.

52. P. Cheng, C.T. Hsu, and A. Chowdhury, Forced Convection in the Entrance Region of a Packed channel with Asymmetric Heating, *J. Heat Transfer*, vol. 110, pp. 946-954, 1988.

53. M. Sozen, and K. Vafai, Longitudinal Heat Dispersion in Porous Beds with Real Gas Flow, *AIAA J. Thermophysics Heat Transfer*, vol. 7, pp. 153-157, 1993.

54. P. Adnani, I. Catton, and M.A. Abdou, Non-Darcian Forced Convection in Porous Media with Anisotropic Dispersion, *J. Heat Transfer*, vol. 117, pp. 447-451, 1995.

55. F. Kuwahara, A. Nakayama, and H. Koyama, Study on Thermal Dispersion in Fluid Flow and Heat Transfer in Porous Media (Numerical Prediction of Thermal Dispersion Using a Two-dimensional Structural Model), *Trans. Japan Soc. Mech. Eng., Part B*, vol. 62, pp. 3118-3124, 1996.

56. W.S. Fu, and H.C. Huang, Effects of a Random Porosity Model on Heat Transfer Performance of Porous Media, *Int. J. Heat Mass Transfer*, vol. 42, pp. 13-25, 1999.

57. A. Amiri, K. Vafai, and T.M. Kuzay, Effects of Boundary Conditions on Non-Darcian Heat Transfer Through Porous Media and Experimental Comparisons, *Num. Heat Transfer, Part A*, vol. 27, pp. 651-664, 1995.

58. A. Amiri and K. Vafai, Transient Analysis of Incompressible Flow Through a Packed Bed, *Int. J. Heat Mass Transfer*, vol. 41, pp. 4259-4279, 1998.

59. D.Y. Lee, and K. Vafai, Analytical Characterization and Conceptual Assessment of Solid and Fluid Temperature Differentials in Porous Media, *Int. J. Heat Mass Transfer*, vol. 42, pp. 423-435, 1999.

60. A. Hadim, Forced Convection in a Porous Channel with Localized Heat Sources, *J. Heat Transfer*, vol. 116, pp. 465-472, 1994.

61. M. Sozen, and T.M. Kuzay, Enhanced Heat Transfer in Round Tubes with Porous Insert, *Int. J. Heat Fluid Flow*, vol. 17, pp. 124-129, 1996.

62. A.R. Martin, C. Saltiel, J. Chai, W. Shyy, Convective and Radiative Internal Heat Transfer Augmentation with Fiber Arrays, *Int. J. Heat Mass Transfer*, vol. 41, pp. 3431-3440, 1998.

63. S. Kuo, and C. Tien, Heat Transfer Augmentation in a Foam-Material Filled Duct With Discrete Heat Sources, *Proc. Thermal Phenomena in the Fabrication and Operation of Electronic Components*, Los Angeles, CA, May 11-13, pp. 87-91, 1988.

64. G.J. Hwang, C.C. Wu, and C.H. Chao, Investigation of Non-Darcian Forced Convection in an Asymmetrically Heated Sintered Porous Channel, *J. Heat Transfer*, vol. 117, pp. 725-732, 1995.

65. J.L. Lage, A.K. Weinert, D.C. Price, and R.M. Weber, Numerical Study of a Low Permeability Microporous Heat Sink for Cooling Phased-Array Radar Systems, *Int. J. Heat Mass Transfer*, vol. 39, pp. 3633-3647, 1996.

66. A.G. Fedorov and R. Viskanta, A Numerical Simulation of Conjugate Heat Transfer in an Electronic Package Formed by Embedded Discrete Heat Sources in Contact with a Porous Heat Sink, *J. Heat Transfer*, vol. 119, pp. 8-16, 1997.

67. P.C. Huang, and K. Vafai, Analysis of Forced Convection Enhancement in a Channel Using Porous Blocks, *AIAA J. Thermophysics Heat Transfer*, vol. 8, pp. 563-573, 1994.

68. S. Chikh, A. Boumedien, K. Bouhadef, and G. Lauriat, Analysis of Fluid Flow and Heat Transfer in a Channel with Intermittent Heated Porous Blocks, *Heat Mass Transfer/ Waerme- und Stoffuebertragung*, vol. 33, pp. 405-413, 1998.

69. Y. Ould-Amer, S. Chikh, K. Bouhadef, and G. Lauriat, Forced Convection Cooling Enhancement by Use of Porous Materials, *Int. J. Heat Fluid Flow*, vol. 19, pp. 251-258, 1998.

70. W. Fu, H. Huang, and W. Lion, Thermal Enhancement in Laminar Channel Flow with a Porous Block, *Int. J. Heat Mass Transfer*, vol. 39, pp. 2165-2175, 1996.

71. T.W. Tong, and M.C. Sharatchandra, Heat Transfer Enhancement Using Porous Inserts in Heat Transfer and Flow in Porous Media, *ASME HTD-Vol. 156*, pp. 41-46, 1990.

72. H.A. Hadim, and A. Bethancourt, Numerical Study of Forced Convection in a Partially Porous Channel With Discrete Heat Sources, *ASME J. Electronic Packaging*, vol. 117, pp. 46-51, 1995.

73. Z. Guo, S.Y. Kim, and H.J. Sung, Pulsating Flow and Heat Transfer in a Pipe Partially Filled With a Porous Medium, *Int. J. Heat Mass Transfer*, vol. 40, pp. 4209-4218, 1997.

74. M.A. Al-Nimr and M.K. Alkam, Unsteady non-Darcian Forced Convection Analysis in an Annulus Partially Filled with a Porous Material, *J. Heat Transfer*, vol. 19, pp. 799-804, 1997.

75. M.K. Alkam and M.A. Al-Nimr, Transient Non-Darcian Forced Convection Flow in a Pipe Partially Filled With a Porous Material, *Int. J. Heat Mass Transfer*, vol. 41, pp. 347-356, 1998.

76. M. Sozen and K. Vafai, Analysis of Oscillating Compressible Flow Through

a Packed Bed, *Int. J. Heat Fluid Flow*, vol. 12, pp. 130-136, 1991.

77. M. Sozen and K. Vafai, Thermal Charging and Discharging of Sensible and Latent Heat Storage Packed Beds, *AIAA J. Thermophysics Heat Transfer*, vol. 5, pp. 623-625, 1991.

78. M.A. Al-Nimr, T. Aldoss, and M.I. Naji, Transient Forced Convection in the Entrance Region of a Porous Tube, *Canadian J. Chem. Eng.*, vol 72, pp. 249-255, 1994.

79. M.A. Al-Nimr, T. Aldoss, and M.I. Naji, Transient Forced Convection in the Entrance Region of Porous Concentric Annuli, *Canadian J. Chem. Eng.*, vol. 72, pp. 1092-1096, 1994.

80. V. Srinivasan, K. Vafai, and R.N. Christensen, Analysis of Heat Transfer and Fluid Flow Through a Spirally Fluted Tube Using a Porous Substrate Approach, *J. Heat Transfer*, vol. 116, pp. 543-551, 1994.

81. A.Nakayama, and A.V. Shenoy, Non-Darcy Forced Convective Heat Transfer in a Channel Embedded in a Non-Newtonian Inelastic Fluid-saturated Porous Medium, *Can. J. Chem. Eng.*, vol. 71, pp. 168-173, 1993.

82. G. Chen and H.A. Hadim, Numerical Study of Non-Darcy Forced Convection in a Packed Bed Saturated With a Power-Law Fluid, *J. Porous Media*, vol. 1, pp. 147-157, 1998.

83. G. Chen, and H.A. Hadim, Forced Convection of a Power-law Fluid in a Porous Channel – Integral Solutions, *J. Porous Media*, vol. 2, pp. 59- 69, 1999.

84. G. Chen, and H.A. Hadim, Forced Convection of a Power-law Fluid in a Porous Channel – Numerical Solutions, *Heat Mass Transfer/Waerme-und Stoff.*, vol. 34, pp. 221-228, 1998.

85. M.K. Alkam, M.A. Al-Nimr, and Z. Mousa, Forced Convection of Non-Newtonian Fluids in Porous Concentric Annuli, *Int. J. Num. Methods Heat Fluid Flow*, vol. 8, pp. 703-716, 1998.

86. F.C. Chou, W.Y. Lien, and S.H. Lin, Analysis and Experiment of Non-Darcian Convection in Horizontal Square Packed-sphere Channels - I. Forced Convection, *Int. J. Heat Mass Transfer*, vol. 35, pp. 195-205, 1992.

87. G. Chen and H.A. Hadim, Numerical Study of Three-Dimensional Non-Darcy Forced Convection in a Square Porous Duct, *Int. J. Num. Methods Heat Fluid Flow*, vol. 9, pp. 151-169, 1999.

88. M. Sozen and K. Vafai, Analysis of Non-Thermal Equilibrium Condensing Flow of a Gas Through a Packed Bed, *Int. J. Heat Mass Transfer*, vol. 32, pp. 1247-1261, 1990.

89. K. Vafai and M. Sozen, An investigation of a Latent Heat Storage Porous Bed and Condensing Flow Through It, *J. Heat Transfer*, vol. 112, pp. 1014-1022, 1990.

90. K.S. Udell and J. Fitch, Heat Transfer in Capillary Porous Media Considering Evaporation, Condensation and Non-Condensable Gas Effects, in *Heat Transfer in Porous Media and Particulate Flows*, ASME HTD-vol. 46, pp. 103-110, 1985.

91. M. Kaviany and Mittal, Funicular State in Drying of a Porous Slab, *Int. J. Heat Mass Transfer*, vol. 30, pp. 1407-1418, 1987.

92. B.W. Gamson, G. Thodos, and O.A. Hougen, Heat Mass and Momentum Transfer in the Flow of Gases Through Granular Solids, *Trans. AIChE*, vol. 39, pp. 1-35.

93. K. Vafai and M. Sozen, A Comparative Analysis of Multiphase Transport Models in Porous Media, *Ann. Rev. Heat Transfer*, vol. 3, pp. 145-162, 1990.

94. V.K. Dhir, Boiling and Two-Phase Flow in Porous Media, *Ann. Rev. Heat Transfer*, vol. 5, pp. 303-350, 1994.

95. C.Y. Wang and P. Cheng, Multiphase Flow and Heat Transfer in Porous Media, *Advances in Heat Transfer*, vol. 30, 1997.

96. C.Y. Wang and C. Beckermann, Two-Phase Mixture Model of Liquid-Gas Flow and Heat Transfer in Capillary Porous Media - I. Formulation, *Int. J. Heat Mass Transfer*, vol. 36, pp. 2747 - 2758, 1993.

97. Y.S. Wu, and K. Pruess, Numerical Method for Simulating Non-Newtonian Fluid Flow and Displacement in Porous Media, 11th Int. Conf. on Comp. Meth. in Water Res., Cancun, Mexico, CMWR'96., Part 1, vol. 1, pp. 109-117, 1996.

98. C.Fyhr, and A. Rasmuson, Mathematical Model of Steam Drying of Wood Chips and Other Hygroscopic Porous Media, *AIChE J.*, vol. 42, pp. 2491-2502, 1996.

99. H.H. Vaziri, Theory and Application of a Fully Coupled Thermo-hydro-mechanical Finite Element Model, *Comp. and Structures*, vol. 61, pp. 131-146, 1996.

100. W. Xicheng and B.A. Schrefler, Multi-frontal Parallel Algorithm for Coupled Thermo-hydro-mechanical Analysis of Deforming Porous Media, *Int. J. Num. Meth. Eng.*, vol. 43, pp. 1069-1083, 1998.

101. Y. Zhou, R.K.N.D. Rajapakse, and J. Graham, Coupled Heat-Moisture-Air Transfer in Deformable Unsaturated Media, *J. Eng. Mech.*, vol. 124, pp. 1090-1099, 1998.

102. N.E. Wijeysundera, B.F. Zheng, M. Iqbal, and E.G. Hauptmann, Numerical Simulation of the Transient Moisture Transfer Through Porous Insulation, *Int. J. Heat Mass Transfer*, vol. 39, pp. 995-1004, 1996.

103. P.M. Adler and J.F. Thovert, Real Porous Media: Local Geometry and Macroscopic Properties, *App. Mech. Rev.*, vol. 51, pp 537-585, 1998.

104. K. Vafai and C.L. Tien, Numerical Investigation of Phase Change Effects in Porous Materials, *Int. J. Heat Mass Transfer*, vol. 32, pp. 1261-1277, 1989.

105. A.Y. Bakier, Aiding and Opposing Mixed Convection Flow in Melting From a Vertical Flat Plate Embedded in a Porous Medium, *Transp. Porous Media*, vol. 29, pp. 127-139, 1997.

106. K. Sasaguchi, K. Kusano, and R. Viskanta, Numerical Analysis of Solid-Liquid Phase Change Heat Transfer Around a Single and Two Horizontal, Vertically Spaced Cylinders in a Rectangular Cavity, *Int. J. Heat Mass Transfer*, vol. 40, pp. 1343-1354, 1997.

107. R. Viswanath, and Y. Jaluria, Numerical Study of Conjugate Transient Solidification in an Enclosed Region, *Num. Heat Transfer, Part A*, vol. 27, pp. 519-536, 1995.

108. W.J. Chang, and D.F. Yang, Natural Convection for the Melting of Ice in Porous Media in a Rectangular Enclosure, *Int. J. Heat Mass Transfer*, vol. 39, pp. 2333-2348, 1996.

109. J. Pak and O.A. Plumb, Melting in a Two-Component Packed Bed, *J. Heat Transfer*, vol. 119, pp. 553-559, 1997.

110. A.K. Stubos, C. Perez Caseiras, J.M. Buchlin, and N.K. Kanellopoulos, Numerical Investigation of Vapor-Liquid Flow and Heat Transfer in Capillary Porous Media, *Num. Heat Transfer; Part A*, vol. 31, pp., 1997.

111. J.V. Daurelle, F. Topin, and R. Occelli, Modeling of Coupled Heat and Mass Transfer With Phase Change in a Porous Medium: Application to Superheated Steam Drying, *Num. Heat Transfer A*, vol. 33, pp. 39-63, 1998.

112. P.S. Ramesh and K.E. Torrance, Numerical Algorithm for Problems Involving Boiling and Natural Convection in Porous Materials, *Num. Heat Transfer B*, vol.17, pp. 1-24, 1990.

113. P.A. Forsyth and R.B. Simpson, Two-Phase, Two-Component Model for Natural Convection in a Porous Medium, *Int. J. Num. Meth. Fluids*, vol. 12, pp. 655-682, 1991.

114. C.Y. Wang and C. Beckermann, Two-phase Mixture Model of Liquid-Gas Flow and Heat Transfer in Capillary Porous Media-II. Application to Pressure-Driven Boiling Flow Adjacent to a Vertical Heated Plate, *Int. J. Heat Mass Transfer*, vol. 36, p 2759-2768, 1993.

115. A.K. Stubos, C. Satik, and Y.C. Yortsos, Effects of Capillary Heterogeneity on Vapor-Liquid Counterflow in Porous Media, *Int. J. Heat Mass Transfer*, vol. 36, pp. 967-976, 1991.

116. C.Y. Wang and C. Beckermann, Boundary Layer Analysis of Buoyancy-Driven Two-Phase Flow in Capillary Porous Media, *J. Heat Transfer*, vol. 117, pp. 1082-1087, 1995.

117. G.R. Schmidt, T.J. Chung, and A. Nadarajah, Thermocapillary Flow With Evaporation and Condensation at Low Gravity, Part 1. Non-Deforming Surface, *J. Fluid Mech.*, vol. 294, pp. 323-347, 1995.

118. A.Majumdar and C.L. Tien, Effects of Surface Tension on Film Condensation in a Porous Medium, *J. Heat Transfer*, vol. 112, pp. 751-757, 1990.

119. J.N. Chung, O.A. Plumb, and W.C. Lee, Condensation in a Porous Region Bounded by a Cold Vertical Surface, *J. Heat Transfer*, vol. 114, pp. 1011-1018, 1992.

120. N.D. Konstantinou, A.K. Stubos, and J.C. Statharas, N.K. Kanellopoulos, and A.Ch. Papaioannou, Enhanced Boiling Heat Transfer in Porous Layers With Application to Electronic Component Cooling, *J. Enhanced Heat Transfer*, vol. 4, pp. 175-186, 1997.

121. A.Bouddour, J.L. Auriault, M. Mhamdi-Alaoui, and J.F. Bloch, Heat and Mass Transfer in Wet Porous Media in Presence of Evaporation-Condensation, *Int. J. Heat Mass Transfer*, vol. 41, pp. 2263-2277, 1998.

122. P.A. Forsyth, Radioactive Waste Disposal Heating Effects in Unsaturated Fractured Rock, *Num. Heat Transfer A*, vol. 17, pp. 29-51, 1990.

123. I.Catton, and M. Chung, Two-Phase Flow in Porous Media With Phase Change: Post Dryout Heat Transfer and Steam Injection, *Nuclear Eng. Design*, vol. 151, pp. 185-202, 1994.

124. W. Kissling, M. McGuinness, A. McNabb, G. Weir, S. White, and R. Young, Analysis of One-Dimensional Horizontal Two-Phase Flow in Geothermal Reservoirs, *Trans. Porous Media*, vol. 7, pp. 223-253, 1992.

125. P. Cheng and C.Y. Wang, Multiphase Mixture Model for Multiphase, Multicomponent Transport in Capillary Porous Media - II. Numerical Simulation of the Transport of Organic Compounds in the Subsurface, *Int. J. Heat Mass Transfer*, vol. 39, pp. 3619-3632, 1996.

126. C.Y. Wang, and P. Cheng, Multidimensional modeling of steam injection into porous media, *J. Heat Transfer*, vol. 120, pp 286-290, 1998.

127. M. Emmert, R. Helmig, and W. Baechle, Numerical Modelling of Steam Injection in a Vertical Sand-Filled Column, Proc. 27th Congress of the International Association of Hydraulic Research, IAHR. Part C, Aug 10-15 1997 vol. C, San Francisco, CA, pp. 107-112, 1997.

OVERVIEW OF CURRENT COMPUTATIONAL STUDIES OF HEAT TRANSFER IN POROUS MEDIA AND THEIR APPLICATIONS – NATURAL AND MIXED CONVECTION

K. Vafai
H. Hadim

1 INTRODUCTION

The topic of buoyancy-driven convection in fluid-saturated porous media has received significant research interest motivated by a wide range of geophysical and engineering applications including geothermal energy extraction, groundwater resource management, nuclear waste disposal, building thermal insulation, grain storage, enhanced oil recovery, metal casting (alloy solidification), and heat transfer in electronic equipment, among many others. Reviews of several topics within this broad research area can be found in earlier books presented by Kakac et al. [1], Nield and Bejan [2], Kaviany [3] and more recently Ingham and Pop [4]. An earlier extensive review of relevant studies related to thermal convection in porous media was also reported by Tien and Vafai [5].

Most of the analytical and numerical work presented in the literature is based on the Darcy-Oberbeck-Boussinesq formulation. The most common reason for the wide use of the Darcian formulation is its simplicity in linearizing the momentum equations, thus removing a considerable amount of difficulty in solving the governing equations. However, it is well established now that Darcy's law is valid only when the pore Reynolds number is of the order of unity. For many applications, Darcy's law is not valid, and the non-Darcy effects of inertia within the porous matrix and viscous shear at the boundary must be taken into consideration. Furthermore, additional important physical phenomena related to variable porosity, thermal dispersion, and local thermal non-equilibrium must be taken into consideration. A detailed review of these phenomena and their related studies on

331

forced convection is presented in a previous chapter of this volume [6] and will not
be repeated here. A large number of recent investigations have been focused on
developing robust numerical models that are able to simulate these various physical
aspects.

In this chapter, an up-to-date review of current computational work
involving most of the important aspects of natural and mixed convection in
porous media and their applications mainly over the past decade is presented. This
chapter is divided into two major sections related to natural and mixed convection
in porous media. Generalized models which include all the important physical
phenomena governing convective flow and energy transport in porous media are
presented. Due to space limitations, detailed numerical simulations of these
models have been omitted, and the reader is referred to related studies which have
been reviewed in this chapter. The literature review in each section is divided into
individual sub-topics related to geometrical configurations of fundamental and
practical interest, important physical phenomena, major computational models,
and practical applications.

2 NATURAL CONVECTION IN POROUS MEDIA

2.1 Mathematical Formulation and Numerical Solution

As reported previously [7, 8], the complexity of a porous structure hampers a
general formulation of the detailed 'microscopic' transport phenomena.
Therefore, the point transport equations are commonly integrated over a
representative elementary volume, which accommodates the fluid and the solid
phases in a porous structure. This technique is known as the volume-averaging
method and is widely considered in formulating the transport processes in porous
media [5-7]. However, this technique requires information on a number of
empirical transport coefficients that are supplemented by well-established
experimental findings and analytical solutions for some rather simplified cases.

2.1.1 The General Model.
The governing equations for natural and mixed
convection in porous media are summarized in this section. The thermophysical
properties of the solid and the fluid are assumed to be constant except for the
density variation in the buoyancy term. Later on, this assumption is relaxed for a
model of natural convection with phase change in porous media. Also, it is
assumed that the solid matrix and the fluid are in local thermodynamic
equilibrium. Viscous dissipation and compressional work are negligible. The
equations governing the conservation of mass, momentum, and energy in their
general form (Cartesian coordinates) can be written as [6-9]

$$\nabla \cdot \vec{u} = 0 \tag{1}$$

$$\frac{\rho_f}{\varepsilon} \frac{D\vec{u}}{Dt} = -\nabla P + \rho_f \vec{g} \beta(T - T_\infty) - \frac{\mu_f}{K}\vec{u} - \frac{\rho_f F \varepsilon}{\sqrt{K}}\left|\vec{u}\right|\vec{u} + \mu_{eff}\nabla^2\vec{u} \tag{2}$$

$$\sigma \frac{\partial T}{\partial t} + \vec{u} \cdot \nabla T = \alpha_{eff} \nabla^2 T \tag{3}$$

where

$$|\vec{u}| = \sqrt{u^2 + v^2}$$

As usual, the Richardson number ($R_i = Gr/Re^2$) characterizes mixed convection where the Grashof number ($Gr = g\beta KL\Delta T/v^2$) and the Reynolds number ($Re = VL/v$) represent the dynamism of the natural and the forced convection effects. The limiting values of $R_i \rightarrow 0$ and $R_i \rightarrow \infty$ correspond to the forced and the natural convection limits, respectively.

In the above equations, $\vec{u}$ is the velocity vector, P the pressure, $\vec{g}$ the gravitational vector, T the temperature, ρ_f the fluid density, K the permeability of the porous medium, α_{eff} the effective thermal diffusivity, μ_f the viscosity of the fluid, μ_{eff} the apparent viscosity of the medium, ε the porosity of the porous medium, β the volume thermal expansion coefficient of the fluid, σ the thermal capacitance of the porous matrix, F a geometric function (empirical) which depends on the Reynolds number and the microstructure of the porous medium, and L is the characteristic length. The effective viscosity of the fluid-saturated porous medium μ_{eff} associated with the Brinkman term in the momentum equation may have a different value than the fluid viscosity, μ_f. However, it is customary to assume that $\mu_{eff} = \mu_f$ [8]. Also, the thermal dispersion effect is accounted for by lumping it with the thermal conductivity [6]. The term on the left-hand side of Eq. 2 is the convective term, which may be negligible for certain cases of convective flow through porous media [8]. On the right-hand side of Eq. 2, the first term is the fluid pressure gradient term, the second one is the buoyancy term, the third one is the linear Darcy term, the fourth one is the Forchheimer term (due to the nonlinear inertia effects within the porous medium), and the last one is the Brinkman term which takes into account viscous shear at the boundary. These various effects have been discussed in more detail in [6] and in the related studies reviewed in this chapter.

In two-dimensional cartesian coordinates, upon cross differentiation, eliminating pressure in the momentum equations, and introducing the stream function and the vorticity, the vorticity transport and stream function equations in their general forms can be written as [10]

$$\frac{\rho_f}{\varepsilon} \frac{D\zeta}{Dt} = \mu_f \nabla^2 \zeta - \frac{\mu_f}{K}\zeta - \frac{\rho_f F\varepsilon}{\sqrt{K}}(\frac{\partial}{\partial y}(|\vec{u}|u) - \frac{\partial}{\partial x}(|\vec{u}|v)) - \rho_f g \beta \frac{\partial T}{\partial x} \tag{4}$$

$$\nabla^2 \Psi = \zeta \tag{5}$$

where

$$\zeta = \frac{\partial u}{\partial y} - \frac{\partial v}{\partial x} \tag{6}$$

and

$$u = \frac{\partial \Psi}{\partial y} \quad , \quad v = -\frac{\partial \Psi}{\partial x} \tag{7}$$

The permeability of the porous structure K and the geometric function F in the momentum equation are inherently tied to the porous structure and are generally based on experimental findings such that no universal representation does exist. For a randomly packed bed of spheres, such transport coefficients as reported by Ergun [11] could be expressed in terms of porosity ε and the particle diameter d_p as follows [12]:

$$F = \frac{1.75}{\sqrt{150\varepsilon^3}} \tag{8}$$

$$K = \frac{\varepsilon^3 d_p^2}{150(1-\varepsilon)^2} \tag{9}$$

The experimental investigations by Benenati and Brosilow [13] show a distinct porosity variation in packed beds. Their results show a high porosity region ($\approx 0.5 d_p$ from the wall), which then decays in an oscillatory fashion while moving away from the wall until an asymptotic free stream porosity value is reached at about $5 d_p$ from the wall. It is common practice to consider an exponential decaying function since the oscillatory behavior is considered to be a secondary effect as shown by Amiri and Vafai [14]. A typical form of such an exponential decay is given by

$$\varepsilon = \varepsilon_\infty [1 + a\, e^{(by/d_p)}] \tag{10}$$

The free stream porosity value ε_∞ is a function of the bed-to-particle D_h/d_p ratio and varies between 0.259 and 0.43, depending on the packing arrangement. In addition, for uniform solid sphere particles, the values of a =1.7 and b=6 were found to yield close approximation to the experimental data.

2.1.2 Numerical Solution. Several numerical techniques are available for solving the nonlinear governing equations and associated boundary conditions for the general model described above. These well-established techniques belong to the three main categories including: finite difference, finite volume, and finite element. Spectral element methods have also been used in a few cases. Due to space limitations, a detailed presentation of all these techniques is outside of the main scope of the present overview and the reader may refer to the studies reviewed in the following sections where the related numerical solution techniques have been identified as much as possible for convenience. In addition, there is a large number of references which discuss these techniques in more detail such as the comprehensive handbook edited by Minkowycz et al. [15].

Some important issues related to numerical grid generation for problems involving natural and mixed convection in porous media are discussed here. Numerical solutions of natural and mixed convection in porous media have shown the existence of viscous sublayers which are adjacent to the solid boundary and whose thickness depends on the Rayleigh (or Richardson) number and the Darcy number. As a result of imposing no-slip boundary conditions in porous media, steep velocity gradients exist near the walls in the convection regimes such that a large number of grid points are required near the solid boundary to adequately resolve the flow field. However, away form the impermeable boundary, such small grid size is not needed as it may generate inaccuracies (due to accumulation of rounding errors) and stability problems. Hence variable grid size is required to predict the flow and temperature fields accurately. To prevent potential numerical instability and inaccuracy, abrupt changes in mesh size should be avoided. Appropriate stretching functions or coordinate transformations can be used to increase the resolution in the viscous sublayers.

2.2 Analysis of Natural Convective Heat and Mass Transfer with Phase Change in Porous Materials

Several applications of natural convection in porous media involve phase change. Such applications include building insulation, heat exchangers, grain storage, drying, and geothermal systems, among others. For example, the performance of a porous insulation may be influenced significantly by condensation of the water vapor when the latter reaches the saturation point at the corresponding temperature. In such cases, the porous medium consists of three phases: the solid matrix, the liquid water, and a binary gas phase composed of air and water vapor. The main transport phenomena include vapor diffusion due to vapor concentration gradients, bulk natural convection due to density gradients induced by temperature gradients, and heat conduction in all three phases. An extensive review of the literature revealed that most of the existing models for this problem involve various simplifying assumptions. A more general model developed by Vafai and Tien [16] is presented in this section.

2.2.1 Generalized Model The governing equations for a generalized model of natural convection in a porous medium with phase change are derived using the local volume-averaging technique. With this technique, two different averages of a quantity are defined [7]. The local volume average $\langle \Psi \rangle$ for a quantity Ψ is defined as

$$\langle \Psi \rangle = \frac{1}{V} \int_V \Psi \, dV \tag{11}$$

where V is an averaging volume which is bounded by a closed surface in the porous medium. The averaging volume V is composed of three phases, the solid phase $V_\sigma(t)$, the liquid phase $V_\beta(t)$, and the gas phase $V_\gamma(t)$. The second averaging quantity is the intrinsic phase average, which is given by

$$\langle \Psi_\alpha \rangle^\alpha = \frac{1}{V_\alpha(t)} \int_{V_\alpha(t)} \Psi_\alpha dV \tag{12}$$

where Ψ_α is a quantity associated with the α phase. Using these definitions, the nondimensional governing equations for natural convection in porous media with phase change are derived as [16]:

Thermal energy equation:

$$\frac{\partial \langle T \rangle}{\partial t} + \frac{P_1 P_2 P_{18}}{P_{19}} \psi_\varepsilon \langle \vec{v}_\beta \rangle \cdot \nabla \langle T \rangle + \frac{P_3 P_4 P_{18} Pe}{P_{19}} \langle \rho_\gamma \rangle^\gamma \langle \vec{v}_\gamma \rangle \cdot \nabla \langle T \rangle$$

$$+ \frac{P_{18}}{P_{19}} \langle \dot{m} \rangle = P_{18} \nabla^2 \langle T \rangle + \frac{P_{18}}{P_{19}} \nabla P_{19} \cdot \nabla \langle T \rangle \tag{13}$$

Liquid phase equation of motion:

$$\langle \vec{v}_\beta \rangle = - K_{r\beta}(\nabla \varepsilon_\beta + \psi_T \nabla \langle T \rangle - \psi_g \vec{g}) \tag{14}$$

Liquid phase continuity equation:

$$\frac{\partial \varepsilon_\beta}{\partial t} + \psi_\varepsilon \nabla \cdot \langle \vec{v}_\beta \rangle + \frac{1}{P_1 P_6} \langle \dot{m} \rangle = 0 \tag{15}$$

Gas phase equation of motion:

$$\langle \vec{v}_\gamma \rangle = P_{20} K_{r\gamma}(- \nabla \langle p_\gamma \rangle^\gamma + P_5 \langle \rho_\gamma \rangle^\gamma \vec{g}) \tag{16}$$

Gas phase continuity equation:

$$\frac{\partial}{\partial t}(\varepsilon_\gamma \langle \rho_\gamma \rangle^\gamma) + Pe \nabla \cdot (\langle \rho_\gamma \rangle^\gamma \langle \vec{v}_\gamma \rangle) - \frac{1}{P_4 P_6} \langle \dot{m} \rangle = 0 \tag{17}$$

Gas phase diffusion equation:

$$\frac{\partial}{\partial t}(\varepsilon_\gamma \langle \rho_v \rangle^\gamma) + Pe \nabla \cdot (\langle \rho_v \rangle^\gamma \langle \vec{v}_\gamma \rangle) - \frac{1}{P_4 P_6 P_{11}} \langle \dot{m} \rangle$$

$$= \frac{1}{Le} \nabla \cdot [\langle \rho_\gamma \rangle^\gamma \nabla(\frac{\langle \rho_v \rangle^\gamma}{\langle \rho_\gamma \rangle^\gamma})] \tag{18}$$

Volume constraint:

$$\varepsilon_\sigma + \varepsilon_\beta + \varepsilon_\gamma = 1 \tag{19}$$

Thermodynamic relations:

$$\left(P_v\right)^\gamma = P_9\left(\rho_v\right)^\gamma \langle T\rangle \tag{20}$$

$$\left(P_a\right)^\gamma = P_9\left(\rho_a\right)^\gamma \langle T\rangle \tag{21}$$

$$\left(\rho_\gamma\right)^\gamma = P_{11}\left(\rho_v\right)^\gamma + P_{12}\left(\rho_a\right)^\gamma \tag{22}$$

$$\left(P_\gamma\right)^\gamma = P_{13}\left(P_v\right)^\gamma + P_{14}\left(P_a\right)^\gamma \tag{23}$$

$$\left(\rho_{v,s}\right)^\gamma = \frac{1}{P_9\langle T\rangle}\exp\left(-\frac{P_{15}+P_{16}}{\langle T\rangle}+\frac{P_{16}}{\langle T_0\rangle}\right) \tag{24}$$

The variable properties in the porous matrix are:

$$\overline{k}_{eff} \cong \varepsilon_\sigma \overline{k}_\sigma + \varepsilon_\beta \overline{k}_\beta + \varepsilon_\gamma \frac{\left(\overline{k}_v\left(\overline{\rho}_v\right)^\gamma + \overline{k}_a\left(\overline{\rho}_a\right)^\gamma\right)}{\left(\left(\overline{\rho}_v\right)^\gamma + \left(\overline{\rho}_a\right)^\gamma\right)} \tag{25}$$

$$\overline{\rho} = \varepsilon_\sigma \overline{\rho}_\sigma + \varepsilon_\beta \overline{\rho}_\beta + \varepsilon_\gamma\left(\left(\overline{\rho}_v\right)^\gamma + \left(\overline{\rho}_a\right)^\gamma\right) \tag{26}$$

$$\overline{C}_p = \frac{\varepsilon_\sigma \overline{\rho}_\sigma \overline{c}_\sigma + \varepsilon_\beta \overline{\rho}_\beta \overline{c}_\beta + \varepsilon_\gamma\left(\left(\overline{\rho}_v\right)^\gamma \overline{c}_v + \left(\overline{\rho}_a\right)^\gamma \overline{c}_a\right)}{\overline{\rho}} \tag{27}$$

$$\overline{\alpha}_{eff} = \frac{\overline{k}_{eff}}{\overline{\rho}\,\overline{C}_p} \tag{28}$$

In the above equations, the main variables of interest are the temperature T, the liquid volume fraction ε_β, the vapor density ρ_v, the gas density ρ_γ, the gas volume fraction ε_γ, and the condensation rate $\dot{m}$. The variables with a bar on top of them refer to dimensional quantities. The dimensionless variables and non-dimensional parameters Pe, Le, ψ_ε, ψ_g, ψ_T and P_j's are defined in [16].

In arriving at the governing equations, three common assumptions were made. These are: (1) the porous insulation material is homogeneous and isotropic; (2) the three phases in the porous medium are in local thermodynamic equilibrium; and (3) Darcy's flow model is valid in describing the motion of the gas phase and of the liquid phase. The first assumption is the common simplifying procedure to rationally tackle problems for heat and mass transport in porous materials. The second assumption can be justified based on the work of Amiri and Vafai [14]. The third assumption is justified as follows: Vasseur et al. [17] used the results of Vafai and Tien [8] to examine the validity of Darcy's law. Based on their results, two conditions which characterize the inertia and the boundary effects should be satisfied if the results obtained from Darcy's law are to be within a 10% error band. These two conditions are

$$\overline{U} < \frac{6 \times 10^{-3}\overline{v}}{(1 - \varepsilon_\sigma)F\sqrt{\overline{K}}} \quad \text{and} \quad \overline{L} > \text{Pr} \sqrt{\overline{K}/(1 - \varepsilon_\sigma)} \quad (29)$$

where $\overline{U}$ is the characteristic fluid velocity, $\overline{v}$ the kinematic viscosity, ε_σ the porosity, F the geometric function described previously, $\overline{L}$ the characteristic length of the porous material, and Pr is the Prandtl number. Based on the typical thermophysical data for applications such as porous insulation, these two conditions are satisfied. The thermal dispersion effects are also neglected for applications of this type based on Amiri and Vafai [14]. Aside from these assumptions, the governing equations are very general and can be applied to a large class of problems in heat and mass transfer in porous materials. Also, unlike most studies of buoyancy-induced flow in porous media, the Boussinesq approximation has not been invoked in the present model.

It is noteworthy that the effective gas permeability $\overline{K}_\gamma$ and the effective liquid permeability $\overline{K}_\beta$ for partially liquid saturated media can be expressed in terms of the permeability $\overline{K}$ and the relative permeabilities, $K_{r\beta}$ and $K_{r\gamma}$, as

$$\begin{aligned} \overline{K}_\gamma &= K_{r\gamma}\overline{K} \\ \overline{K}_\beta &= K_{r\beta}\overline{K} \end{aligned} \quad (30)$$

Based on the relative permeability model suggested by Wyllie [18] that agrees well with previous work [19, 20], the relative permeabilities are taken to have the following forms:

$$\begin{aligned} K_{r\beta} &= s^3 \\ K_{r\gamma} &= (1 - s)^3 \end{aligned} \quad (31)$$

where

$$s = \frac{s_\beta - s_{\beta p}}{1 - s_{\beta p}}, \quad \text{and} \quad s_\beta = \frac{\varepsilon_\beta}{\varepsilon_\beta + \varepsilon_\gamma} \quad (32)$$

The variable $s_{\beta p}$ is the saturation of the liquid in the pendular state in the porous medium. Below this saturation, the liquid is essentially immobile due to no inter-pore connections. There were no concrete experimental data available for $s_{\beta p}$; however, a value of 0.1 was found to be a reasonable one [16].

2.2.2 Numerical Solution. A finite difference technique can be used to solve numerically the governing equations described above and their associated boundary conditions. However, the techniques wich necessitate a conservative form of the governing equations cannot be used in the present formulation whose governing equations are too complicated to be cast in conservative form without a loss of accuracy. Consequently, an explicit scheme is used to obtain high accuracy such that the spatial derivatives are discretized by central differencing, except for the convective terms which are approximated by an upwind differencing scheme. However, based on extensive trial and error, it was found that central differencing is more apropriate for the convective term in the gas phase continuity equation for better stability. Depending whether phase change occurs or not, the numerical scheme exhibits two different formats in time and space. Since the complex physical phenomena for this problem are highly transient, the required time step size must be quite small. In the present case, numerical experimentation with the accuracy and numerical stability of various versions of upwind differencing (e.g. first order, third order, third order combined with fourth order artificial viscosity, etc.) has shown that first order upwind differencing is the most appropriate for the present case. Further details of the numerical scheme can be found in [16].

2.3 Natural Convection over External Boundaries

An important fundamental research topic of thermal convection in porous media is related to natural convection over solid surfaces, which are embedded in porous media. For problems involving standard geometry and boundary conditions, simplified analytical methods based on the Darcy flow model have been used such as similarity solutions and the perturbation technique. For problems dealing with complex geometry or boundary conditions and/or non-Darcy effects, more detailed numerical simulations are required. A summary of the literature related to these numerical simulations is presented in Table 1.

2.3.1 Vertical Plate. The vertical flat plate is the most fundamental configuration used to study natural convection over external surfaces in porous media. Among the initial studies of this problem, Cheng and Minkowycz [21] presented similarity solutions using a boundary layer formulation based on Darcy's law. Hong et al. [22] included both boundary and inertia effects as well as the convective term in the momentum equation. Subsequently, several studies focused on various important physical phenomena such as the non-Darcy effects, transient phenomena, and conjugate natural convection.

Kim and Vafai [9] obtained analytical and numerical solutions for buoyancy-driven fluid flow and heat transfer from the vertical plate considering both the constant wall temperature and the constant wall heat flux cases. They discussed in detail the interplay between the velocity boundary layer (VBL) and the thermal boundary layer (TBL). Theses are illustrated in Fig. 1 based on an order of magnitude analysis for the cases when the thickness of the TBL, ε_T, is much larger than that of the VBL, ε_V (Fig. 1a), and vice versa (Fig. 1b). A typical velocity profile for each case is also shown.

Table 1 Summary of Computational Studies of Natural Convection in Porous Media – External Flows
(UWT = uniform wall temperature; UHF = uniform wall heat flux; VWT = variable wall temperature)

Author	Geometry	Boundary Conditions	Inertia Effects	Boundary Effects	Other Effects	Numerical Solution
Kim and Vafai [9]	vertical plate	UWT, UHF		*		analytical
Ettefagh et al. [10]	cavity	UWT	*	*		finite difference
Bradeen et al. [28]	vertical plate	VWT				"
Bradeen et al.[29]	"	VWT			transient	"
Rees [30]	"	VWT				Keller box
Harris et al. [31]	"	UHF			transient	finite difference
Chamkha [33]	"	UWT		*	magnetic field	"
Bassom and Rees [39]	vertical cylinder	VWT				Keller box
Oosthuizen and Naylor [40]	"	UWT		*	partially porous	finite element
Bradeen et al. [44]	horizontal plate	VWT	*			finite difference
Rees [45]	"	UWT				Keller box
Hossain and Rees [46]	"	VWT	*			
Angirasa and Peterson [47]	"	UWT				finite difference
Pop et al. [49]	horizontal cylinder	UHF	*	*	transient	match. asymp. exp.
Nguyen et al. [50]	"	UWT			"	spectral
Bradeen et al. [51]	"	UWT	*	*	"	finite difference
Yan et al. [53]	sphere	UWT, UHF			"	"
Yang and Wang [55]	axi-symmetric shapes	UWT			power-law fluid	Runge-Kutta
Shu and Pop [58]	line heat source	adiabatic				finite difference

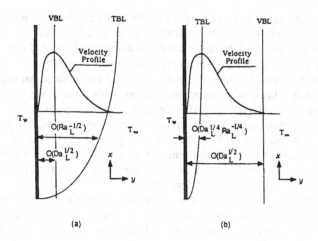

Figure 1 Hydrodynamic and thermal boundary layers for natural convection over a flat plate in a porous medium: (a) $\varepsilon_T \gg \varepsilon_V$; (b) $\varepsilon_T \ll \varepsilon_V$ (From Kim and Vafai [9]; Copyright 1989 Elsevier Science Ltd – Reprinted with permission).

For the same configuration, Chen and Chiou [23] studied the non-Darcian effects and conjugate natural convection. Vynnycky and Kimura [24] also investigated the problem of conjugate natural convection and later [25] considered the transient effects. Four different mathematical models describing transient non-Darcy natural convection from a corrugated plate embedded in an enclosed porous medium were presented by Hsiao [26], and the results obtained using a finite difference technique were compared to those for a flat plate. Non-Darcy boundary layer convective flow over a wavy surface was investigated by Rees and Pop [27].

    Besides the standard thermal boundary conditions of uniform temperature or uniform heat flux, other thermal boundary conditions have been considered for both steady-state and transient cases. Bradean et al. [28, 29] considered the case of a sinusoidally heated or cooled vertical plate. They obtained analytical solutions for small times then they used a finite difference technique based on the fully implicit Crank-Nicholson method to solve the problem for a wide range of the Rayleigh number. Beyond a specified value of the Rayleigh number, they observed a cellular flow separating from the plate. Rees [30] considered the effect of a non-uniform surface temperature distribution along the vertical plate. He solved the governing equations for the three-dimensional boundary layer using a combination of a spectral decomposition in the spanwise direction and the Keller box method which is a well-established technique for studying nonsimilar boundary layer flows [15]. He argued that this method is faster and more accurate than a full finite difference discretization which would require a very fine grid for this type of problem. He found that the effect of nonuniform heating became confined to a thin layer of uniform thickness embedded within the main growing boundary layer. Harris et al. [31] obtained numerical solutions of the full boundary-layer equations governing transient natural convection due to a sudden change in heat flux applied on the vertical plate. They obtained analytical solutions for small times when the transient effects are confined to a thin region near the surface, then they used a finite difference based numerical solution for subsequent times until steady-state is reached.

In several practical applications involving natural convection, the effects of thermal radiation heat transfer often become significant since they are often comparable to or sometimes larger than natural convection effects. Hossain and Pop [32] investigated the effects of thermal radiation on buoyancy-induced Darcy flow of an optically dense, viscous, incompressible fluid along an isothermal flat surface. They implemented a Rosseland diffusion approximation and an implicit finite difference method together with the Keller box technique. Both the streamwise and the normal components of the buoyancy force were retained in the momentum equations. Chamkha [33] analyzed natural convection in a porous medium bounded by a semi-infinite, ideally transparent, vertical flat plate exposed to solar radiation. The resulting approximate nonlinear ordinary differential equations based on the local similarity method were solved by a standard implicit iterative finite difference method.

Various other physical phenomena have been investigated for natural convection near a vertical surface embedded in a porous medium. Telles and Trevisan [34] studied the effects of hydrodynamic dispersion using an enhanced form of the Runge-Kutta algorithm. They obtained similarity solutions for the convective flows promoted by the density variation due to the combination of temperature and concentration. Angirasa and Peterson [35] investigated the case when the vertical plate is exposed to a stably stratified, fluid-saturated, low-porosity medium in which Darcy flow prevailed. Their results indicated that the ambient thermal stratification had a significant effect on the flow and temperature fields. The effects of viscous dissipation were studied by Murthy and Singh [36] using a non-Darcy model. Pascal and Pascal [37] used similarity transformations to investigate free convection of a non-Newtonian power-law fluid with yield stress along a permeable vertical surface (flat plate or circular cylinder) embedded in a porous medium. They derived the governing modified Darcy's law from the non-linear rheology of the non-Newtonian fluid. The problem of natural convection boundary layer flow along an inclined plate embedded in a thermally stratified porous medium in the presence of a nonuniform transverse magnetic field was considered by Chamkha [38]. He conducted finite-difference-based numerical simulations to investigate the influence of various phenomena including non-Darcy effects, the Hartmann and Hall effects of magneto-hydrodynamics, and the thermal stratification in the porous medium.

2.3.2 Vertical Cylinder. The vertical cylinder is another configuration of fundamental as well as practical interest. Bassom and Rees [39] presented asymptotic solutions and numerical simulations using the Keller box method and the Newton-Raphson algorithm for the governing nonsimilar boundary layer equations for convective flow induced by a heated vertical cylinder in a fluid-saturated porous medium. As an application to hot water pipes within structural components of buildings, Oosthuizen and Naylor [40] investigated natural convection from a cylinder placed on the vertical centerline of a square enclosure partly filled with a fluid-saturated porous medium. They used a vorticity-stream function formulation to generate the governing equations which they solved using a Galerkin-based finite element procedure. They analyzed the effects of the size of the fluid gap at the top of the enclosure on the average heat transfer rate from the cylinder. Hossain et al. [41] studied conjugate free convection along a vertical cylindrical fin in a non-Newtonian fluid-saturated porous medium. They used a finite difference method to solve the boundary layer equations based on the modified Darcy's law for power-law fluids. They investigated the effects of the power-law index, conjugate convection-conduction parameter, and surface curvature parameter on heat transfer.

2.3.3 Horizontal Plate. Natural convection from a horizontal plate facing upward or downward in a fluid-saturated porous medium is one of the fundamental heat transfer phenomena which occurs in geothermal reservoirs. Rees and Pop [42] studied the effects of stationary surface waves on the free convection induced by a horizontal uniformly heated surface in a porous medium. Conjugate free convection effects were studied by Lesnic et al. [43], who used the boundary-layer approximation to obtain numerical and asymptotic solutions based on a stream function – temperature formulation. Bradean et al. [44] considered the case of a sinusoidally heated or cooled horizontal surface maintained at a constant temperature. Rees [45] studied the case when the effects of inertia on the free convection boundary layer flow were significant and solved the resulting nonsimilar boundary layer equations using the Keller box method. He found that the inertia effects dominated near the leading edge, but Darcy flow was reestablished further downstream. Using a similar numerical technique, Hossain and Rees [46] analyzed the inertia effects for the case when the surface temperature was assumed to display a power-law variation with distance from the leading edge. For natural convection from a heated plate facing upward, Angirasa and Peterson [47] used the alternating direction implicit (ADI) finite difference technique and successive overrelaxation and determined the range of the Rayleigh number for which two-dimensional flow occurs as being within $40 \leq Ra \leq 600$. They found that for higher Rayleigh number, the flow became three-dimensional with multiple plume formation, while at lower Rayleigh number the transport was dominated by conduction.

2.3.4 Horizontal Cylinder. Natural convection from a horizontal cylinder in a fluid-saturated porous medium is another problem of fundamental as well as practical interest which occurs in thermal insulation, underground cable systems, and steam and water distribution lines. For example, cylindrical shapes have been used in catalytic beds and for nuclear waste disposal in sub-sea beds.

Using a Darcy-flow model, Facas [48] presented numerical solutions for natural convection from a pipe with two baffles attached along its surface while it is buried beneath a semi-infinite-fluid-saturated porous medium. This complicated geometry was handled through the use of a body-fitted curvilinear coordinate system. Pop et al. [49] used the method of matched asymptotic expansions to investigate transient free convection boundary-layer flow on a horizontal circular cylinder exposed to a uniform surface heat flux. Bradean et al. [50] analyzed unsteady natural convection adjacent to an impulsively heated horizontal circular cylinder in a porous medium. Their results showed that as convection became increasingly more dominant, a single hot cell of fluid formed vertically above the cylinder and rapidly moved away from the cylinder as time increased. Nguyen et al. [51] considered the case when a surface reaction occurs on the cylinder. The convective phenomena driven by temperature gradients via heat release from an nth-order irreversible reaction were modeled using the Brinkman-Forchheimer equation combined with the Boussinesq-Oberbeck approximation. They obtained numerical solutions using a Fourier spectral element method in conjunction with a semi-implicit method combining the second-order Adams-Bashforth and first-order backward Euler schemes for advancing the solution in time. Their results showed a strong influence of natural convection on the physicochemical process.

2.3.5 Other External Surfaces.   Among studies involving various other geometries, Kimura and Pop [52] studied conjugate natural convection about a sphere in a fluid-saturated porous medium, and later Yan et al. [53] studied the transient effects for the same problem. Pop and Na [54] studied natural convection along an isothermal frustum of a wavy cone embedded in a fluid-saturated porous medium. They solved the boundary layer equations using the Keller-box method. Yang and Wang [55] solved the problem of natural convection of a non-Newtonian power-law fluid with or without yield stress about a two-dimensional or axi-symmetric body of arbitrary shape in a fluid-saturated porous medium. They used the boundary layer approximation assuming a high modified Rayleigh number and obtained similarity solutions using the fourth-order Runge-Kutta scheme and a shooting method. Kimura et al. [56] studied conjugate natural convection from plates and other bodies in a fluid-saturated porous medium. They considered several different configurations including slender bodies, rectangular slabs, horizontal cylinders and spheres.

2.3.6 Heat Sources. Configurations involving natural convection in the presence of heat sources are of practical interest in nuclear waste disposal. Masuoka et al. [57] studied the development of a buoyant plume arising from a line heat source. Their detailed numerical analysis indicated that the lateral flow induced near the interface caused deviation from the similarity solutions. Shu and Pop [58] investigated natural convection from inclined wall plumes arising from a line thermal source embedded at the leading edge of an adiabatic plate with arbitrary tilt angle in a fluid-saturated porous medium. They solved the governing nonsimilar equations using the finite difference technique and they studied the effect of the tilt angle on the velocity and temperature profiles. By employing local similarity and the modified Keller Box method, Leu and Jang [59] studied natural convection from a point heat source embedded in a non-Darcian porous medium. Their results indicated that the non-Darcy effects decreased the centerline velocity and temperature and increased the velocity and thermal boundary layer thicknesses. Facas [60] used the Darcy model to study natural convection from an elliptic heat source buried beneath a semi-infinite saturated porous medium. The complicated geometry for this problem was handled using the finite-difference method and a body-fitted curvilinear coordinate system. The numerical simulations indicated that the boundary-layer approximations could not be employed for low ellipse aspect ratio and that the slender orientation resulted in higher heat transfer rates than the blunt orientation.

2.4 Natural Convection in Porous Enclosures

The topic of natural convection in porous enclosures is divided into two main categories: enclosures heated from the side and enclosures heated from below. Various flow and heat transfer patterns are possible depending on the imposed thermal boundary condition, aspect ratio, and geometry of the enclosure. A previous review of existing experimental and numerical studies of this topic was presented by Combarnous and Bernard [61] while Cheng et al. [62] reported a review of numerical studies of variable porosity and thermal dispersion effects. The latter included configurations consisting of rectangular enclosures heated from below or from the side as well as heated horizontal cylinders embedded in an enclosed porous cavity. A summary of the recent literature related to natural convection in porous enclosures is presented in Table 2.

2.4.1 Vertical an Inclined Enclosures.
Earlier studies of natural convection in vertical porous enclosures were based on Darcy's law for modeling flow within the enclosure [1,2]. However, during the past two decades, several studies have shown that the non-Darcy effects of inertia within the porous matrix and viscous shear at the boundary, as well as variable porosity and thermal dispersion play significant roles in natural convection inside the enclosure. Earlier studies of inertia and boundary effects were presented by Beckermann et al. [63] and Lauriat and Prasad [64] for the case of a vertical rectangular enclosure with isothermal vertical walls. They used the Brinkman-Forchheimer-extended Darcy model and clearly demonstrated the importance of non-Darcy effects, especially for high Darcy numbers (Da > 10-7). Prasad and Tuntomo [65] used the Forchheimer-extended Darcy equation to study in detail the inertia effects. Recently, Satya Sai et al. [66] used the finite element technique based on a semi-implicit, operator-splitting method to investigate in detail non-Darcy effects in vertical enclosures of axisymmetric geometry.

Table 2 Summary of Computational Studies of Natural Convection in Porous Media – Internal Flows
(UWT = uniform wall temperature; UHF = uniform wall heat flux; VWT = variable wall temperature)

Author	Geometry	Boundary Conditions	Inertia Effects	Boundary Effects	Other Effects	Numerical Solution
Lauriat and Prasad [64]	rectangular cavity	UWT	•	•		finite difference
Satya Sai et al. [66]	rect./annular cavity	UWT, UHF	•	•		finite element
Degan and Vasseur [70]	rectangular cavity	UWT			anisotropic	finite difference
Degan and Vasseur [71]	"	UHF		•		"
Beckermann et al. [73]	"	UWT	•		partially porous	finite volume
Antohe and Lage [78]	"	VHF			transient	finite element
Nithiarasu et al. [80]	"	convective				finite element
Bian et al. [81]	"	UWT	•		power-law fluid	Runge-Kutta
Getachew et al. [82]	"	UWT			"	finite volume
Hadim and Chen [83]	"	UWT	•		"	"
Hsiao et al. [84]	inclined cavity	heat source		•	thermal dispersion	finite difference
Bian et al. [85]	"	UWT		•	elect/magnetic field	finite volume
Ettefagh and Vafai [86]	open-ended cavity	UWT				"
Nguyen et al. [88]	spherical enclosure	UWT			transient	spectral element
Mamou et al. [95]	horizontal layer	VWT				finite difference
Ozoe et al. [111]	rectangular corner	UWT			three-dimensional	finite difference
Dawood and Burns [112]	rectangular box	convective		•		finite difference
Goyeau et al. [114]	rectangular cavity	UWT	•		double diffusive	finite volume
Nithiarasu et al. [115]	"	UWT		•	"	finite element

In some porous media, thermophysical properties such as porosity, permeability and thermal conductivity may vary directionally or randomly within the porous medium. Anisotropic porous media are encountered in various industrial and geophysical applications including fibrous materials (e.g. insulation, filters), sedimentary soils, rock formations, some biological materials, dendritic solidification of multi-component mixtures, and preforms of aligned ceramic or graphite fibers used in casting of metal matrix composites. Ni and Beckermann [67] studied natural convection in a vertical enclosure filled with a homogeneous porous medium that is both hydrodynamically and thermally anisotropic, while Chang and Hsiao [68] considered a similar problem for a vertical cylinder. As a fundamental study towards understanding the interacting flows between the melt pool and the mushy zone of solidifying alloys, Song and Viskanta [69] considered the case when the vertical enclosure is partially filled with an anisotropic porous medium. Degan and Vasseur [70, 71] considered a porous medium in which the principal axes of permeability are oriented in a direction that is oblique to the gravity vector. They used a Darcy flow model and developed a stream function-temperature formulation. Their numerical results obtained using the ADI scheme and a successive over-relaxation (SOR) method confirmed the flow features anticipated by an approximate boundary layer solution and showed that the anisotropic properties of the porous medium modified the flow pattern and the heat transfer rate from the ones expected under isotropic conditions. Murthy et al. [72] studied the effects of surface undulations on natural convection from an isothermal surface in a Darcian fluid-saturated porous enclosure. They used the finite element method on a graded nonuniform mesh system. The flow-driving Rayleigh number as well as the geometrical parameters of wave amplitude, wave phase, and the number of waves considered were found to influence the flow and heat transfer process in the enclosure.

The problem of natural convection in enclosures partially filled with porous media occurs in several practical applications such as porous insulation, geothermal systems, and metal casting. This problem involves coupling of the governing momentum and energy equations in the nonporous (fluid) region with the corresponding ones in the porous region through an appropriate set of matching conditions at the fluid/porous-layer interface. This is discussed in detail by Beckermann et al. [73] who used a control volume formulation based on the SIMPLER algorithm to study various configurations of the composite (fluid-porous) system in a vertical enclosure. Sathe et al. [74] reported a comparative numerical study of various insulation schemes for reducing natural convection in vertical rectangular enclosures. The schemes included a fully porous enclosure and a partially porous enclosure with or without an impermeable partition. The flow in the porous region was modeled using the Brinkman-extended Darcy model to account for no-slip at the walls and at the interface. Du and Bilgen [75] used the Brinkman-extended Darcy model to study the case when the enclosure is partially filled with a uniform heat-generating saturated porous medium. They analyzed in detail the interactions between the flow and temperature fields depending on various parameters including Rayleigh number, Darcy number, aspect ratio, porous layer location, asymmetric cooling, and filling factor.

Several studies of natural convection in vertical porous enclosures considered boundary conditions other than the standard ones involving constant temperature or constant heat flux at the wall. Storesletten and Pop [76] presented similarity solutions of the two-dimensional free convection flow in a porous medium bounded by two vertical walls maintained at non-uniform temperatures. The temperature variation of the walls was selected such that similarity solutions for the flow and temperature fields were possible. They considered both symmetric and asymmetric heating at the walls. Using both numerical and asymptotic methods, Rees and Lage [77] considered the case of natural convection flow in a rectangular porous container where the impermeable bounding walls are held at a temperature which is a linearly decreasing function of height. Antohe and Lage [78] considered a rectangular enclosure filled with a fluid-saturated porous medium exposed to time-periodic heating from the side in the horizontal direction. The same authors [79] also studied the Prandtl number effects for a similar problem. Their numerical simulations confirmed the existence of a preferred heat-pulsating frequency that compared well with the theoretical estimates. Nithiarasu et al. [80] used a Galerkin finite element method combined with a Eulerian velocity correction procedure to investigate non-Darcy natural convection in a vertical enclosure exposed to a convective boundary condition on one wall. They performed the study for a wide range of Rayleigh number (10^2 - 10^9), Darcy number (10^{-7} - 10^{-2}), and aspect ratio (1-10).

Non-Newtonian fluids in porous media can be found in several engineering applications including enhanced oil recovery, ceramic engineering, polymer composite molding, filtration, biomechanics, and food technology. Various studies of natural convection in a vertical enclosure filled with a porous medium saturated by a non-Newtonian (power-law) fluid have been reported [81-83]. Bian et al. [81] used a modified Darcy model for power-law fluids. Getachew et al. [82] derived a similar model using the volume-averaging technique. Hadim and Chen [83] presented a detailed numerical study using a modified Brinkman-Forchheimer-extended Darcy model for power-law fluids. They used a finite volume numerical technique based on the SIMPLER algorithm and presented the results for a wide range of Rayleigh number, Darcy number, and power-law index of the fluid.

Vertical enclosures are a special case of the more general category of inclined enclosures. Hsiao et al. [84] considered the case of an inclined porous cavity with a discrete heat source on one wall. They included non-Darcy and thermal dispersion effects as well as the effects of variable porosity. The governing equations expressed in a body-fitted coordinate system were solved numerically by a finite difference method. Bian et al. [85] studied the effect of an electromagnetic field on free convection in an inclined rectangular porous cavity saturated with an electrically-conducting fluid.

Several studies dealt with other configurations of enclosures filled with a fluid-saturated porous medium. Among them, buoyancy-induced flows in open-ended cavities can be found in various applications such as connections between reservoirs, nuclear waste repositories, solar thermal receiver systems, and brake housings commonly used in various aircraft models. This configuration was considered by Ettefagh and Vafai [86] who studied the transient natural convection due to a temperature differential in the open-ended cavity for the case when the Darcy-Rayleigh number Ra ≤ 350. They investigated the effects of important variables including the aspect ratio, the temperature difference, and the Darcy-Rayleigh number on the flow and temperature fields and the Nusselt number in the cavity. Their numerical results indicated that the presence of external corners augments flow instability within the cavity while decreasing the aspect ratio has a stabilizing effect on the flow field. For the same configuration, Ettefagh et al. [10] analyzed in detail non-Darcy effects within the open-ended cavity.

Among other configurations, Dharma Rao et al. [87] studied non-Darcy effects on natural convection heat transfer in a vertical porous annulus. Nguyen et al. [88] studied transient natural convection in a spherical enclosure containing a central core fluid and a porous shell fully saturated with the same fluid. They used the Darcy-Brinkman model and obtained numerical solutions using a hybrid spectral method, which combined the Galerkin and collocation techniques. Time advancement was accomplished by combined Adams-Bashforth and backward Euler schemes. The numerical results showed significant variations along the porous-fluid interface in terms of various governing parameters. However, the overall heat flux was only sensitive to the ratio of the solid thermal conductivity to that of the fluid.

Various other applications of natural convection in porous enclosures have been investigated. Vasseur et al. [89] studied the effects of a transverse magnetic field on buoyancy-driven convection in an inclined two-dimensional cavity, and Alchaar et al. [90] considered a similar problem involving an electrically-conducting fluid in the presence of a magnetic field. Ozaki et al. [91] investigated characteristics of a small heat storage vessel packed with phase-change material (PCM) encapsulated into a spherical hollow. The heat storage vessel was dealt with as a porous medium, and the flow of the working fluid was analyzed using a modified Darcy model which takes into account both the buoyancy effects and the channeling effects.

2.4.2 Horizontal Porous Layer. Natural convection through horizontal porous layers, fully or partially heated from below, occurs in applications such as geologic repository for the storage of nuclear waste and in some geothermal operations. Caltagirone and Fabrie [92] used a Darcy flow model and a pseudo-spectral numerical method based on Fourier functions to investigate the case when the horizontal porous layer is heated from below at high Rayleigh number. For a similar configuration, Prasad and Kladias [93] used the Brinkman-Forchheimer-extended Darcy model to investigate in detail the non-Darcy effects including the effects of variable porosity and conductivity ratio. Their model predicted the existence of an asymptotic convection regime where the flow and heat transfer solutions were independent of permeability of the porous matrix. Vadasz et al. [94] investigated the case when the porous medium is heated from below or from above while it is bounded by perfectly conducting isothermal side walls. Numerical solutions for identical uniform temperatures imposed on both side walls showed that when heating from below, a sub-critical flow resulted mainly near the side walls, which amplified and extended over the entire domain under supercritical conditions. Mamou et al. [95] considered a horizontal saturated porous layer exposed to a sinusoidal temperature distribution on the lower boundary, which simulates a moving wave. Their results showed that, for a specified Rayleigh number, the Benard cells moved with the imposed wave if the velocity of the latter remained below a critical value. Beyond this critical value, there was still an entrainment of the cells, but at a much lower rate. Further, the cell motion was irregular and time-periodic. The effects of anisotropy of the effective thermal conductivity tensor were investigated by Howle and Georgiadis [96]. They found that an increase in the degree of anisotropy leads to reduced heat transfer.

Some studies of natural convection in a horizontal porous layer considered fluids other than water. For a non-Newtonian fluid, Bian et al. [97] conducted their study using a modified Darcy model for power-law fluids. Bian et al. [98] studied natural convection in a shallow horizontal porous layer saturated by an electrically-conducting fluid, to which a transverse magnetic field was applied. They used matched asymptotic expansions for the limiting cases and performed a more detailed numerical study for a wide range of governing parameters including the Darcy-Rayleigh number, cavity aspect ratio, and Hartmann number.

A horizontal cylinder or annulus is another common configuration used to study natural convection in porous enclosures. Barbosa Mota and Saatdjian [99] investigated natural convection in a horizontal eccentric porous annulus. Using a very fine grid for this configuration, they solved the parabolic-elliptic system using a second-order finite difference scheme based on the implicit alternating-direction method coupled with successive under-relaxation. Chen et al. [100] used a vorticity-stream function formulation in a body-fitted coordinate system to investigate a similar problem including the transient effects. Pan and Lai [101] considered layered porous annuli and investigated the appropriateness of using an effective permeability for layered porous media. Zhao et al. [102] studied natural convection in a fluid-saturated porous medium confined in a horizontal circular cylinder which is rotating about its axis. They considered the case of isothermal boundary conditions and uniform internal heat sink. Both numerical and perturbation results indicated that rotation significantly decreased the amplitude of fluid particle trajectories in the radial direction and thus reduced the overall heat transfer. For a horizontal porous annulus heated from the inner surface, Charrier-Mojtabi [103] studied two-dimensional and three-dimensional natural convection using a Fourier-Galerkin approximation for the periodic azimuthal and axial directions and a collocation-Chebyshev approximation in the confined radial direction.

2.4.3 Other Applications. Several computational studies have been reported for various other applications of natural convection in porous enclosures. Pop et al. [104] presented numerical solutions for natural convection in a Darcian fluid confined in the region of a horizontal, L-shaped corner formed by a heated isothermal vertical plate connected to a horizontal surface which was either adiabatic or held at ambient temperature. Debeda et al. [105] studied the problem of natural convection in a fissured porous layer with permeable top and heated from below. They used a local multigrid method in order to take local variations induced by the fissure into account. The method was based on a correction of the fluxes at the interface between the different grids. Mullis [106] presented numerical modeling of natural convection in porous, permeable media including sheets, wedges and lenses to simulate more closely geological conditions. The model was also extended to cover other simple geometry for which analytical solutions were not available. Chiorean and Chiorean [107] presented a parallel-computing algorithm based on the Gauss-Seidel iterative method for the analysis of free convection in an inclined porous enclosure bounded by four rigid isothermal walls. Using linear stability theory, Zhao et al. [108] presented a progressive asymptotic approach procedure associated with the finite element method for the analysis of steady-state natural convection problems in fluid-saturated porous media heated from below. With this method, they considered a series of modified Horton-Rogers-Lapwood problems in which gravity was assumed to tilt a small angle away from the vertical. An extensive review of the fundamental theory of rotating flows in porous media was reported by Vadasz [109]. He identified two main classes of problems: (1) rotating flows in isothermal heterogeneous porous systems and (2) natural convection in homogeneous non-isothermal porous systems. He presented a few examples of solutions to selected problems.

2.4.4 Three-dimensional Studies. Few three-dimensional investigations of natural convection in porous enclosures have been reported, and most of them were based on a Darcy flow model. Rao et al. [110] used a Galerkin scheme for the numerical study of three-dimensional natural convection in a fluid-saturated porous annulus heated from the inner surface. They showed that the secondary cells consisting of three-dimensional, closed, co-axial double helices extended along the axial direction in the top region of the annulus which led to an increase in the overall heat transfer compared with that for the two-dimensional unicellular flow. Ozoe et al. [111] used a vorticity-stream function formulation based on a Darcy flow model to study three-dimensional natural convection at a rectangular corner. In their work, four different extrapolation schemes for the average Nusselt number were implemented. Dawood and Burns [112] investigated three-dimensional natural convection in a rectangular parallelepiped filled with a fluid-saturated porous medium. They used a multigrid method to enhance the convergence of their numerical model. Yamaguchi et al. [113] studied three-dimensional combined natural convection and thermal radiation in a long and wide vertical porous layer with a hexagonal honeycomb core. They used Darcy's law for fluid flow and Rosseland's approximation for radiation heat transfer. The numerical methodology was based on an algebraic coordinate transformation technique and the SIMPLER algorithm.

2.4.5 Double-Diffusive Convection. Double-diffusive convection occurs when heat and mass transfer take place within a fluid layer such that a combined convection phenomenon results from both temperature and concentration gradients. Heat and mass transfer in porous media can be found in many applications such as packed bed chemical reactors, enhanced oil recovery and nuclear waste disposal. Several studies have been reported for the problem of double-diffusive natural convection in a porous cavity. Goyeau et al. [114] studied this problem using the Darcy-Brinkman model and a finite volume approach. Nithiarasu et al. [115] used a generalized model and a finite-element-based numerical method to study both Darcy and non-Darcy flow and heat transfer in an axi-symmetric cavity. A comparative study between several different flow models for double diffusion in a porous cavity was performed by Karimi-Fard et al. [116]. These models include Darcian flow, Forchheimer's extension, Brinkman's extension, and the generalized flow model which was described earlier. The coupled equations were solved using a finite volume approach with a projection algorithm for the momentum equation. A comparison between the different flow models is illustrated in Fig. 2 where it is shown clearly how the divergence between the models increases with increasing Darcy number. However, for the case of double-diffusive convection, it is shown that the Brinkman model and the generalized model are very close, indicating that the inertia effects are negligible. The authors also investigated the effects of the Lewis number on heat transfer for a wide range of Prandtl number. As illustrated in Fig. 3, the generalized model and Brinkman's extension produce results that are very close. For double-diffusive convection, it is shown in Fig. 3 that there exists a critical value of the Lewis number at which heat transfer is maximized.

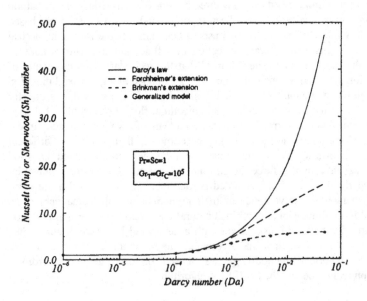

Figure 2 Variation of Nusselt or Sherwood numbers as a function of Darcy number for different flow models (From Karimi-Fard et al. [116]; Copyright 1997 Taylor & Francis – Reprinted with permission).

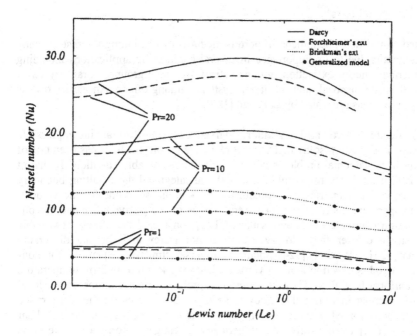

Figure 3 Variation of Nusselt Number as a function of Lewis and Prandtl numbers for $Gr_T = Gr_C = 10^5$, $Da = 10^{-3}$, and $\Lambda = 2.34$. From Karimi-Fard et al. [116]; Copyright 1997 Taylor & Francis – Reprinted with permission).

Nguyen et al. [117] investigated double-diffusive natural convection in an enclosure filled with two layers of anisotropic porous media saturated with a two-component fluid. They used domain decomposition and a pseudo-spectral method with Chebyshev polynomials as the basis functions. Casada and Young [118] developed a numerical model for double-diffusive natural convection in arbitrarily shaped two-dimensional porous media. They used a finite difference method in a generalized coordinate system and a stream function formulation. They found that the energy and moisture transport equations were best solved using a modified Crank-Nicolson method that was developed to control the tendency for instability caused by the source terms in the governing equations. More recently, Bera et al. [119] used the spectral element method to study double-diffusive convection within a rectangular porous enclosure exposed to simultaneous heating and cooling on the sides. They used a Brinkman-extended Darcy model in which the principal direction of the permeability tensor was considered to be oblique to the gravity vector.

3 MIXED CONVECTION IN POROUS MEDIA

3.1 External Flows

Mixed convection phenomena in porous media is of continuing interest to many investigators due to its importance in several engineering applications including geothermal energy extraction, groundwater flow, underground storage systems, nuclear waste disposal, and insulation systems, among others. An earlier review of this topic was presented by Lai et al. [120].

3.1.1 Vertical Plate and Cylinder. Mixed convection of an incompressible viscous fluid past an isothermal vertical plate in a porous medium has been one of the most fundamental problems which received considerable attention. Takhar et al. [121] used the Darcy-Brinkman model and integrated the resulting boundary layer equations numerically to obtain the non-similar solutions for the velocity and temperature distributions for several values of the permeability and viscous dissipation parameters. Lai and Kulacki [122] considered the effects of variable viscosity and later they closely examined non-Darcy and thermal dispersion effects [123]. Yu et al. [124] obtained universal similarity equations for non-Darcy mixed convection along a vertical plate exposed to a uniform temperature or a uniform heat flux. They presented comprehensive correlations of the local Nusselt number including the Darcy and non-Darcy regimes for the entire range from the pure forced convection limit to the pure natural convection limit. Takhar and Beg [125] theoretically and numerically investigated non-Darcy effects as well as the effects of viscosity, porosity, buoyancy parameter, and thermal conductivity ratio. The Brinkman Forchheimer-extended Darcy model was implemented in the boundary layer equations which they solved using Keller's implicit finite difference technique and a double-shooting Runge-Kutta method. Recently Hassanien et al. [126] presented a boundary layer analysis using the local similarity approach. They investigated the effects of thermal dispersion and stratification on the flow and temperature fields.

Besides the standard configuration in which the vertical plate is exposed to a uniform temperature or a uniform heat flux, other types of boundary conditions have been investigated. Chen et al. [127] considered the case when the vertical surface is subjected to a power-law variation of wall temperature or wall heat flux, and they included non-Darcy and thermal dispersion effects. The inertia effects were studied by Kodah and Duwairi [128] for a vertical plate exposed to variable wall temperature. Recently, Kodah and Al-Gasem [129] presented the non-similarity solutions for non-Darcy mixed convection from a vertical plate exposed to a variable surface heat flux of the power-law form. They solved the governing equations using the Keller box method.

Due to its occurrence in many industrial applications, magneto-hydrodynamic mixed convection flow of an electrically-conducting fluid in different porous geometries has been considered by many investigators. Among them, Chamkha [130] derived the boundary layer equations for hydromagnetic mixed convection in the presence of heat generation and magnetic dissipation within the porous medium. He defined a mixed convection parameter covering the entire range from pure forced convection to pure free convection. He performed a non-similarity transformation of the governing equations, which were solved using an implicit finite difference method. The effects of transverse magnetic field, viscous dissipation, variable viscosity, and plate transpiration (lateral mass flux) on mixed convection were investigated by Takhar and Beg [131]. They solved the governing equations based on a non-Darcy model using Keller's implicit finite difference scheme.

The vertical cylinder is another configuration of fundamental interest in studying mixed convection heat transfer over external surfaces in porous media. Chen and Chen [132] and Chen et al. [133] analyzed the effects of conjugate heat transfer, solid boundary shear, inertia forces, variable porosity distribution, and transverse thermal dispersion. They found that the boundary and inertia effects decreased the heat transfer rate, while the near-wall porosity variation and thermal dispersion increased it. Gill et al. [134] analyzed a similar problem including conjugate conduction and non-Darcy effects in a high-porosity medium and obtained numerical solutions using a local non-similarity method. Aldoss et al. [135] obtained non-similarity solutions for the case of variable wall temperature and variable surface heat flux.

3.1.2 Horizontal flat plat and cylinder. Kumari et al. [136] studied the non-similar, non-Darcy mixed convection about a heated horizontal surface in a saturated porous medium when the surface temperature is a power function of distance. The analysis was performed for the cases of parallel and stagnation flows with favorable induced pressure gradient. The governing equations were solved numerically using the Keller box method. They found that the heat transfer was enhanced due to the buoyancy parameter, but the non-Darcy parameter reduced it. For non-Darcy flow, the similarity solution existed only for the case of parallel flow. The influence of surface mass transfer on mixed convection over horizontal plates in saturated porous media was studied by Lai and Kulacki [137]. In this study, the similarity solutions were reported for the special cases in which the wall temperature, free stream velocity, and the injection or withdrawal velocity were a prescribed power function of distance. Non-uniform boundary conditions were also studied including variable wall temperature [138, 139] and variable surface heat flux [140, 141]. Chen [142] analyzed both types of boundary conditions using the Brinkman-Forchheimer-extended Darcy model including porosity variation and thermal dispersion effects.

3.1.3 Other Configurations. Mixed convective heat transfer about bodies of various other geometry has also attracted a great deal of attention. Kumari and Nath [143] studied non-Darcy mixed convection over a non-isothermal horizontal cylinder and sphere embedded in a saturated porous medium. The governing boundary-layer equations were solved numerically using the Keller box method. It was found that the heat transfer was reduced by increasing the non-Darcy parameter, but it was enhanced with increasing buoyancy forces or increasing wall heating. Using a Chebyshev-Legendre spectral method, Nguyen and Paik

[144] investigated unsteady mixed convection about a sphere. They used the Darcy flow model and they included the effects of nonlinear dependence of density on temperature. The problem of mixed convection along wedges embedded in porous media was investigated by Vargas et al. [145] using the Darcy flow formulation and a non-similarity solution technique.

3.1.4 Heat Sources. Mixed convection in the presence of a heat source/sink is another problem of fundamental as well as practical interest. Jang and Shiang [146] studied mixed convection from a line heat source embedded at the leading edge of an adiabatic vertical surface which is immersed in a non-Darcian porous medium. They employed an implicit finite difference technique and the Keller box method, and they analyzed both buoyancy-assisting and buoyancy-opposing flow conditions. Similarity solutions were also obtained by Merkin and Pop [147] for the same problem. Yih [148] used the Keller box method to analyze the effect of heat source/sink on laminar MHD mixed convection owing to the stagnation flow against a vertical permeable flat plate with linear wall temperature variation in a fluid-saturated porous medium.

3.1.5 Non-Newtonian Fluids. As indicated earlier, non-Newtonian fluids in porous media occur in various applications. Shenoy [149] derived the governing equation for Darcy-Forchheimer flow of a non-Newtonian (power-law), inelastic fluid through a porous medium. They obtained a similarity solution using an approximate integral method. Gorla et al. [150] presented a non-similar boundary layer analysis of mixed convection of a power-law fluid along a horizontal surface embedded in a porous medium and exposed to variable surface heat flux distribution. Mansour et al. [151] considered an isothermal vertical cylinder and solved the governing equations using a finite difference method. Recently, Mansour et al. [152] presented a boundary layer analysis for the interaction of mixed convection with thermal radiation in laminar boundary layer flow from a vertical wedge in a porous medium saturated with a power-law-type non-Newtonian fluid. The fluid was considered as a gray medium, and the Rosseland approximation was used to describe the radiative heat flux in the energy equation. The transformed conservation laws were solved numerically for the case of variable surface temperature conditions.

3.2 Internal Flows

3.2.1 Vertical Channels. Among the initial studies of internal flow mixed convection in two-dimensional vertical channels, Hadim [153] used the Brinkman-Forchheimer-extended Darcy model and a stream-function-vorticity formulation to analyze in detail the evolution of mixed convection in the entrance region as illustrated in Fig. 4. Their results showed that in the Darcy regime, at high values of the Reynolds number, the describing parameter is the Gr^*/Re ratio (where Gr^* is the modified Grashof number). At relatively low Re, axial conduction effects could be significant especially in the entrance region (Fig. 4). The effects of the governing parameters on the hydrodynamic and thermal entrance lengths were discussed in detail. For a similar configuration, Hadim and Chen [154] compared heat transfer enhancement in the mixed convection region between the two cases of uniform wall temperature and uniform wall heat flux. For the case when discrete heat sources are mounted on one wall of the channel, Hadim [155] found that the Nusselt number increased with decreasing Darcy number, and the effect of Darcy number was more pronounced over the upstream heat sources and in the non-Darcy regime.

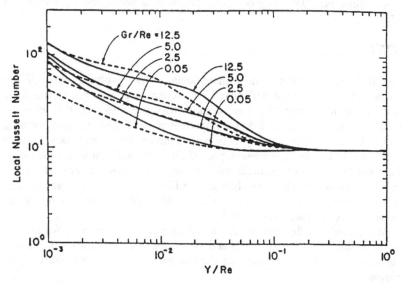

Figure 4 Local Nusselt number variation for non-Darcy mixed convection in a vertical channel for $Da = 10^{-6}$, $Re = 20$ (——) and $Re = 100$ (- - - -) (From Hadim [153]; Copyright 1993 American Institute of Aeronautics and Astronautics – Reprinted with permission).

Chang and Chang [156] studied developing non-Darcy mixed convection in a vertical tube partially filled with a porous medium. They used the SIMPLEC numerical algorithm combining the alternating directions implicit (ADI) and successive over-relaxation techniques (SOR), and the diffusion and convection terms were handled by a hybrid scheme. Later, the same authors solved a similar problem for the configuration consisting of a vertical parallel-plate channel partially filled with a porous medium [157].

Fully developed hydromagnetic non-Darcy mixed convection of an electrically-conducting heat-generating/absorbing fluid in a channel embedded in a uniform porous medium was investigated by Chamkha [158]. Various combinations of thermal boundary conditions at the walls of the channel were prescribed. The fully nonlinear governing equations were solved numerically by an implicit finite difference method.

3.2.2 Horizontal Channels. Hwang and Chao [159] investigated the effects of peripheral wall conduction and Darcy number on laminar mixed convection in the fully developed region of a horizontal square porous channel with a uniform heat input. They studied the effects of non-axisymmetric channel configuration, buoyancy-induced secondary flow, and wall conduction parameter on the flow and heat transfer characteristics. They used a modified Darcy-Forchheimer-Ergun flow model in the porous medium and a finite-difference method. Chou and Chung [160] presented a numerical study of stagnant conductivity effects on the fully developed non-Darcian mixed convection in a horizontal packed-sphere channel.

4 CONCLUSIONS

An extensive review of current computational work involving most of the important aspects of natural and mixed convection in porous media and their applications mainly over the past decade was presented in this chapter. Generalized models which include the important physical phenomena governing convective flow and energy transport in porous media have been presented. Due to space limitations, detailed presentation of the computational techniques used for these models has been omitted, and the reader is referred to related studies which have been reviewed in this chapter. The literature review was presented in a convenient format allowing the reader to identify the latest progress in individual areas of research in this field.

It is hoped that the present review will provide a convenient reference for the current research related to natural and mixed convection in porous media and their applications and a better understanding of the current research directions in these areas.

5 REFERENCES

1. S. Kakac, S. Kilkis, F.A. Kulacki, and F. Arinc, *Convective Heat and Mass Transfer in Porous Media*, Kluwer Academic, Dordrecht, 1991.
2. D.A. Nield and A. Bejan, *Convection in Porous Media*, Springer-Verlag, New York, 1992.
3. M. Kaviany, *Principles of Heat Transfer in Porous Media*, Springer-Verlag, New York, 1995.
4. D.B. Ingham and I. Pop, *Transport Phenomena in Porous Media*, Pergamon, Elsevier Science Ltd., UK, 1998.
5. C.L. Tien and K. Vafai, Convective and Radiative Heat Transfer in Porous Media, in *Adv. Appl. Mech.*, vol. 27, pp. 225-281, 1990.
6. H. Hadim and K. Vafai, Overview of Current Computational Studies of Heat Transfer in Porous Media and Their Applications – Forced Convection and Multiphase Heat Transfer, in *Advances in Numerical Heat Transfer*, vol. 2 (eds. W.J. Minkowycz and E.M. Sparrow), Taylor & Francis, Washington, D.C., 1999.

7. S. Whitaker, Simultaneous Heat, Mass and Momentum Transfer in Porous Media: a Theory of Drying, *Adv. Heat Transfer*, vol. 13, pp. 119-203, 1977.
8. K. Vafai and C.L. Tien, Boundary and Inertia Effects on Flow and Heat Transfer in Porous Media, *Int. J. Heat Mass Transfer*, vol. 24, pp. 195-203, 1981.
9. S.J. Kim and K. Vafai, Analysis of Natural Convection About a Vertical Plate Embedded in a Porous Medium, *Int. J Heat Mass Transfer*, vol. 32, pp. 665-677, 1989.
10. Ettefagh, K. Vafai, and S.J. Kim, Non-Darcian Effects in Open-Ended Cavities Filled With a Porous Medium, *J. Heat Transfer*, vol. 113, pp. 747-756, 1991.
11. S. Ergun, Fluid Flow Through Packed Columns, *Chem. Eng. Prog.*, vol. 48, pp. 89-94, 1952.
12. K. Vafai, Convective Flow and Heat Transfer in Variable Porosity Media, *J. Fluid Mechanics*, vol. 147, pp. 233-259, 1984.
13. R.F. Benenati and C.B. Brosilow, Void Fraction Distribution in Beds of Spheres, *AIChE J.*, vol. 8, pp. 359-361, 1962.
14. A. Amiri and K. Vafai, Analysis of Dispersion Effects and Non-thermal Equilibrium, Non-Darcian, Variable Porosity Incompressible Flow through Porous Media, *Int. J. Heat Mass Transfer*, vol. 37, pp. 939-954, 1994.
15. W.J. Minkowycz, E.M. Sparrow, G.E. Schneider, R.H. Pletcher, *Handbook of Numerical Heat Transfer*, John Wiley, New York, 1988.
16. K. Vafai and H. C. Tien, A Numerical Investigation of Phase Change Effects in Porous Materials, *Int. J. Heat Mass Transfer*, vol. 32, pp. 1261-1277, 1989.
17. P. Vasseur, T. H. Nguyen, L. Robillard and V.K.T. Thi, Natural Convection between Horizontal Concentric Cylinders Filled with a Porous Layer with Internal Heat Generation, *Int. J. Heat Mass Transfer*, vol. 27, pp. 337-349, 1984.
18. M.R.J. Wyllie, Relative Permeability. In *Petroleum Production Handbook* (Edited by Frick), vol. 2, Chap. 25. McGraw-Hill, New York (1962).
19. I. Fatt and W.A. Klikoff, Effect of Fractional Wettability on Multiphase Flow Through Porous Media, AIME Technical Note No. 2043, *AIME trans.* 216, pp. 426-432, 1959.
20. K.S. Udell, Heat Transfer in Porous Media Considering Phase Change and Capillarity-The Heat Pipe Effect, *Int. J. Heat Mass Transfer*, vol. 28, pp. 485-495, 1985.
21. P. Cheng and W.J. Minkowycz, Free Convection about a Vertical Flat Plate Embedded in a Porous Medium with Application to Heat Transfer from a Dike, *J. Geophys Res*, vol. 82, pp. 2040-2044, 1977.
22. J.T. Hong, C.L. Tien, and M. Kaviany, Non-Darcian Effects on Vertical-Plate Natural Convection in Porous Media with High Porosities, *Int. J. Heat Mass Transfer*, vol. 28, pp. 2149-2157, 1985.
23. C.H. Chen and J.S. Chiou, Conjugate Free Convection Heat Transfer Analysis of a Vertical Plate Fin Embedded in Non-Darcian Porous Media, *Int. J. Eng. Sci.*, vol. 32, pp. 1703-1716, 1994.

24. M. Vynnycky and S. Kimura, Conjugate Free Convection due to a Vertical Plate in a Porous Medium, *Int. J. Heat Mass Transfer*, vol. 37, pp. 229-236, 1994.

25. M. Vynnycky and S. Kimura, Transient Conjugate Free Convection due to a Vertical Plate in a Porous Medium, *Int. J. Heat Mass Transfer*, vol. 38, pp. 219-231, 1995.

26. S.W. Hsiao, Numerical Study of Transient Natural Convection About a Corrugated Plate Embedded in an Enclosed Porous Medium, *Int. J. Num. Meth. Heat Fluid Flow*, vol. 5, pp. 629-645, 1995.

27. D.A.S. Rees and I. Pop, Non-Darcy Natural Convection form a Vertical Wavy Surface in a Porous Medium, *Transport Porous Media*, vol. 20, pp. 223-234, 1995.

28. R. Bradean, D.B. Ingham, P.J. Heggs, and I. Pop, Free Convection Fluid Flow Due to a Periodically Heated and Cooled Vertical Flat Plate Embedded in a Porous Media, *Int. J. Heat Mass Transfer*, vol. 39, pp. 2545-2557, 1996.

29. R. Bradean, D.B. Ingham, P.J. Heggs, and I. Pop, The Unsteady Penetration of Free Convection Flows Caused by Heating and Cooling Flat Surfaces in a Porous Media, *Int. J. Heat Mass Transfer*, vol. 40, pp. 665-687, 1997.

30. D.A.S. Rees, Three-dimensional Free Convection Boundary Layers in Porous Media Induced by a Heated Surface With Spanwise Temperature Variations, *J. Heat Transfer,* vol. 119, pp. 792-798, 1997.

31. S.D. Harris, D.B. Ingham, and I. Pop, Free Convection From a Vertical Plate in a Porous Medium Subjected to a Sudden Change in Surface Heat Flux, *Transport Porous Media*, vol. 26, pp. 205-224, 1997.

32. M.A. Hossain and I. Pop, Radiation Effect on Darcy Free Convection Flow Along an Inclined Surface Placed in Porous Media, *Heat Mass Transfer/ Waerme-Stoffuebertr.*, vol. 32, pp. 223-227, 1997.

33. A.J. Chamkha, Hydromagnetic Free Convection Flow Over an Inclined Plate Caused by Solar Radiation, *AIAA J. Thermophys. Heat Transfer*, vol. 11, pp. 312-315, 1997.

34. R.S. Telles and O.V. Trevisan, Dispersion in Heat and Mass Transfer Natural Convection Along Vertical Boundaries in Porous Media, *Int. J. Heat Mass Transfer*, vol. 36, pp. 1357-1365, 1993.

35. D. Angirasa and G.P. Peterson, Natural Convection Heat Transfer From an Isothermal Vertical Surface in a Fluid-Saturated Thermally-Stratified Porous Medium, *Int. J. Heat Mass Transfer*, vol. 40, pp. 4329-4335, 1997.

36. P.V.S.N. Murthy and P. Singh, Effect of Viscous Dissipation on a Non-Darcy Natural Convection Regime, *Int. J. Heat Mass Transfer*, vol. 40, pp. 1251-1260, 1997.

37. J.P. Pascal and H. Pascal, Free Convection in a Non-Newtonian Fluid-Saturated Porous Medium With Lateral Mass Flux, *Int. J. Non-Linear Mechanics,* vol. 32, pp. 471-482, 1997.

38. A.J. Chamkha, Hydromagnetic Natural Convection From an Isothermal Inclined Surface Adjacent to a Thermally Stratified Porous Medium, *Int. J. Eng. Science*, vol. 35, pp. 975-986, 1997.

39. A.P. Bassom and D.A.S. Rees, Free Convection From a Heated Vertical Cylinder Embedded in a Fluid-Saturated Porous Medium, *Acta Mechanica*, vol. 116, pp. 139-151, 1996.

40. P.H. Oosthuizen and D. Naylor, Natural Convective Heat Transfer From a Cylinder in an Enclosure Partly Filled with a Porous Medium, *Int. J. Num. Methods Heat Fluid Flow*, vol.6, pp. 51-63, 1996.

41. M.A. Hossain, A. Nakayama, and I. Pop, Conjugate Free Convection of Non-Newtonian Fluids About a Vertical Cylindrical Fin in Porous Media, *Heat Mass Transfer/Warme-Stoffuebertr.*, vol. 30, pp. 149-153, 1995.

42. D.A.S. Rees and I. Pop, Free Convection Induced by a Horizontal Wavy Surface in a Porous Medium, *Fluid Dyn. Res.*, vol. 14, pp. 151-166, 1994.

43. D. Lesnic, D.B. Ingham, and I. Pop, Conjugate Free Convection From a Horizontal Surface in a Porous Medium, *Z. Angew Math. Mech.*, vol. 75, pp. 715-722, 1995.

44. R. Bradean, D.B. Ingham, P.J. Heggs, and I. Pop, Buoyancy-Induced Flow Adjacent to a Periodically Heated and Cooled Horizontal Surface in Porous Media, *Int. J. Heat Mass Transfer*, vol. 39, pp. 615-630, 1996.

45. D.A.S. Rees, Effect of Inertia on Free Convection From a Horizontal Surface Embedded in a Porous Medium, *Int. J. Heat Mass Transfer*, vol. 39, pp. 3425-3430, 1996.

46. M.A. Hossain and D.A.S. Rees, Non-Darcy Free Convection Along a Horizontal Heated Surface, *Transport Porous Media*, vol. 29, pp. 309-321, 1997.

47. D. Angirasa and G.P. Peterson, Upper and Lower Rayleigh Number Bounds for Two-Dimensional Natural Convection Over a Finite Horizontal Surface Situated in a Fluid-Saturated Porous Medium, *Num. Heat Transfer; Part A*, vol. 33, pp. 477-493, 1998.

48. G.N. Facas, Natural Convection from a Buried Pipe With External Baffles, *Num. Heat Transfer, Part A*, vol. 27, pp. 595-609, 1995.

49. I. Pop, D.B. Ingham, and R. Bradean, Transient Free Convection About a Horizontal Circular Cylinder in a Porous Medium with Constant Surface Flux Heating, *Acta Mechanica*, vol. 119, pp. 79-91, 1996.

50. R. Bradean, D.B. Ingham, P.J. Heggs, and I. Pop, Unsteady Free Convection Adjacent to an Impulsively Heated Horizontal Circular Cylinder in Porous Media, *Num. Heat Transfer, Part A*, vol. 32, pp. 325-346, 1997.

51. H.D. Nguyen, S. Paik, R.W. Douglass, and I. Pop, Unsteady Non-Darcy Reaction-Driven Flow From an Anisotropic Cylinder in Porous Media, *Chem. Eng. Sci.*, vol. 51, pp. 4963-4977, 1996.

52. S. Kimura and I. Pop, Conjugate Convection From a Sphere in a Porous Medium, *Int. J. Heat Mass Transfer*, vol. 37, pp. 2187-2192, 1994.

53. B. Yan, I. Pop, and D.B. Ingham, Numerical Study of Unsteady Free Convection From a Sphere in a Porous Medium, *Int. J. Heat Mass Transfer*, vol. 40, pp. 893-903, 1997.

54. I. Pop and T.Y. Na, Natural Convection Over a Frustum of a Wavy Cone in a Porous Medium, *Mech. Res. Comm.*, vol. 22, pp. 181-190, 1995.

55. Y.T. Yang and S.J. Wang, Free Convection Heat Transfer of Non-Newtonian

Fluids over Axisymmetric and Two-dimensional Bodies of Arbitrary Shape Embedded in a Fluid-saturated Porous Medium, *Int. J. Heat Mass Transfer*, vol. 39, pp. 203-210, 1996.

56. S. Kimura, T. Kiwata, A. Okajima, and I. Pop, Conjugate Natural Convection in Porous Media, *Adv. Water Resources*, vol. 20, pp. 111-126, 1997.

57. T. Masuoka, Y. Takatsu, S. Kawamoto, H. Koshino, and T. Tsuruta, Buoyant Plume Through a Permeable Porous Layer Located Above a Line Heat Source in an Infinite Fluid Space, *JSME Int. J., Series B: Fluids Thermal Eng.*, vol. 38, pp. 79-85, 1995.

58. J.J. Shu and I. Pop, Inclined Wall Plumes in Porous Media, *Fluid Dyn. Res.*, vol. 21, pp. 303-317, 1997.

59. J.S. Leu and J.Y. Jang, Natural Convection From a Point Heat Source Embedded in a Non-Darcian Porous Medium, *Int. J. Heat Mass Transfer*, vol. 38, pp. 1097-1104, 1995.

60. G.N. Facas, Natural Convection From an Elliptic Heat Source Buried in a Semi-Infinite, Saturated, Porous Medium, *Proc. 1995 ASME/JSME Thermal Engineering Joint Conference*, Part 3 (of 4), pp. 423-430, 1995.

61. M. Combarnous and D. Bernard, Modeling of Free Convection in Porous Media: From Academic Cases to Real Configurations, *Proc. 1988 National Heat Transfer Conference*, vol. 96, pp. 735-745, 1988.

62. P. Cheng, S.W. Hsiao, and C.K. Chen, Natural Convection in Porous Media With Variable Porosity and Thermal Dispersion, in *Convective Heat and Mass Transfer in Porous Media* (eds. S. Kakac, B. Kilkis, F.A. Kulacki and F. Arinc), Kluwer Academic Pub., Dordrecht, pp. 543-563, 1991.

63. C. Beckermann, R. Viskanta, and S. Ramadhyani, Numerical Study of Non-Darcy Natural Convection in a Vertical Enclosure Filled with a Porous Medium, *Num. Heat Transfer*, vol. 10, pp. 557-570, 1986.

64. G. Lauriat and V. Prasad, Non-Darcian Effects on Natural Convection in a Vertical Porous Layer, *Int. J. Heat Mass Transfer*, vol. 32, pp. 2135-2148, 1989.

65. V. Prasad and A. Tuntomo, Inertia Effects on Natural Convection in a Vertical Porous Cavity, *Num. Heat Transfer*, vol. 11, pp. 295-320, 1987.

66. B.V.K. Satya Sai, K.N. Seetharamu, and P.A. Aswathanarayana, Finite Element Analysis of Heat Transfer by Natural Convection in Porous Media in Vertical Enclosures: Investigations in Darcy and non-Darcy regimes, *Int. J. Num. Meth. Heat Fluid Flow*, vol. 7, pp. 367-400, 1997.

67. J. Ni and C. Beckermann, Natural Convection in a Vertical Enclosure Filled with Anisotropic Porous Media, *J. Heat Transfer*, vol. 113, 1033-1037, 1991.

68. W.J. Chang and C.F. Hsiao, Natural Convection in a Vertical Cylinder Filled with Anisotropic Porous Media, *Int. J. Heat Mass Transfer*, vol. 36, pp. 3361-3367, 1993.

69. M. Song and R. Viskanta, Flow Characteristics of Anisotropic Structures Constructed with Porous Layers, *Transport Porous Media*, vol. 15, pp. 151-173, 1994.

70. G. Degan and V. Vasseur, Natural Convection in a Vertical Slot Filled with an Anisotropic Porous Medium with Oblique Principal Axes, *Num. Heat Transfer, Part A,* vol. 30, pp. 397-412, 1996.

71. G. Degan and V. Vasseur, Boundary-Layer Regime in a Vertical Porous Layer with Anisotropic Permeability and Boundary Effects, *Int. J. Heat Fluid Flow,* vol. 18, pp. 334-343, 1997.

72. P.V.S.N. Murthy, B.V.R. Kumar, and P. Singh, Natural Convection Heat Transfer From a Horizontal Wavy Surface in a Porous Enclosure, *Num. Heat Transfer; Part A,* vol. 31, pp. 207-221, 1997.

73. C. Beckermann, R. Viskanta, and S. Ramadhyani, Natural Convection in Vertical Enclosures Containing Simultaneous Fluid and Porous Layers, *J. Fluid Mech.,* vol. 186, pp. 557-570, 1988.

74. S.B. Sathe, W.-Q. Lin, and T.W. Tong, Natural Convection in Enclosures Containing an Insulation with a Permeable Fluid-Porous Interface, *Int. J. Heat Fluid Flow,* vol. 9, pp. 389-395, 1988.

75. Z.G. Du, and E. Bilgen, Natural Convection in Vertical Cavities With Partially Filled Heat-Generating Porous Media, *Num. Heat Transfer, Part A,* vol. 18, pp. 371-386, 1990.

76. L. Storesletten, and I. Pop, Free Convection in a Vertical Porous Layer with Walls at Non-uniform Temperature, *Fluid Dyn. Res.,* vol. 17, pp. 107-119, 1996.

77. D.A.S. Rees and J.L. Lage, Effect of Thermal Stratification on Natural Convection in a Vertical Porous Insulation Layer, *Int. J. Heat Mass Transfer,* vol. 40, pp. 111-121, 1997.

78. B.V. Antohe and J.L. Lage, Amplitude Effect on Convection Induced by Time-periodic Horizontal Heating, *Int. J. Heat Mass Transfer,* vol. 39, pp. 1121-1133, 1996.

79. B.V. Antohe, and J.L. Lage, Prandtl Number Effect on the Optimum Heating Frequency of an Enclosure Filled With Fluid or With a Saturated Porous Medium, *Int. J. Heat Mass Transfer,* vol. 40, pp. 1313-1323, 1997.

80. P. Nithiarasu, K.N. Seetharamu, T. Sundarajan, Numerical Investigation of Buoyancy Driven Flow in a Fluid-Saturated non-Darcian Porous Medium, *Int. J. Heat Mass Transfer,* vol. 42, pp. 1205-1215, 1999.

81. W. Bian, P. Vasseur, and E. Bilgen, Boundary-Layer Analysis for Natural Convection in a Vertical Porous Layer Filled With a Non-Newtonian Fluid, *Int. J. Heat Fluid Flow,* vol. 15, pp. 384-391, 1994.

82. D. Getachew, W.J. Minkowycz, and D. Poulikakos, Natural Convection in a Porous Cavity Saturated With a Non-Newtonian Fluid, *J. Thermophys. Heat Transfer,* vol. 10, pp. 640-651, 1996.

83. H.A. Hadim and G. Chen, Numerical Study of Non-Darcy Natural Convection of a Power-law Fluid in a Porous Cavity, *ASME-HTD vol. 317,* pp. 301-307, 1995.

84. S.W. Hsiao, C.K. Chen, and P. Cheng, Natural Convection an Inclined Porous Cavity With a Discrete Heat Source on One Wall, *Int. J. Heat Mass Transfer,* vol. 39, pp. 1121-1133, 1994.

85. W. Bian, P. Vasseur, E. Bilgen, and F. Meng, Effect of an Electromagnetic Field on Natural Convection in an Inclined Porous Layer, *Int. J. Heat Fluid Flow*, vol. 17, pp. 36-44, 1996.

86. J. Ettefagh and K. Vafai, Natural Convection in Open-Ended Cavities with a Porous Obstructing Medium, *Int. J. Heat Mass Transfer*, vol. 31, pp. 673-693, 1988.

87. V. Dharma Rao, S.V. Naidu, and P.K. Sarma, Non-Darcy Effects in Natural Convection Heat Transfer in a Vertical Porous Annulus, *J. Heat Transfer*, vol. 118, pp. 502-505, 1996.

88. H.D. Nguyen, S. Paik, and I. Pop, Transient Thermal Convection in a Spherical Enclosure Containing a Fluid Core and a Porous Shell, *Int. J. Heat Mass Transfer*, vol. 40, pp. 379-392, 1997.

89. P. Vasseur, M. Hasnaoui, E. Bilgen, and L. Robillard, Natural Convection in an Inclined Fluid Layer With a Transverse Magnetic Field, Analogy With a Porous Medium, *J. Heat Transfer*, vol. 117, pp. 121-129, 1995.

90. S. Alchaar, P. Vasseur, and E. Bilgen, Effects of a Magnetic Field on the Onset of Convection in a Porous Medium, *Heat Mass Transfer/Waerme-Stoffuebertr.*, vol. 30, pp. 259-267, 1995.

91. K. Ozaki, H. Inaba, and A. Shigemori, Numerical Analysis of Heat Storage Characteristics of a Small Heat Storage Vessel Packed With Phase-Change Material Encapsulated into Spherical Hollow, *Trans. Japan Soc. Mech. Eng., Part B*, vol. 63, pp. 1762-1769, 1997.

92. J.P. Caltagirone and P. Fabrie, Natural Convection in a Porous Medium at High Rayleigh Numbers. Part 1 – Darcy's Model, *European J. Mech. B*, vol. 8, pp. 207-227, 1989.

93. Prasad and Kladias, Non-Darcy Natural Convection in Saturated Porous Media, *in Convective Heat and Mass Transfer in Porous Media* (eds. S. Kakac, B. Kilkis, F.A. Kulacki and F. Arinc), Kluwer Academic Pub., Dordrecht, pp. 173-224, 1991.

94. P. Vadasz, C. Braester, and J. Bear, Effect of Perfectly Conducting Side Walls on Natural Convection in Porous Media, *Int. J. Heat Mass Transfer*, vol. 36, pp. 1159-1170, 1993.

95. M. Mamou, L. Robillard, E. Bilgen, and P. Vasseur, Effects of a Moving Thermal Wave on Benard Convection in a Horizontal Saturated Porous Layer, *Int. J. Heat Mass Transfer*, vol. 39, pp. 347-354, 1996.

96. L.E. Howle and J.G. Georgiadis, Natural Convection in Porous Media With Anisotropic Dispersive Thermal Conductivity, *Int. J. Heat Mass Transfer*, vol. 37, pp. 1081-1094, 1994.

97. W. Bian, P. Vasseur, and E. Bilgen, Natural Convection of Non-Newtonian Fluids in an Inclined Porous Layer, *Chem. Eng. Comm.*, vol. 129, pp. 79-97, 1994.

98. W. Bian, P. Vasseur, E. Bilgen, and F. Meng, Effect of an Electromagnetic Field on Natural Convection in an Inclined Porous Layer, *Int. J. Heat Fluid Flow*, vol. 17, pp. 36-44, 1996.

99. J.P. Barbosa Mota and E. Saatdjian, Natural Convection in Porous Cylindrical Annuli, *Int. J. Num. Meth. Heat Fluid Flow*, vol. 5, pp. 3-12, 1995.

100. C.K. Chen, S.W. Hsiao, and P. Cheng, Transient Natural Convection in an Eccentric Porous Annulus Between Horizontal Cylinders, *Num. Heat Transfer, Part A*, vol. 17, pp. 431-448, 1990.

101. C.P. Pan, and F.C. Lai, Natural Convection in Horizontal-Layered Porous Annuli, *J. Thermophys. Heat Transfer*, vol. 9, pp. 792-795, 1996.

102. M. Zhao, L. Robillard, and M. Prud'homme, Effect of Weak Rotation on Natural Convection in a Horizontal Porous Cylinder, *Heat Mass Transfer/ Warme-Stoffuebertr.*, vol. 31, pp. 403-409, 1996.

103. M.C. Charrier-Mojtabi, Numerical Simulation of Two- and Three-Dimensional Free Convection Flows in a Horizontal Porous Annulus Using a Pressure and Temperature Formulation, *Int. J. Heat Mass Transfer*, vol. 40, pp. 1521-1533, 1997.

104. I. Pop, D. Angirasa, and G.P. Peterson, Natural Convection in Porous Media Near L-shaped Corners, *Int. J. Heat Mass Transfer*, vol. 40, pp. 485-490, 1997.

105. V. Debeda, J.P. Caltagirone, and P. Watremez, Local Multigrid Refinement Method for Natural Convection in Fissured Porous Media, *Num. Heat Transfer, Part B*, vol. 28, pp. 455-467, 1995.

106. A.M. Mullis, Natural Convection in Porous, Permeable Media: Sheets, Wedges and Lenses, *Marine Petroleum Geology*, vol. 12, pp. 17-25, 1995.

107. I. Chiorean and I.D. Chiorean, Parallel Algorithm for Solving a Problem of Convection in Porous Medium. *Adv. Eng. Software*, vol. 28, pp. 463-467, 1997.

108. Zhao, H.B. Muhlhaus, and B.E. Hobbs, Finite Element Analysis of Steady-State Natural Convection Problems in Fluid-saturated Porous Media Heated From Below, *Int. J. Num. Analyt. Meth. Geomech.*, vol. 21, pp. 863-881, 1997.

109. P. Vadasz, Flow in Rotating Porous Media, in *Advances in Fluid Mechanics, Fluid Transport in Porous Media*, vol. 13, pp. 161-214, 1997.

110. Y.F. Rao, K. Fukuda, and S. Hasegawa, Numerical Study of Three-Dimensional Natural Convection in a Horizontal Porous Annulus With Galerkin Method, *Int. J. Heat Mass Transfer*, vol. 31, pp. 698-707, 1988.

111. H. Ozoe, H. Matsumoto, T. Nishimura, and Y. Kawamura, Three-Dimensional Natural Convection in Porous Media at a Rectangular Corner, *Num. Heat Transfer, Part A*, vol. 17, pp. 249-268, 1990.

112. A.S. Dawood and P.J. Burns, Steady Three-Dimensional Convective Heat Transfer in a Porous Box via Multigrid, *Num. Heat Transfer, Part A*, vol. 22, pp. 167-198, 1992.

113. Y. Yamaguchi, Y. Asako, and M. Faghri, Natural Convection and Radiation Heat Transfer in a Vertical Porous Layer With a Hexagonal Honeycomb Core (1st Report, Numerical Analysis), *Trans. Japan Soc. Mech. Eng., Part B*, vol. 63, pp. 2119-2126, 1997.

114. Goyeau, J.-P. Songbe, and D. Gobin, Numerical Study of Double-Diffusive Natural Convection in a Porous Cavity Using the Darcy-Brinkman Formulation, *Int. J. Heat Mass Transfer*, vol. 39, pp. 1363-1378, 1996.

115. P. Nithiarasu, K.N. Seetharamu, and T. Sundararajan, Double-Diffusive Natural Convection in an Enclosure Filled with a Fluid-Saturated Porous Medium: a Generalized Non-Darcy Approach, *Num. Heat Transfer; Part A*, vol. 30, pp. 413-426, 1996.

116. M. Karimi-Fard, M.C. Charrier-Mojtabi, and K. Vafai, Non-Darcian Effects on Double-Diffusive Convection Within a Porous Medium, *Num. Heat Transfer; Part A*, vol. 31, pp. 837-852, 1997.

117. H.D. Nguyen, S. Paik, and R.W. Douglass, Study of Double-Diffusive Convection in Layered Anisotropic Porous Media, *Num. Heat Transfer, Part B*, vol. 26, pp. 489-505, 1994.

118. M.E. Casada and J.H. Young, Model for Heat and Moisture Transfer in Arbitrarily Shaped Two-dimensional Porous Media, *Transactions of the ASAE*, vol. 37, pp. 1927-1938, 1994.

119. P. Bera, V. Eswaran, and P. Singh, Numerical Study of Heat and Mass Transfer in an Anisotropic Porous Enclosure due to Constant Heating and Cooling, *Num. Heat Transfer*, vol. 34, pp. 887-905, 1998.

120. F.C. Lai, F.A. Kulacki, and V. Prasad, Mixed Convection in Porous Media, *in Convective Heat and Mass Transfer in Porous Media* (eds. S. Kakac, B. Kilkis, F.A. Kulacki and F. Arinc), Kluwer Academic Pub., Dordrecht, pp. 225-288, 1991

121. H.S. Takhar, V.M. Soundalgekar, and A.S. Gupta, Mixed Convection of an Incompressible Viscous Fluid in a Porous Medium Past a Hot Vertical Plate, *Int. J. Non-Linear Mech.*, vol. 25, pp. 723-728, 1990.

122. F.C. Lai, and F.A. Kulacki, Effect of Variable Viscosity on Convective Heat Transfer Along a Vertical Surface in a Saturated Porous Medium, *Int. J. Heat Mass Transfer*, vol. 33, pp. 1028-1031, 1990.

123. F.C. Lai and F.A. Kulacki, Non-Darcy Mixed Convection Along a Vertical Wall in a Saturated Porous Medium, *J. Heat Transfer*, vol. 113, pp. 252-255, 1991.

124. W.S. Yu, H.T. Lin, and C.S. Lu, Universal Formulations and Comprehensive Correlations for Non-Darcy Natural Convection and Mixed Convection in Porous Media, *Int. J. Heat Mass Transfer*, vol. 34, pp. 2859-2868, 1991.

125. H.S. Takhar and O.A. Beg, Non-Darcy Effects on Convective Boundary Layer Flow Past a Semi-infinite Vertical Plate in Saturated Porous Media, *Heat Mass Transfer/ Waerme- Stoffuebertr.*, vol. 32, pp. 33-44, 1996.

126. I.A. Hassanien, A.Y. Bakier, and R.S.R. Gorla, Effects of Thermal Dispersion and Stratification on Non-Darcy Mixed Convection From a Vertical Plate in a Porous Medium, *Heat Mass Transfer/Warme-Stoffuebertr.*, vol. 34, pp. 209-212, 1998.

127. C.H. Chen, T.S. Chen, and C.K. Chen, Non-Darcy Mixed Convection in Porous Media, *Int. J. Heat Mass Transfer*, vol. 39, pp. 1157-1164, 1996.

128. Z.H. Kodah and H.M. Duwairi, Inertia Effects on Mixed Convection for Vertical Plates With Variable Wall Temperature in Saturated Porous Media, *Heat Mass Transfer/Waerme-Stoffuebertr.*, vol. 31, pp. 333-338, 1996.

129. Z.H. Kodah and A.M. Al-Gasem, Non-Darcy Mixed Convection From a Vertical Plate in Saturated Porous Media-Variable Surface Heat Flux, *Heat Mass Transfer/Waerme-Stoffuebertr.*, vol. 33, pp. 377-382, 1996.

130. A.J. Chamkha, Mixed Convection Flow Along a Vertical Permeable Plate Embedded in a Porous Medium in the Presence of a Transverse Magnetic Field, *Num. Heat Transfer A*, vol. 34, pp. 93-103, 1998.

131. H.S. Takhar, O.A. Beg, Effects of Transverse Magnetic Field, Prandtl Number and Reynolds Number on Non-Darcy Mixed Convective Flow of an Incompressible Viscous Fluid Past a Porous Vertical Flat Plate in a Saturated Porous Medium, *Int. J. Energy Research*, vol. 21, pp. 87-100, 1997.

132. C.K. Chen and C.H. Chen, Non-Darcian Effects on Conjugate Mixed Convection About a Vertical Circular Pin in a Porous Medium, *Computers Structures*, vol. 38, pp. 529-535, 1991.

133. C.K. Chen, C.H. Chen, U.S. Gill, and W.J. Minkowycz, Non-Darcian Effects on Mixed Convection About a Vertical Cylinder Embedded in a Saturated Porous Medium, *Int. J. Heat Mass Transfer*, vol. 35, pp. 3041-3046, 1992.

134. U.S. Gill, W.J. Minkowycz, C.K. Chen, and C.H. Chen, Boundary and Inertia Effects on Conjugate Mixed Convection-Conduction Heat Transfer From a Vertical Cylindrical Fin Embedded in a Porous Medium, *Num. Heat Transfer A*, vol. 21, pp. 423-441, 1992.

135. T.K. Aldoss, M.A. Jarrah, B.J. Al-Sha'er, Mixed Convection From a Vertical Cylinder Embedded in a Porous Medium: Non-Darcy Model, *Int. J. Heat Mass Transfer*, vol. 39, pp. 1141-1148, 1996.

136. M. Kumari, I. Pop, and G. Nath, Nonsimilar Boundary Layers for Non-Darcy Mixed Convection Flow About a Horizontal Surface in a Saturated Porous Medium, *Int. J. Eng. Sci.*, vol. 28, pp. 253-263, 1990.

137. F.C. Lai and F.A. Kulacki, Thermal Dispersion Effects on Non-Darcy Convection Over Horizontal Surface in Saturated Porous Media, *Int. J. Heat Mass Transfer*, vol. 32, pp. 971-976, 1989.

138. T.K. Aldoss, T.S. Chen, and B.F. Armaly, Nonsimilarity Solutions for Mixed Convection From Horizontal Surfaces in a Porous Medium - Variable Wall Temperature, *Int. J. Heat Mass Transfer*, vol. 36, pp. 471-477, 1993.

139. C.H. Chen, Analysis of non-Darcian Mixed Convection From Impermeable Horizontal Surfaces in Porous Media: The Entire Regime, *Int. J. Heat Mass Transfer*, vol. 40, pp. 2993-2997, 1997.

140. T.K. Aldoss, T.S. Chen, and B.F. Armaly, Nonsimilarity Solutions for Mixed Convection From Horizontal Surfaces in a Porous Medium - Variable Surface Heat Flux, *Int. J. Heat Mass Transfer*, vol. 36, pp. 463-470, 1993.

141. C.H. Chen, Non-Darcy Mixed Convection From a Horizontal Surface With Variable Surface Heat Flux in a Porous Medium, *Num. Heat Transfer A*, vol. 30, pp. 859-869, 1996.

142. C.H. Chen, Nonsimilar Solutions for Non-Darcy Mixed Convection From a Non-isothermal Horizontal Surface in a Porous Medium, *Int. J. Eng. Sci.*, vol. 36, pp. 251-263, 1998.

143. M. Kumari and G. Nath, Non-Darcy Mixed Convection Flow Over a Non-isothermal Cylinder and Sphere Embedded in a Saturated Porous Medium, *J. Heat Transfer*, vol. 112, pp. 518-521, 1990.

144. H.D. Nguyen and S. Paik, Unsteady Mixed Convection From a Sphere in Water-Saturated Porous Media With Variable Surface Temperature/Heat Flux, *Int. J. Heat Mass Transfer*, vol. 37, pp. 1783-1793, 1994.

145. J.V.C. Vargas, T.A. Laursen, and A. Bejan, Nonsimilar Solutions for Mixed Convection on a Wedge Embedded in a Porous Medium, *Int. J. Heat Fluid Flow*, vol. 16, pp. 211-216, 1995.

146. J.Y. Jang and C.T. Shiang, 1997, The Mixed Convection Plume Along a Vertical Adiabatic Surface Embedded in a non-Darcian Porous Medium, *Int. J. Heat Mass Transfer*, vol. 40, pp. 1693-1699, 1997.

147. J.H. Merkin, and I. Pop, Mixed Convection on a Horizontal Surface Embedded in a Porous Medium: The Structure of a Singularity, *Transp. Porous Media*, vol. 29, pp. 355-364, 1997.

148. K.A. Yih, Heat source/sink effect on MHD Mixed Convection in Stagnation Flow on a Vertical Permeable Plate in Porous Media, *Int. Comm. Heat Mass Transfer*, vol. 25, pp. 427-442, 1998.

149. A.V. Shenoy, Darcy-Forchheimer Natural, Forced and Mixed Convection Heat Transfer in Non-Newtonian Power-Law Fluid-Saturated Porous Media, *Transport Porous Media*, vol. 11, pp. 219-241, 1993.

150. R. Gorla, R. Subba, K. Shanmugam, and M. Kumari, Nonsimilar Solutions for Mixed Convection in Non-Newtonian Fluids Along Horizontal Surfaces in Porous Media, *Transport Porous Media*, vol. 28, pp. 319-334, 1997.

151. M.A. Mansour, M. Abd El-Hakiem, S. Abd El-Gaid, R. Subba, and R. Gorla, Mixed Convection in Non-Newtonian Fluids Along an Isothermal Vertical Cylinder in a Porous Medium, *Transport Porous Media*, vol. 28, pp. 307-317, 1997.

152. M.A. Mansour, R. Subba, and R. Gorla, Mixed Convection-Radiation Interaction in Power-Law Fluids along a Non-isothermal Wedge Embedded in a Porous Medium, *Transport Porous Media*, vol. 30, pp. 113-124, 1998.

153. A. Hadim, Numerical Study of Non-Darcy Mixed Convection in a Vertical Porous Channel, *J. Thermophys. Heat Transfer*, vol. 8, pp. 371-373, 1993.

154. H.A. Hadim and G. Chen, Non-Darcy Mixed Convection in a Vertical Porous Channel With Asymmetric Wall Heating, *J. Thermophys. Heat Transfer*, vol. 8, pp. 805-808, 1994.

155. H.A. Hadim, Non-Darcy Mixed Convection in a Vertical Porous Channel With Discrete Heat Sources at the Walls, *Int. Comm. Heat Mass Transfer*, vol. 21, pp. 377-387, 1994.

156. W.J. Chang and W.L. Chang, Mixed Convection in a Vertical Tube Partially Filled With a Porous Medium, *Num. Heat Transfer A*, vol. 28, pp. 739-754, 1995.

157. W.J. Chang and W.L. Chang, Mixed Convection in a Vertical Parallel-Plate Channel Partially Filled With Porous Media of High Permeability, *Int. J. Heat Mass Transfer*, vol. 39, pp. 1331-1342, 1996.
158. A.J. Chamkha, Non-Darcy Fully Developed Mixed Convection in a Porous Medium Channel With Heat Generation/Absorption and Hydromagnetic Effects, *Num. Heat Transfer A*, vol. 32, pp. 653-675, 1997.
159. G.J. Hwang, C.H. Chao, Effects of Wall Conduction and Darcy Number on Laminar Mixed Convection in a Horizontal Square Porous Channel, *J. Heat Transfer*, vol. 114, pp. 614-621, 1992.
160. F.C. Chou and P.Y. Chung, Effect of Stagnant Conductivity on Non-Darcian Mixed Convection in Horizontal Square Packed Channels, *Num. Heat Transfer A*, vol. 27, pp. 195-209, 1995.

ELEVEN

RECENT PROGRESS AND SOME CHALLENGES IN THERMAL MODELING OF ELECTRONIC SYSTEMS

Y. Joshi

1 INTRODUCTION

Electronic systems are occupying an increasingly large part of our lives. Over the past decade, one of the most noticeable changes in our day-to-day activities is the ubiquitous presence of these, from ever smaller and more powerful computers to hand-held communication devices to "smart" automobiles. New product models are being aggressively introduced by companies at dizzying frequencies to capture consumer attention and stay competitive. For some products, such as laptop computers and cellular telephones, these cycle times are on the order of a few months. In order to ensure the products' success and companies' survival, their design must be right the first time, leaving little room for trial and error. In many of these new system developments, thermal design supported by computational modeling is playing a key role.

For example, a cellular telephone (Fig. 1) has almost become a modern necessity for many people. The electronics in these systems are housed in a compact enclosure and may dissipate 1-3 W [1]. In order to conserve precious battery power, all cooling must be passive, including a combination of conductive spreading, natural convection and radiation. The design must satisfy two key thermal limits. The chip operating temperature in typical commercial silicon devices is limited to about 125 °C. This limit is often a result of properties of materials used in packaging, such as the glass transition temperature of the printed wiring board, as well as degradation in circuit electrical performance. It can be increased to about 200 °C for silicon devices, with appropriate choice of materials and circuit technologies. Another crucial operating limit is the touch temperature of the casing of the enclosure. This must be maintained below about 40 °C in order to avoid skin injury. Thermal management schemes to maintain these limits are currently an active topic of research, as overall system sizes continue to shrink and volumetric heat dissipation rates continue to increase. A related activity that assesses the thermal fields within electronic systems is thermal characterization through modeling and measurements. Thermal modeling of electronic systems is the focus of this chapter.

In addition to being a tool to assess the adequacy of thermal management, thermal modeling is employed in support of a number of other activities associated with the development of electronic systems. Since electronic systems employ a number of packaging materials with differences in the coefficients of thermal expansion, thermal gradients inevitably result in thermo-mechanical stresses. The determination of these, and their impact on the life of the system, is a key design activity. Thermal fields in the form of temperatures and their spatial and/or temporal gradients become inputs to many reliability and performance assessment activities. For example, several failure mechanisms at the semiconductor chip and package level such as electromigration and fatigue of solder joints are thermally influenced. Additionally, key chip electrical parameters such as thermal noise degrade performance at elevated temperatures [2].

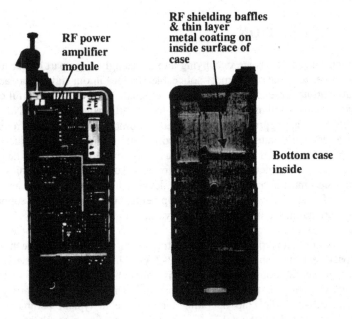

Figure 1 A typical cellular telephone [1]. The view on the left shows the printed wiring board containing the power amplifier, the highest power dissipation component. The view on the right displays the bottom case which conducts the heat to the outer surface, from which it is convected and radiated to the surroundings.

2 THE HIERARCHICAL NATURE OF ELECTRONIC SYSTEMS AND THEIR THERMAL MODELING

As seen in Fig. 2, electronic systems include a large variation in geometric length scales and thermophysical properties.   This arises naturally due to the multiple levels of assembly involved in the manufacturing of electronic products. Semiconductor chips containing the electronic circuits are usually encapsulated within various types of plastic or ceramic packages, with peripheral or area array interconnects for attachment to an organic or ceramic substrate, or printed wiring assembly or board (PWA or PWB). The PWA in turn is attached to the next level of packaging (typically a "mother board" or module) usually with edge connectors.   A plastic or metal enclosure serves as the final housing of the electronic system.

At the small scales associated with the semiconductor chips, in addition to an average chip or "junction" temperature, there is also an interest in knowing the temperature variations within localized regions (from few μm to tens of μm). These are required for the assessments of chip reliability and electrical performance. At the package and PWA levels, the interest is in the temperature field and its spatial gradients near the interconnects, which directly impact the long-term reliability of the solder joints. With the availability of computational fluid dynamics/heat transfer (CFD/HT) tools, there is considerable interest currently in simulating the fluid flows and temperature variations within electronic system enclosures.   As mentioned earlier, these allow for the assessment of the adequacy of the thermal management schemes. As system sizes

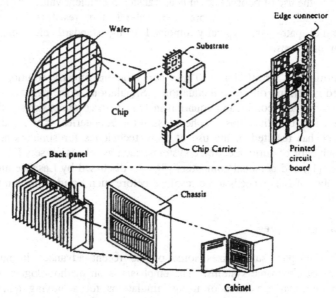

Figure 2 Widely varying length scales in electronic systems [3].

continue to shrink, the effect of enclosure boundaries on the thermal fields is becoming increasingly significant.

2.1 Thermal Modeling Approaches for Electronic Systems

Traditionally, thermal modeling of electronic systems has focused on one level of the packaging hierarchy. The schemes employed can be broadly categorized into three types:

2.1.1 Analytical Solutions. For a class of idealized situations involving the chip, package or the PWA, exact solutions of linear sets of governing equations have been obtained. The heat input is prescribed as a heat flux boundary condition or a uniform volumetric heat generation rate. For the chip or package, two-or three-dimensional heat conduction solutions have been carried out, e.g. Kennedy [4], Albers [5], Lee et al. [6]. PWA simulations include an extended surface type formulation, with one-or two-dimensional heat conduction.

2.1.2 Resistor Networks. Thermal resistor networks have been used extensively for simulation of electronic systems. The basic building blocks for these are one- and two-dimensional heat conduction solutions in rectangular and radial geometries.   Such networks have been used for simulations at the package, module, and system levels. Their greatest use continues to be for back-of-the-envelope calculations of package or module effective thermal resistances. Their conceptual simplicity and rapid solution times are offset by the potential for large errors due to the use of poorly known heat transfer coefficient values. Solutions to network models containing more than about ten resistors in multiple series/parallel paths are typically obtained using standard electrical circuit simulation programs.

2.1.3 Numerical Simulations. These computations employ solutions of the discretized governing differential equations. Simulations at the package or PWA level solve the heat conduction equation in two or three dimensions, subject to heat generation from the chips, and heat loss at the boundaries. The rapid growth in CFD/HT has resulted in the use of these techniques for simulations of the thermal behavior of entire electronic systems, such as television sets [7], cellular telephones [1], and personal computers [8]. As discussed by Lasance and Joshi [9], a number of issues needs to be resolved before such simulations can become truly predictive.

2.2 Outline of Chapter

This chapter summarizes some of the recent advances in numerical simulations of electronic systems. The emphasis is on methodologies that will enable the ultimate objective of using simulations for achieving quantitative thermal predictions during system design. We start by considering the complex interactions between the modes of heat transfer in typical electronic systems.

These interactions must be modeled carefully in order to achieve accurate thermal predictions. The large range in the relevant length scales and thermophysical properties is currently one of the biggest challenges in the thermal modeling of electronic systems. The determination of thermal information at different length scales can be facilitated by the use of a modeling approach that provides just the adequate level of detail at each level of system hierarchy. Information from system-level coarse grid models is fed into more detailed "zoomed-in" models of selected subsystems or components. This integrated approach is discussed with examples. The chapter ends with a discussion of some of the key barriers that make system-level predictive modeling currently unfeasible. Major efforts are currently under way to meet these challenges.

3 CONJUGATE TRANSPORT MODELING

Heat is generated in electronic systems within the semiconductor chips. It is transported through the packaging architecture and is ultimately rejected to the ambient environment from the boundaries of the system enclosure. Multiple heat transfer modes that are often comparable in magnitude are involved both in the transmission of heat within the system and its loss to the environment. The ratios of thermal conductivity of many electronic packaging materials (k_p) to common coolant fluids, air ($k_f = 0.027$ W/mK) and dielectric liquid fluorinert FC 72 ($k_f = 0.058$ W/mK), are collected in Table 1 at 300 K. It is seen that in all cases these values are considerably larger than one, signifying the importance of conduction within the various solid materials. Simulations of convection thus inevitably require a coupled accounting of heat spreading through conduction effects.

In air-cooled systems, radiation effects also need careful consideration. While radiative heat transfer in forced convection-dominated systems is often negligible, it becomes a comparable mode under natural and mixed convection conditions. Improper accounting of it, or its neglect, can result in large errors (over 20% in some cases) in predicted temperatures compared to measurements. In order to develop accurate computational thermal models of electronic systems, it is crucial to understand the interplay between these heat transfer modes. With the availability of three-dimensional CFD/HT capabilities, it has become possible to analyze these interactions. The most commonly employed formulation is based on finite-volume discretization of the primitive variables form of the governing equations in a single domain. It is relatively straightforward to include the large variations in thermophysical properties of the packaging materials by using the harmonic mean formulation [10].

The conjugate conduction and natural convection from small heat sources mounted on conducting substrates have been studied extensively in air [11-13] and dielectric liquids [14,15]. The temperature responses of the heat sources under these conditions are quite different from those obtained by prescribing convection coefficients from conventional correlations as surface boundary conditions.   The package material-to-fluid thermal conductivity ratio k_p/k_f determines the conductive spreading in the solid and the resulting thermal

Table 1 Thermal conductivity ratios of selected packaging materials employed in electronic systems to air and dielectric liquid FC 72 coolants at 300K.

Material	Thermal Conductivity Ratio (k_p/k_f)	
	Air Cooling	Liquid Cooling
Chip/Package Materials		
• Silicon	5692	11931
• Silicon Carbide	8991	18845
• Plastic Encapsulant	8	3.5
• Ceramic (Al$_2$O$_3$)	1385	2903
PWA Materials		
• Epoxy Fiberglass	12	24
• Ceramic (AlN)		
Enclosure Materials		
• Plastic	8	3.5
• Aluminum	9115	19105
• Magnesium	6000	12576
Attachment Materials		
• Solder (eutectic lead-tin)	1385	2903
• Polymer Epoxy	11	23

response of the heat source. Most of the earlier studies have focused on the responses of individual discrete heat sources, and not the effects of component interactions through the substrate, and the thermal interactions with the enclosure boundaries. As electronic systems continue to shrink in size, these effects are receiving greater attention. The complexity of conjugate interactions in compact electronic systems is illustrated through an example case study next.

3.1 Interactions Between Thermal Modes in a Laptop Computer Enclosure

Portable computing and communications devices are some of the fastest growing segments of the consumer electronics products market. The ever-reducing design cycle times of these products has resulted in increasing interest in developing predictive thermal modeling approaches. Due to the battery power and weight limitations, passive cooling including combined conduction, natural convection and radiation is usually employed for their thermal management. The compact sizes result in strong coupling between the various transport modes. Typically, models of entire systems include approximate representations of many of the constituent sub-systems such as electronic packages and heat sinks. Once the modeling approach is validated, parametric studies to evaluate influences of design changes such as increasing chip power, different component locations, alternate choice of materials and ambient environment conditions can be easily carried out.

As an example of the combined heat transfer effects, consider the transport in a small height-to-width aspect ratio enclosure in Fig. 3, investigated computationally and experimentally by Adams et al. [16,17]. The enclosure dimensions (0.1524 m x 0.1524 m x 0.0318 m on the inside) are representative of a laptop computer. The enclosure is constructed of 0.0127-m-thick plexiglas side walls, a 0.0063-m-thick plexiglas bottom wall, and a 0.0095-m-thick nominally isothermal aluminum top plate. The internal power dissipation is simulated by a three-by-three array of 88-lead plastic quad flat packages (PQFP) containing silicon chip or die, seen in Fig. 4. The 22 package leads on each of the four sides of the chips are in a "gull wing" configuration and are soldered to an epoxy fiberglass printed wiring board (PWB). An air gap of 0.064 m exists between the PWB and bottom plexiglas plate of the enclosure.

Adams et al. [16-18] considered the combined natural convection, conduction and radiation effects within the enclosure in Fig. 3. The entire enclosure, including the PWB and electronic packages, is modeled using a single computational domain. Time-dependent, laminar governing conservation equations, assuming negligible viscous dissipation and pressure stress effects, are employed and Boussinesq approximations are invoked. The resulting governing equations are [19]:

continuity:

$$\frac{\partial u}{\partial x} + \frac{\partial v}{\partial y} + \frac{\partial w}{\partial z} = 0 \qquad (1)$$

x momentum:

$$\rho\left[\frac{\partial u}{\partial t} + \frac{\partial}{\partial x}(uu) + \frac{\partial}{\partial y}(vu) + \frac{\partial}{\partial z}(wu)\right] = -\frac{\partial p}{\partial x} + \mu\left[\frac{\partial^2 u}{\partial x^2} + \frac{\partial^2 u}{\partial y^2} + \frac{\partial^2 u}{\partial z^2}\right] \qquad (2)$$

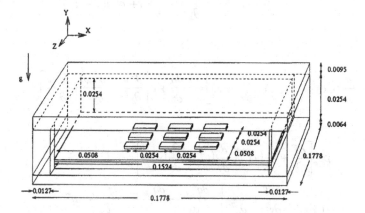

Figure 3 Laptop enclosure with electronic package array [19]. All dimensions are in m.

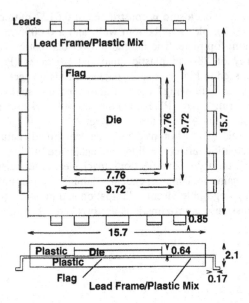

Figure 4 Details of the plastic quad flat (PQFP) electronic package [19]. All dimensions are in mm.

y-momentum:

$$\rho\left[\frac{\partial v}{\partial t}+\frac{\partial}{\partial x}(uv)+\frac{\partial}{\partial y}(vv)+\frac{\partial}{\partial z}(wv)\right]=$$

$$-\frac{\partial p}{\partial y}+\mu\left[\frac{\partial^2 v}{\partial x^2}+\frac{\partial^2 v}{\partial y^2}+\frac{\partial^2 v}{\partial z^2}\right]+\rho g\beta(T-T_r) \tag{3}$$

z-momentum:

$$\rho\left[\frac{\partial w}{\partial t}+\frac{\partial}{\partial x}(uw)+\frac{\partial}{\partial y}(vw)+\frac{\partial}{\partial z}(ww)\right]=-\frac{\partial p}{\partial z}+\mu\left[\frac{\partial^2 w}{\partial x^2}+\frac{\partial^2 w}{\partial y^2}+\frac{\partial^2 w}{\partial z^2}\right] \tag{4}$$

energy:

$$\rho c_p\left[\frac{\partial T}{\partial t}+\frac{\partial}{\partial x}(uT)+\frac{\partial}{\partial y}(vT)+\frac{\partial}{\partial z}(wT)\right]=$$

$$k\left[\frac{\partial^2 T}{\partial x^2}+\frac{\partial^2 T}{\partial y^2}+\frac{\partial^2 T}{\partial z^2}\right]-\frac{\partial q''_{rad,x}}{\partial x}-\frac{\partial q''_{rad,y}}{\partial y}-\frac{\partial q''_{rad,z}}{\partial z}+q''' \tag{5}$$

The modified Rayleigh number (Ra_H^*) and Prandtl number (Pr) are defined by:

$$Ra_H^* = \frac{g\beta q_T H^2}{k_f \alpha_f \nu_f}, \quad \Pr = \frac{\nu_f}{\alpha_f} \qquad (6)$$

It is noted that the above formulation applies for laminar conditions. For the highest power level experiments by Adams [19], thermocouple measurements in the air above the packages revealed significant time-wise fluctuations indicating departures from laminar flow. However, the maximum chip temperatures for these power levels were larger than the conventionally acceptable limits, and hence modeling of transport is confined to laminar conditions in this chapter.

3.2 Radiation Formulation

Radiative heat transfer is included by using an irradiation/radiosity method which assumes that the air is non-participating, the radiating surfaces are opaque, and that radiation exchange between surfaces is gray and diffuse. The net radiative flux from surface i is given by

$$(7)$$

$$q''_{rad,i} = \frac{\sigma T_i^4 - J_i}{(1-\varepsilon_i)/\varepsilon_i A_i} = \sum_{j=1}^{N} \frac{J_i - J_j}{(A_i F_{ij})^{-1}},$$

where σ is the Stefan-Boltzmann constant, $q''_{rad,i}$, T_i, J_i, ε_i, and A_i are the net radiative heat flux, temperature, radiosity, emissivity, and area of the ith surface, respectively, and F_{ij} is the view factor from the ith to jth surface. View factors from surface i to surface j are given by:

$$(8)$$

$$F_{ij} = \frac{1}{A_i} \int_{A_i} \int_{A_j} \frac{\cos\theta_i \cos\theta_j}{\pi R^2} dA_j \, dA_i$$

where As are the surface areas, θs are the angles between the two surfaces and their respective normals, and R is the distance between the two surfaces. Analytical expressions for the view factors for this configuration are available from Howell [20]. Solving eq. (3.21) for the radiosity, J_i, results in a system of algebraic equations given by:

$$(9)$$

$$J_i = \varepsilon_i \sigma T_i^4 + (1-\varepsilon_i)\sum_{j=1}^{N} F_{ij} J_j,$$

which are to be solved simultaneously. From the solution for radiosity, the radiative heat flux from each surface is determined. It is assumed that this heat flux is uniform over the surface. The radiative heat fluxes are applied uniformly to their respective control volume faces at the air-solid boundary such that the conduction heat flux on the solid side of the interfaces match the combined radiative and convective heat fluxes on the air side of the interface.

3.3 Boundary Conditions

A constant temperature of 20 $^\circ$C is specified for the top surface of the enclosure, and adiabatic boundary conditions are specified for all other outer boundaries. The resulting boundary conditions are:

(10)

$$x = 0., \qquad \frac{\partial T}{\partial x} = 0., \qquad u = v = w = 0.$$

$$x = X_L, \qquad \frac{\partial T}{\partial x} = 0., \qquad u = v = w = 0.$$

$$z = 0., \qquad \frac{\partial T}{\partial z} = 0., \qquad u = v = w = 0.$$

$$z = Z_L, \qquad \frac{\partial T}{\partial z} = 0., \qquad u = v = w = 0.$$

$$y = 0., \qquad \frac{\partial T}{\partial y} = 0., \qquad u = v = w = 0.$$

$$y = Y_L, \qquad T = T_r, \qquad u = v = w = 0.$$

The origin of the coordinate system is at the bottom left corner of the box on the back face in Figure 3. For the radiation transfer, the boundary conditions are built into the radiosity equations themselves.

3.4 Additional Modeling Considerations

Before a numerical simulation is attempted, a number of modeling issues needs resolution. The PWB is an anisotropic composite consisting of a fiber matrix within an epoxy resin. Copper metalization traces on its upper face also need to be accounted for. It is modeled as a homogeneous material with uniform effective thermophysical properties, estimated by considering thermal resistor networks and validated experimentally [19].

For simplifying the calculations of gray diffuse radiative heat transfer within the enclosure, the electronic packages are considered flush with the lower surface of the enclosure. The walls of the enclosure are divided into isothermal surfaces, based on area-averaged temperatures determined from the CFD/HT computations. These temperatures, along with the appropriate view factors and emissivities, are used in the radiation calculations.

3.5 Methodology for Numerical Computations

A single-domain approach using the SIMPLER algorithm to handle the pressure-velocity coupling as described by Patankar [10] is used to solve the governing equations. A non-uniform, staggered grid is used for the velocities and pressure/temperature. A power law scheme is used to discretize the convection and diffusion terms. Non-uniformities in material properties are handled by using harmonic mean values at the interfaces. In solid materials (package, substrate, enclosure walls), large values of the viscosity and zero volumetric expansion coefficients are prescribed. A fully implicit scheme is employed in which the steady-state results are obtained by using a single large time step.

3.5.1 Incorporation of radiation in computations. Consider a one-dimensional geometry with a control volume of unit width and thickness located at the surface of an object in Figure 5. From eq. (5), the steady-state energy equation with uniform heat generation is given by:

$$\frac{\partial}{\partial x}\left(\rho c_p u T\right)=\frac{\partial}{\partial x}\left(k\frac{\partial T}{\partial x}\right)-\frac{\partial q_{rad,x}}{\partial x}+q''' \tag{11}$$

Integrating the energy equation over the control volume results in

$$\int_x^{x+\Delta x}\frac{\partial J}{\partial x}dx = \int_x^{x+\Delta x}\left(q'''-\frac{\partial q'''_{rad,x}}{\partial x}\right)dx \tag{12}$$

Where $J = -k\frac{\partial T}{\partial x}+\rho c_p u T$ is the heat flux due to convection and conduction. The resulting expression is

$$J_e - J_w = q''\Delta x - \left[\left(q''_{rad,x}\right)_e-\left(q''_{rad,x}\right)_w\right] \tag{13}$$

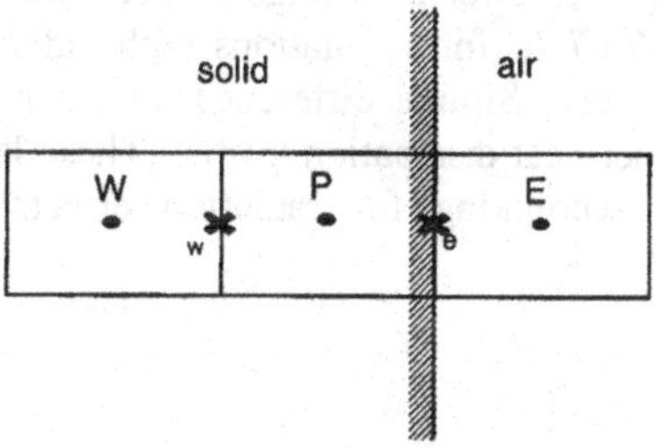

Figure 5 One-dimensional example used in describing the inclusion of thermal radiation in the computational methodology.

where the lower case subscripts represent the condition at the interfaces of the control volume. Since side w is internal to the solid surface and $q''_{rad,i}$ is a prescribed value from the radiation subroutine, the radiative heat flux term is combined with the heat generation term to form the source term, S, in the governing equation. In this manner, the net radiative heat flux on a surface is included in the CFD computations. A negative value of $q''_{rad,i}$ would indicate radiative heating of the control volume while a positive value would result in radiative cooling of the control volume.

3.6 Computations of Combined Transport Modes

Computations were performed with and without the radiation effects for uniform package power inputs in the range of 0.047 W to 0.75 W (total enclosure heat generation of 0.423 W to 6.75 W). The influence of heat conduction in the enclosure walls was also evaluated. The natural convection flow patterns were qualitatively similar for the various cases, although the maximum velocities in the enclosure central plane are lowered by about 10% for a power input of 0.5 W per component, when radiation is accounted for. Significant differences were found in the computed temperatures at the heat sources depending upon the treatment of conduction and radiation effects. Accurate thermal characterization required the inclusion of enclosure wall conduction, conduction across the air layer under the PWB and radiative transfer between the enclosure surfaces.

The computed temperature patterns without and with radiation effects are seen in Figs. 6a and 6b in a vertical plane through the enclosure center (z=0.0889 m). Near the enclosure interior walls, the effects of radiative heating result in significant differences in the shapes of the temperature contours. The lower air flow velocities in the presence of radiation also result in greater necking and reduced definition of the individual plumes above the heat sources.

The predicted temperatures with and without radiation are compared to experimental measurements for the same configuration in Tables 2 and 3 as junction (chip)-to-ambient (upper cold plate), and board (PWB)-to-ambient temperature differences, ΔT_{ja} and ΔT_{ba} respectively. The temperature difference ΔT_{ja} for the central heat source for a package power input of 0.5 W/package was 77.7 °C, 142.1 °C, and 75.7 °C for simulations with radiation, without radiation, and experiment, respectively. Similar differences are seen for the corner and side heat sources, and for other heat dissipation levels. These large differences clearly indicate the need for accounting for radiation effects for accurate thermal predictions.

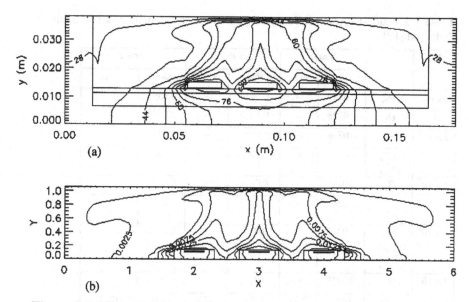

Figure 6 Isotherms in a vertical plane through the enclosure centerline; (a) without radiation, (b) with radiation.

Table 2 Comparison of Measured and Simulated Package Junction-to-Ambient and PWB Board-to-Ambient Temperature Differences [19]

Configuration	Package ΔT_{ja} (°C)			PWB ΔT_{ba} (°C)					
	corner	side	center	corner	center	side	loc. 1	loc. 2	loc. 3
P = 0.047 W/pkg, exp.	8.0	8.8	10.0	6.9	9.0	7.8	5.9	3.4	2.3
P = 0.047 W/pkg, sim.	8.1	8.9	10.2	7.5	9.6	8.3	6.3	3.2	1.9
P = 0.25 W/pkg, exp.	33.3	37.4	42.5	29.5	38.5	33.3	24.4	13.6	8.5
P = 0.25 W/pkg, sim.	36.1	39.3	44.3	33.1	41.3	36.4	26.7	13.3	8.0
P = 0.50 W/pkg, exp.	60.4	67.3	75.7	53.1	67.8	59.4	43.1	23.9	15.1
P = 0.50 W/pkg, sim.	64.7	69.7	77.7	59.0	72.1	64.2	47.0	23.2	14.2
P = 0.75 W/pkg, exp.	83.7	92.3	102.9	72.2	90.3	79.9	58.1	33.3	22.1
P = 0.75 W/pkg, sim.	90.9	97.3	107.7	82.7	99.5	89.3	65.4	32.3	19.9

Table 3 Junction-to-Ambient Temperature Difference for Simulations Without Radiation [19]

Ra_H^*		center		side		corner	
		sim.	exp.	sim.	exp.	sim.	exp.
cond.	Θ_{ja}	0.0729	--	0.0657	--	0.0595	--
1	Θ_{ja}	0.0728	--	0.0656	--	0.0593	--
1×10^4	Θ_{ja}	0.0738	--	0.0653	--	0.0584	--
	ΔT_{ja} (°C)	0.5	--	0.4	--	0.4	--
1×10^5	Θ_{ja}	0.0596	--	0.0464	--	0.0398	--
	ΔT_{ja} (°C)	3.9	--	3.0	--	2.6	--
1×10^6	Θ_{ja}	0.0370	--	0.0284	--	0.0252	--
	ΔT_{ja} (°C)	24.1	10.0	18.5	8.8	16.4	8.0
1×10^7	Θ_{ja}	0.0218	--	0.0187	--	0.0178	--
	ΔT_{ja} (°C)	142.1	75.7	121.9	67.3	116.0	60.4

3.6.1 Influence of Model Assumptions. The effects of various modeling assumptions on the computed temperatures were systematically evaluated by comparing the results with measurements. These comparisons for the junction and board temperature increases are presented in Table 4. The baseline case neglects heat conduction in the walls and air gap under the PWB, as well as radiation. The mean Nusselt numbers over the packages and the PWB are collected in Table 5 for the various cases. These include the values for the individual packages, a value for the entire array treated as one large effective source, and the substrate (excluding regions under the packages). Due to the symmetry in computations, only the center, side and corner packages represent results for the entire array.

Inclusion of wall conduction, Case 2, reduces the magnitudes of the air velocities and temperatures. Cases 3-5 examine the effect of the cold plate emissivity. As it increases, the reflected irradiation from the cold plate to the enclosure walls and substrate is reduced. A higher cold plate emissivity results in effective radiative cooling of the PWB. The PWB thermal conductivity is varied between 0.35 W/mK (based on handbook values for epoxy fiberglass) and 0.8 W/mK (to account for substrate metalization).

The effects of interactions between the substrate conduction, convection and radiation are examined through three comparisons: (i) cases 1 and 8, without enclosure wall conduction and radiation, (ii) cases 3 and 9, with enclosure wall conduction and radiation ($\varepsilon_{Al} = 0.07$), and (iii) cases 4 and 6, similar to cases 3 and 9 respectively, but with $\varepsilon_{Al} = 0.2$. The substrate acts as an extended surface whose heat transfer effectiveness depends not only on its thermal conductivity but also its convective and radiative interactions.

Table 4 Effect of Model Assumptions on Chip Junction-to-Ambient and PWB Board-to-Ambient Temperature Differences [19]

Case	Configuration	Heater ΔT_{ja} (°C)			Substrate ΔT_{ba} (°C)					
		corner	side	center	corner	center	side	Loc. 1	Loc. 2	Loc. 3
exp.	Pwr = 0.5 W/htr (Ra_H^* = 1 ×10^7)	60.4	67.3	75.7	53.1	67.8	59.4	43.1	23.9	15.1
1	Base = no wall or radiation, k_{PWB} = 0.35 W/m-K	108.6	115.6	135.7	103.2	130.4	110.7	66.5	31.5	21.2
2	Base + encl. wall and air gap under PWB	92.4	98.7	114.8	83.8	106.3	90.7	54.4	24.7	14.2
3	Base + wall, air gap, and rad. (ε_{Al} = 0.07)	77.2	79.7	86.5	69.6	79.2	72.4	45.5	24.1	15.4
4	Base + wall, air gap, and rad. (ε_{Al} = 0.2)	76.4	78.9	85.8	68.6	78.2	71.5	43.5	20.6	13.1
5	Base + wall, air gap, and rad. (ε_{Al} = 0.9)	68.2	70.7	77.3	60.2	79.6	63.0	34.1	11.0	4.5
6	Base + wall, air gap, rad. (ε_{Al} = 0.2), and k_{PWB} = 0.8 W/m-K	69.7	73.5	81.2	60.2	72.0	64.5	43.0	22.8	14.4
7	Base + wall, air gap, rad. (ε_{Al} = 0.2), k_{PWB} = 0.8 W/m-K, and lead footprint	64.7	69.7	77.7	59.0	72.1	64.2	47.0	23.2	14.2
8	Base + k_{PWB} = 0.8 W/m-K	101.3	111.4	131.3	94.3	124.1	104.6	64.6	33.3	23.0
9	Case 3 + k_{PWB} = 0.8 W/m-K	73.2	77.1	84.9	63.7	75.5	68.0	46.4	25.8	17.1

Table 5 Effect of Modeling Assumptions on Package and PWB Mean Nu for P_d = 0.5 W/package [19]

Package Mean Nu			Array Mean Nu	PWB Mean Nu
corner	side	center		
7.1	7.2	5.5	7.0	1.4
7.3	7.4	5.6	7.2	2.8
6.6	6.4	4.8	6.3	2.3
6.7	6.6	5.0	6.5	2.2
7.2	7.0	5.4	6.9	2.2
6.3	6.0	4.3	6.0	2.4
5.8	5.5	3.8	5.4	2.6
7.1	7.1	5.1	6.9	2.8
6.3	6.1	4.3	6.0	2.5

For enclosures with low emissivity walls (e.g. $\varepsilon_{Al} = 0.07$), radiative heating of the outer regions of the substrate and the enclosure walls increases the overall temperature of the substrate and reduces the effectiveness of the substrate to conduct heat away from the heat sources. Increasing ε_{Al} results in increased radiative heat transfer from the PWB, lower PWB temperatures, and increased heat conduction from the heat sources. Increasing the PWA thermal conductivity under these conditions has a greater effect on thermal performance of the enclosure.

It is clear from this detailed example that complex interactions exist between the various transport modes. The influence of each mode must be assessed in developing system-level models. This example also brings out the diversity of length scales and thermophysical properties present in electronic systems. The number of grid points required to resolve these simultaneously is prohibitively large in most cases. In order to make system-level simulations computationally efficient, some of the smaller length scales must be eliminated, with their thermal effects appropriately modeled. Next we describe the accomplishment of this objective through the development of compact or reduced models.

4 REDUCED OR COMPACT MODELS

The internally complex thermal transport within electronic packages has traditionally been represented as single effective thermal resistances. The primary appeal of these single resistance representations is their easy incorporation into thermal network analyses. These single resistances have the following form:

$$\Theta_{jx} = (T_j - T_x) / P \qquad (14)$$

where T_j and T_x are the temperatures at the chip junction and a specified reference location x, and P is the power dissipation. Commonly used reference locations are the external surface ("case") of the package and the ambient. However, these thermal resistance measures do not provide unique values and have often been used incorrectly. These single-resistance reduced or compact models are incapable of accounting for factors such as cooling method, power dissipation, and package to PWB thermal coupling.

More accurate representations of the package have been developed using thermal resistance networks and block models of effective, homogeneous materials. Examples of these are seen in Figs. 7 and 8. Both types of models can be incorporated within CFD/HT simulations and can substantially reduce computation times for system-level simulations. Before its use, any compact model must be validated to ascertain the correctness of assumptions made in the development of the models. Comparison of a detailed package model with a compact model in a simulated system can be used for this purpose, provided that the system and details of the package are accurately modeled, and any uncertainties in material properties are bounded.

Validated compact models of individual components can be used in simulating multi-component systems where conjugate thermal interactions between the components are significant. The three-by-three array laptop enclosure in Fig. 3 has been simulated by Adams et al. [16-18] with compact representations of individual packages shown in Figs. 7a and 8a-8c. The various reduced representations were found to provide accurate system simulations, with considerable savings in computational time.

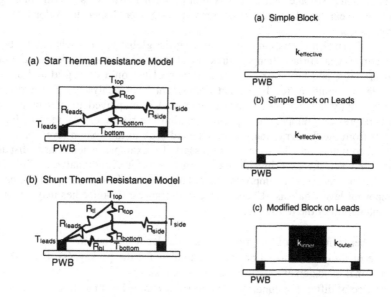

Figure 7 Compact thermal resistance modeling strategies: (a) four resistor, star topology thermal resistance model and (b) six resistor, shunt thermal resistance model.

Figure 8 Compact block modeling strategies: (a) simple block model, (b) simple block on lead (BOL) model, and (c) compound (modified) BOL.

5   INTEGRATED MULTI-SCALE MODELING

While the concept of compact or reduced models is useful for reducing the computational effort associated with coarse-grid system-level simulations, a need still remains for detailed component level thermal modeling for determining local hot spots and for electrical performance and reliability assessments. Investigations of such multi-scale simulations of electronic systems have recently been undertaken by several investigators [21-25]. One approach is to carry out successive levels of computations over increasingly localized computational domains, with just the adequate detail at each level of modeling [26]. As illustrated in Fig. 9 [27], at the system level, a compact representation of the individual components is utilized. At the package level, however, a more detailed representation is required. This zoomed-in model utilizes information from computations performed at the system level.

The first step in the integrated modeling process in Fig. 9 is the development of a model of the entire system on a coarse mesh. The various packages, PWBs, heat sinks, and other significant sub-systems are simulated with reduced or compact models. Details such as individual copper traces on PWBs, individual leads of packages, and individual fins on heat sinks are ignored or specified as modified effective thermal properties at this stage to avoid excessive computational storage and time. A non-uniform mesh is used with finer grid spacings near solid surfaces to resolve large gradients in velocities and temperatures.

Thermal information extracted using the global model includes the board and component surface temperatures, local heat transfer coefficients, reference temperatures, and heat flux. These are interpolated on a finer grid and used as boundary conditions for board-and component-level analyses. For a detailed thermal analysis of a component, a finer grid is employed to predict junction temperatures and temperature gradients within the package. As seen in Fig. 9, results from the fine grid analysis may be fed back to the system level for a more refined computation. This may be necessary, for example, when a large disparity exists between the experimental measurements and thermal analysis. This may also be the case for an improper modeling choice of boundary conditions for component level analysis. Under such circumstances, the remedies may be one or a combination of the following:

- further refinement of the grid in the system-level simulation.
- use of different reduced or compact models for the package(s), PWB(s), and thermal management hardware such as heat sinks.
- choice of different boundary conditions for zoomed-in modeling.

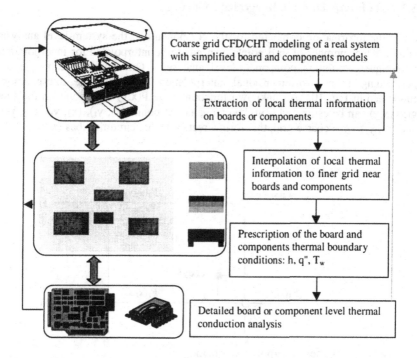

Figure 9 Multi-scale thermal modeling [27].

5.1 Role of Reduced or Compact Thermal Models in Multi-Scale Simulations

As illustrated in Fig. 9, the purpose of using such a model is to reduce the computational effort for the global simulations. The selected model should provide a realistic representation of the actual heat flow paths without using a very fine grid. Some choices of compact models have been shown in Figs. 7 and 8. In layered or block-type models (Fig. 8), the components and PWBs are described with multiple homogeneous layers to represent the key internal structures [16-19, 26, 28]. These layers extend throughout the package or PWB in question. Homogenization techniques can be employed to model effective thermophysical properties for each layer. Package to PWB interconnects can be represented as a peripheral ring (Fig. 8b) for leaded configurations and as one solid layer under the package for area array type interconnects. A considerable reduction in the number of grid points deployed for individual components is obtained along with a significant reduction in the total number of grid points required for system-level modeling. This process also smooths the spatial variations in the thermal properties.

5.2 Data Interpolation in Integrated Modeling

Since the information computed for the components in the system-level analysis is on a fewer number of grid points, it needs to be interpolated for incorporation into higher resolution component-level analyses. One possible technique for performing such two-dimensional interpolation is a bilinear interpolation, illustrated in Fig. 10. An unknown function value $F(x,y)$ on a point on the fine grid (x,y) can be estimated from known values at points (x_1, y_1), (x_2, y_1), (x_2, y_2), and (x_1, y_2), on a coarse grid, that form a rectangle that circumscribes (x,y) :

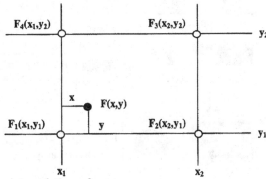

Figure 10   Bilinear interpolation scheme.

$$F = (1-\alpha)(1-\beta)F_1 + \alpha(1-\beta)F_2 + \alpha\beta F_3 + (1-\alpha)\beta F_4 \qquad (15)$$

with $\alpha = \dfrac{x - x_1}{x_2 - x_1}, \ \beta = \dfrac{y - y_1}{y_2 - y_1}$

5.3 Example of an Integrated Methodology Computation

To illustrate the methodology outlined in Fig. 9, we consider its application to a natural-convection-cooled enclosure [26] with electronic components of different size mounted on a PWB (see Figs. 11a and 11b). The specific component analyzed is a leadless ceramic chip carrier (LCCC), seen in Fig. 11a. Figure 12 shows the lumped and multi-layered compact or reduced models of the LCCC, employed in the global (or system-level) simulations. The integrated-analysis approach was validated by comparing the results with those obtained from a fine-grid thermal analysis, where the detailed LCCC representation was employed. The effects of boundary conditions obtained from global analysis, such as heat transfer coefficients, surface temperatures and heat fluxes, on the accuracy of component-level solution were evaluated.

In the *lumped model*, the LCCC is represented as a solid block of the same external dimensions, with uniform effective properties and volumetric heat generation. The effective thermal conductivity of the block was taken as the volume-fraction-averaged value of the constituent packaging materials. The effective density and specific heat were taken as the mass-fraction-averaged values. In the *multi-layered model*, the layers from top to bottom represent: lid, material layer above chip, uniform heat generation layer, and material layer below chip. In the *detailed model*, except for the input/output connections between the chip and the substrate, the details illustrated in Fig. 10c were implemented.

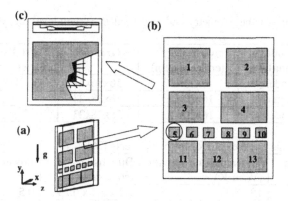

Figure 11 (a) Natural convection enclosure, (b) PWB layout, and (c) Leadless Ceramic Chip Carrier (LCCC) [26].

5.3.1 System-Level Computations. Three sets of computations were performed at the system level corresponding to the three models outlined above for component 5. The geometrical dimensions associated with the enclosure and the components, as well as the grid sizes, are summarized in Tables 6 and 7. Outside of the component, grid point placement was identical for the three sets of computations.

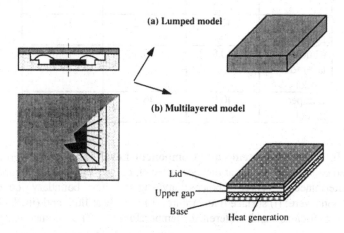

Figure 12 Compact or Reduced Models of LCCC [26]

While 3x3x4 internal grid points were used in representing multi-layered and lumped models, 21x21x12 points were used in the detailed modeling of component 5. Both types of reduced models gave good predictions of the heat flow paths from the component, as summarized in Table 8. Table 8 shows that conduction to the PWB through the bottom surface and convection from the top surface are dominant heat flow paths from the component, with less than 13% of the net power lost through the four side walls by convection.

Table 6 Summary of the geometry and grid used in system-level analysis for the enclosure [26]

Component model in system-level analysis	Geometry (cmx cmx cm)	Grid size	Grid distribution for component 5
Detailed model	16x23x2.3	56x56x31	21x21x12
Multi-layered model	16x23x2.3	38x38x23	3x3x4
Lumped model	16x23x2.3	38x38x23	3x3x4

Table 7 Summary of the data for the components [26]

Components	Thermal conductivity (W/m-K)	Dimension (cm x cm x cm)	Power (W)
1-2	18	5.2 x 5.2 x 0.23	0.5
3	18	4.64x 4.2x 0.23	0.25
4	18	6.04x 4.2x 0.23	0.25
5-10	18	1.4x 1.143x 0.23	0.5
11-13	18	3.72x 4x 0.23	0.5
PWB	0.35	16x 23x 0.2	0.0

Table 8 Thermal budget from the system-level solution for component 5 [26]

	Top surface (%)	Bottom surface (%)	Side walls (%)
Detailed geometry	37.9	50.3	11.8
Multi-layered model	40.8	46.4	12.8
Lumped model	40.9	46.3	12.8

5.3.2   Integrating System-and Component-Level Analyses. From the converged coarse grid results at the system level, component boundary conditions were determined. Three choices investigated for boundary conditions specifications were: (i) surface temperature, (ii) local heat flux, and (iii) local heat transfer coefficient and reference temperature. The component level computations are for component 5 in Fig. 11. The thermophysical properties of the constituent materials are provided in Table 9. The three-dimensional heat conduction equation is solved for the component with the specification of boundary conditions extracted from the system level solution. A number of combinations of boundary conditions is possible and six choices were examined by Tang and Joshi [26], as described in Table 10. Their effectiveness was evaluated by comparing the resulting temperature distributions with those obtained from the detailed conjugate system-level model.

Table 9 Material properties used for various parts of component 5 [26]

	Thermal conductivity (W/m-K)	Density (kg/m^3)	Specific heat (J/kg-K)
Chip: silicon	148	2330	712
Lid: Kovar	18	8360	439
Base: Al$_2$O$_3$	18	3970	765
Air	0.026	1.177	1005

Table 10 Possible B.C. combinations for component level thermal analysis [26]

B.C. types	Bottom surface	Top surface
BFTF	heat flux	heat flux
BFTH	heat flux	convection
BFTT	heat flux	prescribed temperature
BTTF	prescribed temperature	heat flux
BTTH	prescribed temperature	convection
BTTT	prescribed temperature	prescribed temperature

Figure 13 shows the computed temperature contours for component 5, obtained from the detailed model. This is used as a basis for comparison of the results from the integrated modeling approach for the various boundary condition choices for component-level modeling. Some of the key features are the maximum temperature of 125.1 °C located on the chip, symmetric temperature contours above the centerline below the lid and non symmetric distribution on the lid due to thermal interactions with neighboring components.

An example of the results obtained using the integrated formulation is provided in Fig. 14 for the prescribed wall heat flux boundary condition for the side walls. Comparison of Fig. 13a with Fig. 12a reveals an excellent replication of results when heat flux boundary condition (BC) is assigned to the bottom surface and convection BC to the top surface. However, a very large number (nearly 50,000) of iterations was needed to achieve convergence. For prescribed temperature of the bottom surface and prescribed heat flux or convection coefficient on the top surface, an acceptable replication of the component temperature pattern is found in Figs. 13c and 13d. However, when a prescribed temperature BC was used for the top wall, the temperature distribution was not properly replicated, as seen by comparing Figs. 13b and 13e with Fig. 12a.

The above example illustrates the application of the integrated thermal analysis approach described in this section. This methodology allows the use of computed information extracted from coarse grid CFD/HT simulations for fine grid analysis of zoomed regions. The two types of reduced or compact models (Fig. 11) were found to provide very similar results. It has been found by Adams et al. [18] that if the chip size is much smaller than the package size, additional layers to describe the heat generation may have to be incorporated, as illustrated in Fig. 7c. It is also seen that for the component-level computations, specified heat flux or convection BC for the top wall and heat flux BC for the bottom wall result in more realistic temperature patterns, compared to a prescribed temperature boundary condition for the wall facing the fluid (top wall). In the next section, we address the issue of computational efficiency enhancement of the integrated approach through advanced solution methods.

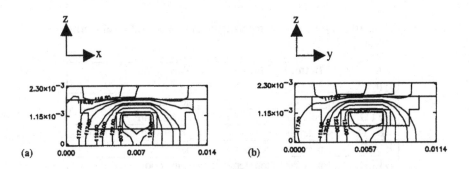

Figure 13 Computed isotherms (°C) for component 5 in two vertical planes, using the detailed model. Distances are in m.

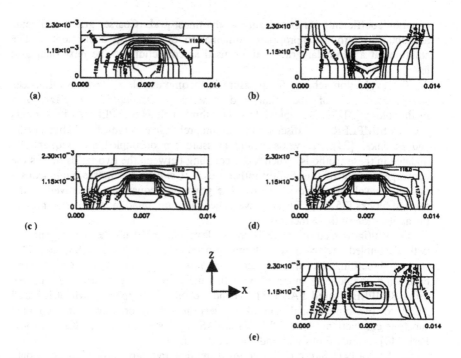

Figure 14 Temperature contours for component 5 with multi-layered model at system level, with B.C.s: heat flux for all the side walls; top and bottom walls, BFTH for (a), BFTT for (b), BTTF for (c), BTTH for (d), and BTTT for (e). (All for x-z plane at y = 5.07 mm)

5.4 A Faster Integrated Modeling Approach Using Multigrid Methods

The computational efficiency advantages of the integrated modeling approach described in Section 5.3 can be further enhanced through the use of multigrid methods.   It is well known that iterative methods used for the solution of discretized equations can effectively smooth errors of wavelength comparable to the mesh size. Their performance deteriorates with increase in the number of grid points. Also, single-level grids, as in Section 5.3, tend to be prohibitively slow to converge for problems involving a wide range of length scales.

Multigrid methods employ a hierarchy of grids to efficiently smooth errors of different wavelength. They differ in the scheme used for obtaining coarser grid discretization coefficients, and the communication between the coarse and fine grid solutions. *Full Approximation Storage (FAS)* scheme [29] computes and stores all discretization coefficients and solutions at every grid level. Solutions at a certain grid are interpolated or extrapolated, depending upon whether the algorithm is switching to or from a coarser grid. *Additive Correction Multigrid (ACM) or Block-Correction-Based Multigrid* method [30] uses discretization equation coefficients computed only on the finest grid. Coarser grid coefficients are obtained through appropriate combination of fine grid coefficients. The solutions at a particular grid level are used as corrections to the next finer grid level.

The evaluation scheme for pressure in incompressible flows also influences the performance of the multi-grid scheme. *Decoupled or Segregated* methodology [31-33] is employed in algorithms such as SIMPLE and its variants such as SIMPLER. As discussed in a comprehensive survey by Sathyamurthy and Patankar [34], the velocity and pressure are decoupled and sequentially updated in these methods. Such loose coupling between the two fields makes the convergence rate of this family rather slow. *Coupled* schemes simultaneously update the velocities and pressure. The governing equations are solved in the form of blocks [35, 36]. Faster convergence rates are found in most cases through the application of these methods.

Multigrid methods have been employed to speed up the convergence of both decoupled and coupled solution procedures. Fourka and Saulnier [37] studied three-dimensional conjugate transport during forced convection from protruding components mounted on a thermally conducting board in a channel, using SIMPLEC-based FAS [38]. Heindel et al. [39] applied SIMPLER-based ACM to two-dimensional natural convection simulations involving coupled substrate conduction. A SIMPLE based FAS approach was used by Kadinski and Peric [40] to study flows with radiative heat transfer.

Vanka [41] introduced a coupled implicit multigrid procedure to update velocities and pressures simultaneously which can be adopted with either ACM or FAS. This symmetrically coupled Gauss-Seidel (SCGS) scheme was used to study the viscous flow in a square cavity for a range of Reynolds numbers. With the combination of ACM-based multigrid method and SCGS (or MG-SCGS), 5 to 15 times increase in convergence rates, compared to the single grid SIMPLER method, was reported by Sathyamurthy and Patankar [34] for laminar flow in lid-driven square and cubic cavities and for flow over a 2-D backward-facing step. Similar large enhancements in computational efficiency were also confirmed in three-dimensional configurations of interest in the modeling of electronic systems by Tang and Joshi [42]. In the following, we present examples illustrating these.

5.5 Application of the MG-SCGS Approach to Electronic Systems

We consider two different configurations involving cooling of substrate-mounted discrete heat sources. These include a natural-convection-cooled rectangular enclosure (Fig. 15) and a mixed/forced-convection-cooled channel with a discrete heat source on the bottom wall (Fig. 16). The convergence rates using SIMPLER, SCGS, and MG-SCGS were compared for both configurations. It was found that convergence rates could be accelerated by applying multigrid technique with only two grid levels. With MG-SCGS, significant performance enhancements over SIMPLER and SCGS were obtained for both configurations.

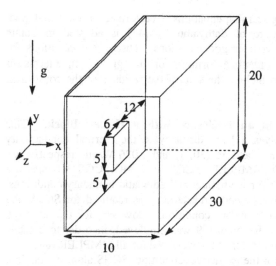

Figure 15 Natural-convection-cooled enclosure with a discrete heat source [42]. All units are in mm.

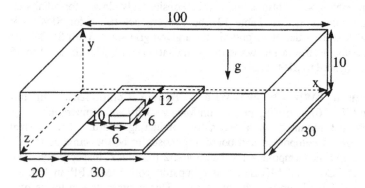

Figure 16 Mixed-and forced-convection cooled component in an enclosure [42]. All units are in mm.

The general governing equations for the fluid and solid regions have been presented in Section 3.1. The procedure to combine fine-grid equations to obtain the coarser block equations has been documented by Hutchinson and Raithby [43] for two-dimensional problems. An extension of this formulation to three dimensions was used by Tang and Joshi [42]. The detailed equations are available in Tang [27]. Coarser blocks containing an integral number of contiguous finer control volumes were first selected such that interfaces at material discontinuities were represented as control volume faces at all grid levels. Every coarser grid control volume was composed of eight finer grid control volumes.

Discretization of the governing equations is performed on the finest grid, and the combination procedure of Sathyamurthy [44] is adapted to obtain coefficients for successively coarser grid equations. Discretized equations for coarser grids are obtained with identical forms as for finer grid. Source terms for the coarse-grid equations also include the sum of the residues of the constituent fine-grid equations.

5.5.1 Natural Convection in an Enclosure with a Heated Block.   This configuration (Fig. 15) involves a large disparity in the thermal conductivity values of the component, board, and air (30, 1, and 0.026 W/mK respectively). The computational times for SIMPLER, SCGS, and MG-SCGS schemes are compared in Table 11 and 12 for isothermal and adiabatic boundary conditions, respectively, on the left wall. No under-relaxation was required for SCGS and MG-SCGS for the isothermal boundary condition. However, for the adiabatic condition, an under-relaxation factor of 0.9 was employed, in order to achieve convergence. Under-relaxation factor of 0.9 was used for all SIMPLER runs.

As seen in Table 11, for the isothermal condition, SCGS alone results in a 20-120% decrease in CPU time. The combination of MG-CGS results in a convergence rate 2 to 14 times faster compared to SIMPLER. As pointed out by Hutchinson and Raithby [43], the convergence rate of iterative solvers is dependent on boundary conditions and can be considerably slower for adiabatic-type boundary conditions. In Table 12, the convergence times for SIMPLER increase by almost an order of magnitude for a given grid size. With SCGS, the CPU times are about 40% faster, while the combination of MG-SCGS is 14 to 18 times faster compared to SIMPLER.

5.5.2 Flow in a PWB Channel With a Heated Component.   In this configuration (Fig. 16) incoming air with uniform velocity and temperature flows over a component mounted on a thermally conducting board. The thermal conductivities of the component and board are chosen to be the same as in Fig. 15. The board and the component are 1 mm and 2 mm thick, respectively. The performance of SCGS and MG-SCGS is compared with SIMPLER in Fig. 17 under mixed and forced convection conditions. This is presented in terms of a performance factor:

$$F_{CPU}= \text{CPU time with SIMPLER/CPU time with SCGS or MG-SCGS} \qquad (16)$$

When $Re/(Gr^*)^{0.5}$ is less than 0.06 in mixed convection, F_{CPU} is less than or slightly larger than 1 for all grids tested with SCGS. The tighter pressure-velocity coupling of SCGS does not result in performance enhancement compared to SIMPLER for this flow regime. With MG-SCGS, for the same flow regime, the range of F_{CPU} is between 6 and 14. Since SCGS does not contribute to this performance enhancement, it is attributed to the use of the multigrid method.

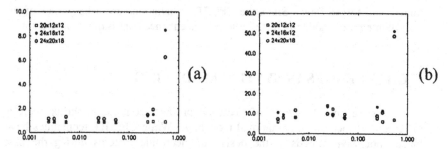

Figure 17 Performance enhancement factor for mixed/forced convection test problem using (a) SCGS and (b) MG-SCGS approach

Table 11 DEC Alpha CPU seconds for natural convection in an enclosure with one single heating block [42]. Base case study with isothermal B. C. for the left wall of the enclosure. $Q=0.1W$, $Ra=1.3\times10^5$

Grid	SIMPLER	SCGS	MG-SCGS
14 x 16 x 14	160	78	21
18 x 20 x 14	635	352	78
22 x 24 x 20	5005	2610	332

Table 12 DEC Alpha CPU seconds/iteration for natural convection in an enclosure with one single heating block, with adiabatic B.C. for the left wall of the enclosure. (18x20x14 grids) [42]

Power	SIMPLER	SCGS	MG-SCGS
0.1 W	6172	3241	318
0.2 W	5639	3173	339
0.3 W	5239	2990	346

When $Re/(Gr^*)^{0.5}$ is above 0.2, F_{CPU} of SCGS is larger than 1 for higher resolution grids, and for MG-SCGS it ranges between 6 and 14. For this regime, both SCGS and multi-grid method contribute to the performance enhancement of MG-SCGS. For $Re/(Gr^*)^{0.5} = 0.5$, SIMPLER did not yield converged results, while SCGS and MG-SCGS were able to meet the convergence criteria effectively.

The above examples clearly demonstrate the use of the MG-SCGS method in accelerating the computational performance of system-level simulations. Large reductions in computational time compared to SIMPLER are found for natural, mixed, and forced flow. Since the CFD/HT simulations are a large part of the overall integrated modeling methodology, this enhancement is quite significant.

6 CHALLENGES IN SYSTEM MODELING

As the times-to-market of electronic products continue to shrink, there is increased reliance on computational modeling. While CFD/HT techniques have made impressive inroads in the design of electronic systems during the past decade, a number of significant roadblocks remain before they can become truly predictive tools. A new generation of research advances must be made in order to address these. Many of the current challenges are common to other application areas, but the shortened product development cycles make their solution perhaps most critical for electronics applications.

Development of compact or reduced models needs to be done for various sub-systems, of interest in electronic systems. These include heat sinks, fans, and liquid flow through heat exchanger plates. Detailed CFD/HT computations of new heat sink designs such as impingement heat sinks (Sathe et al [45]) and micro-channel heat exchangers (Yin and Bau [46], Rujano and Rahman [47]) have recently been made. The computationally intensive nature of these requires development of simplified models for system-level modeling. A promising approach is the use of porous media modeling, employed for liquid cooled compact heat exchanger plates [48]. Development of heat sink optimization approaches and thermal modeling of optimized geometries is also of interest (Ledezma, et al. [49], Fisher and Torrance [50]). Validation of modeling approaches is a key activity that must be undertaken in concert with the development of reduced models. This may take the form of companion experimental studies, as well as extremely detailed computational studies.

Uncertainties in thermophysical properties of electronic packaging materials invariably prevent detailed critical comparisons between experimental measurements and thermal models of electronic systems. This situation must be rectified before significant needed progress in computational modeling can be justified. Electronic packaging materials such as PWB materials with and without metalization are often designed to display highly anisotropic behavior, requiring the determination of directional properties. Also, thin layers of highly conducting materials such as diamond are sometimes employed to get effective heat spreading (Lu et al. [51]). In conjugate models, techniques for estimating uniform effective thermal properties of these materials must be developed.

The modeling of transients in electronic systems is currently in its infancy. Environmental conditions are invariably time dependent. Also, systems such as phased-array radars dissipate power in a time periodic fashion. Extension of the integrated modeling approach outlined in this chapter to account for these effects needs to be undertaken. Current modeling approaches based on lumped capacitance analyses (Zimmerman and Colwell [52]) need to be extended to include spatial variations.

Transitional and turbulent conditions are encountered in large air-cooled systems. Only limited guidance on modeling these flows for geometries of interest in electronics cooling is currently available. Recent studies have investigated jet impingement cooling (Morris and Garimella [53, 54]) and impingement-cooled heat sinks [45]. Computationally efficient yet sufficiently accurate approaches remain as challenges ahead for the modeling community.

Acknowledgements: The author acknowledges the help of Drs. V.H. Adams (Motorola, Inc.), L. Tang (Fore Systems, Inc.) and Mr. K.A. Moores (University of Maryland) in the preparation of this chapter.

7 NOMENCLATURE

A	area $[m^2]$
c_p	specific heat at constant pressure $[J/(kgK)]$
F_{ij}	radiation view factor [non-dimensional]
g	gravitational acceleration $[m/s^2]$
h	heat transfer coefficient at surface $[W/(m^2K)]$
H	fluid region height [m]
J	radiosity $[W/m^2]$
k	thermal conductivity $[W/(mK)]$
p	pressure $[N/m^2]$
Pr	Prandtl number
q_T	total heat generation rate in enclosure [W]
q''_{rad}	radiative heat flux $[W/m^2]$
q'''	volumetric heat generation rate $[W/m^3]$
Ra_H^*	modified Rayleigh number [non-dimensional]
t	time [s]
T	temperature $[°C]$
ΔT_{ja}	heat source to cold plate temperature difference $[°C]$
ΔT_{ba}	substrate to cold plate temperature difference $[°C]$
u, v, w	velocity components [m/s]
x,y,z	coordinates [m]
X_L, Y_L, Z_L	enclosure dimensions [m]
α	thermal diffusivity $[m^2/s]$
β	coefficient of thermal expansion $[K^{-1}]$
ε	emissivity [non-dimensional]
μ	viscosity [kg/m-s]

ν   kinematic viscosity [m^2/s]

ρ   density [kg/m^3]

σ   Stefan-Boltzman constant [$W/(m^2 K^4)$]

θ   Angle between two surfaces

Θ   Non-dimensional temperature [non-dimensional]

Subscript

f   fluid

i,j   ith/jth package

p   package

PWB   substrate

r   reference

8 REFERENCES

1. K. Ramakrishna, T. Y. Lee, J. V. Hause, B. C. Chambers, and M. Mahalingam, Experimental Evaluation of Thermal Performance of And Cooling Enhancements to A Hand-Held Portable Electronic System, in *Process, Enhanced and Multiphase Heat Transfer: A Festschrift for A. E. Bergles*. Proc. of A. E. Bergles Symposium held at Georgia Institute of Technology, Atlanta, GA (November 16, 1996), pp. 217-226, Begell House, Inc., New York, NY, 1997.

2. F.P. McCluskey, R. Grzybowski, and T. Podlesak, Eds., *High Temperature Electronics*, Selection and Use of Silicon Devices at High Temperatures, Chapter 2, pp. 7-52, CRC Press, Boca Raton, Florida, 1997.

3. J.W. Dally, *Packaging of Electronic Systems*, McGraw-Hill Publishing Company, New York, 1990.

4. D.P. Kennedy, Spreading Resistance in Cylindrical Semiconductor Devices, *J. Appl. Physics*, Vol. 31, pp. 1490-1497, 1960.

5. J.H. Albers, An Exact Recursion Relation Solution for the Steady State Surface Temperature of a General Multi-Layer Structure, IEEE Transactions on CPMT, Vol. 18, pp. 31-38, 1995.

6. S. Lee, V.A.S. Song, and K.P. Moran, Constriction/Spreading Resistance Model for Electronics Packaging, *Proceedings of the ASME/JSME Thermal Engineering Conference*, Vol. 4, pp. 199-206, ASME, 1995.

7. C. Lasance, Numerical Simulation of Natural Convection in a Television Set, ASME HTD-Vol. 171, pp. 101-108, 1991.

8. R.L. Linton and D. Agonafer, Thermal Model of a PC, *ASME J. Electronic Packaging*, Vol. 116, pp.134-137, 1994.

9. C. Lasance, and Y. Joshi, Thermal Analysis of Natural Convection Electronic Systems: Status and Challenges, *Advances in Thermal Modeling of Electronic Components and Systems*, Bar-Cohen, A., and

Kraus, A. D., eds., Vol. 4, Chapter 1, ASME Press/IEEE Press, New York, 1998.

10.   S.V. Patankar, *Numerical Heat Transfer and Fluid Flow*, Hemisphere, New York, 1980.

11.   J. Davalath and Y. Bayazitoglu, Forced Convection Cooling Across Rectangular Blocks, ASME J. Heat Transfer, Vol. 109, pp. 321-328, 1987.

12.   S.H. Kim and N.K. Anand, Turbulent Heat Transfer Between a Series of Parallel Plates With Surface-Mounted Discrete Heat Sources, *ASME J. Heat Transfer*, Vol. 116, pp. 577-587, 1994.

13.   C.Y. Choi, S.J. Kim, and A. Ortega, Effects of Substrate Conductivity on Convective Cooling of Electronic Components, *ASME Transactions J. Electronic Packaging*, Vol. 116, pp. 198, 1994.

14.   D. Wroblewski and Y. Joshi, Computations of Liquid Immersion Cooling For a Protruding Heat Source in a Cubical Enclosure, *Int. J. Heat Mass Transfer*, Vol. 36, pp. 1201, 1993.

15.   T.J. Heindel and F.P. Incropera, Laminar Natural Convection in a Discretely Heated Cavity: I Assessment of Three-Dimensional Effects, *ASME J. Heat Transfer*, Vol. 117, pp. 902, 1995.

16.   V.H. Adams, D.L. Blackburn, Y. Joshi, and D.W. Berning, Issues in Validating Package Compact Thermal Models for Natural Convection Cooled Electronic Systems, *IEEE Trans. on CPMT*, Part A., Vol. 20, pp. 420-431, 1997.

17.   V.H. Adams, D.L. Blackburn, and Y. Joshi, Application of Compact Model Methodologies to Natural Convection Cooling of an Array of Electronic Packages in a Low Profile Enclosure, *Advances in Electronic Packaging*, Vol. 2, pp. 1967-1974, ASME, New York, 1997.

18.   V.H. Adams, D. Blackburn, and Y. Joshi, Package Geometry Considerations in Thermal Compact Modeling Strategies, *Proceedings of the Eurotherm Seminar No. 58 on Thermal Management of Electronic Systems*, Nantes, France, 1997.

19.   V.H. Adams, Three-Dimensional Study of Combined Conduction, Radiation, and Natural Convection From an Array of Discrete Heat Sources in a Horizontal Narrow-Aspect-Ratio Enclosure, Ph.D. thesis, University of Maryland, College Park, Maryland, 1998.

20.   J.R. Howell, 1982, *A Catalog of Radiation Configuration Factors*, McGraw-Hill, New York, 1982.

21.   R.L. Linton and D. Agonafer, Coarse and Detailed CFD Modeling of a Finned Heat Sink, *Proceedings of I-THERM IV, InterSociety Conference on Thermal Phenomena in Electronic Systems*, Washington, D.C., pp. 156-161, 1994.

22.   M.K. Klein, Multi-Level Thermal Analysis Applied to the Design of Electronic Products, *CAE/CAD Application to Electronic Packaging*, ASME EEP-Vol. 9, pp. 59-65, 1994.

23.   A.J. Przekwas, Y.J. Lai, and D. Agonafer, Multi-Scale Thermal Analysis of Electronics Packaging, *CAE/CAD Application to Electronic Packaging*, ASME EEP-Vol. 9, pp.7-14, 1994.

24.   A.J. Przekwas, M.M. Athavale, and Y. Ho, 1996, High Resolution Thermal Design Tools for Air-Cooled Electronics Components and Systems, ASME EEP-Vol. 18, Application of CAE/CAD Electronic Systems, pp. 27-32, 1996.

25.   L. Tang and Y. Joshi, Integrated Thermal Analysis of Indirect Air-Cooled Electronic Chassis, *IEEE Transactions on Components, Packaging and Manufacturing Technology*, Vol. 20, pp. 103-110, 1997.

26.   L. Tang and Y. Joshi, Integrated Thermal Analysis of Natural Convection Air Cooled Electronic Enclosure, *HTD-Vol. 329, National Heat Transfer Conference*, Houston, Texas, Vol. 7, pp.211-220, 1996. Also to appear in *ASME Transactions J. Electronic Packaging*, 1999.

27.   L. Tang, A Multi-Scale Conjugate Thermal Analysis Methodology for Convectively Cooled Electronic Enclosures, Ph.D. thesis, University of Maryland, College Park, Maryland, 1998.

28.   I. Ewes, Modeling of IC-Packages Based on Thermal Characteristics, EUROTHERM Seminar No. 45, *Thermal Management of Electronic Systems*, Leuven, Belgium, 1995.

29.   A. Brandt, Multi-Level Adaptive Solutions to Boundary Valued Problems, *Math. Comput.*, Vol. 31, pp. 333-390, 1977.

30.   A. Settari and K. Aziz, A Generalization of Additive Correction Methods for the Iterative Solution of Matrix Equations, *SIAM J. Numer. Anal.*, Vol. 10, pp. 501-521, 1973.

31.   F.H. Harlow and J.E. Welch, Numerical Calculation of Time-Dependent Viscous Incompressible Flow of Fluid With Free Surface, *Phys. Fluids*, Vol. 8, pp. 2182, 1965.

32.   F.H. Harlow and A.A. Amsden, A Numerical Fluid Dynamics Calculation Method For All Flow Speeds, *J. Comput. Phys.*, Vol. 8, pp. 197-213, 1971.

33.   S.V. Patankar and D.B. Spalding, A Calculation Procedure For Heat, Mass, and Momentum Transfer in Three-Dimensional Parabolic Flows, *Int. J. Heat Mass Transfer*, Vol. 15, pp. 1787, 1972.

34.   P.S. Sathyamurthy and S.V. Patankar, Block-Correction-Based Multigrid Method For Fluid Flow Problems, *Numerical Heat Transfer*, Part B, Vol. 25, pp.375-394, 1994.

35.   S.P. Vanka and G.K. Leaf, An Efficient Finite-Difference Calculation Procedure For Multi-Dimensional Fluid Flows, *Proceedings of the AIAA/SAE/ASME 20th Joint Propulsion Conference*, 1984.

36.   G.E. Schneider and M. Zedan, A Coupled Modified Strongly Implicit Procedure for the Numerical Solution of Coupled Continuum Problems, *Proceedings of the AIAA/SAE/ASME 20th Joint Propulsion Conference*, 1984.

37.   B. Fourka and J.B. Saulnier, Analysis of 3D Conjugate Heat Transfer in Electronic Boards: Interaction Between Three Integrated Circuits, *Eurotherm Seminar No. 45: Thermal Management of Electronic Systems*, Leuven, Belgium, 1995.

38.   J.P. Van Doormaal and G. D. Raithby, Enhancements of the SIMPLE Method for Predicting Incompressible Fluid Flow, *Numerical Heat Transfer*, Vol. 7, No. 2, pp.147-163, 1984.
39.   T.J. Heindel, F.P. Incropera, and S. Ramadhyani, S., 1996, Conjugate Natural Convection From an Array of Protruding Heat Sources, *Numerical Heat Transfer*, Part A, Vol. 29, pp. 1-18.
40.   L. Kadinski and M. Peric, Numerical Study of Grey-Body Surface Radiation Coupled With Fluid Flow For General Geometries Using a Finite Volume Multigrid Solver, *Int. J. Heat Fluid Flow*, Vol. 6, pp.3-18, 1996.
41.   S.P. Vanka, Block-Implicit Multigrid Solution of Navier-Stokes Equations in Primitive Variables, *J. Computational Physics*, **65**, pp. 138-158, 1986.
42.   L. Tang and Y. Joshi, Application of Block-Implicit Multigrid Approach to Three-Dimensional Heat Transfer Problems Involving Discrete Heating, *Numerical Heat Transfer* (To Appear), 1999.
43.   B.R. Hutchinson and G.D. Raithby, A Multigrid Method Based on the Additive Correction Strategy, *Numerical Heat Transfer*, Vol. 9, pp. 511-537, 1986.
44.   P.S. Sathyamurthy, Development and Evaluation of Efficient Solution Procedure For Fluid Flow and Heat Transfer Problems in Complex Geometries, Ph.D. thesis, University of Minnesota, Minneapolis, 1991.
45.   S.B. Sathe, K.M. Kelkar, K.C. Karki, C. Tai, C.R. Lamb, and S.V. Patankar, Numerical Prediction of Flow and Heat Transfer in an Impingement Heat Sink, *ASME Journal of Electronic Packaging*, Vol. 119, pp. 58-63, 1997.
46.   X. Yin and H.H. Bau, Uniform Channel Micro Heat Exchangers, *ASME Journal of Electronic Packaging*, Vol. 119, pp. 89-94, 1997.
47.   J.R. Rujano and M.M. Rahman, Transient Response of Microchannel Heat Sinks in a Silicon Wafer, *ASME Journal of Electronic Packaging*, Vol. 119, pp. 239-246, 1997.
48.   L. Tang, K. Moores, R. Chandrashekhar, and Y. Joshi, Characterizing the Thermal Performance of a Flow Through Electronic Module (SEM-E Format) Using a Porous Media Model, *Proceedings of IEEE SEMITHERM XIV Conference*, San Diego, California, pp.68-77, 1998.
49.   G. Ledezma, A.M. Morega, and A. Bejan, Optimal Spacing Between Pin Fins with Impinging Flow, *ASME Journal of Heat Transfer*, Vol. 118, pp. 570-577, 1996.
50.   T.S. Fisher and K.E. Torrance, Free Convection Limits for Pin-Fin Cooling, *ASME Journal of Heat Transfer*, Vol. 120, pp. 633-640, 1998.
51.   T.J. Liu, A.G. Evans, and J.W. Hutchinson, The Effects of Material Properties on Heat Dissipation in High Power Electronics, *ASME Journal of Electronic Packaging*, Vol. 120, pp. 280-289, 1998.
52.   E.B. Zimmerman and G.T. Colwell, Installation Effects on Air Temperatures Within Outdoor Electronic Cabinets, *ASME Journal of Electronic Packaging*, Vol. 120, pp. 201-206, 1998.

53.   G.K. Morris, S.V. Garimella, and R.S. Amano, Prediction of Jet Impingement Heat Transfer Using a Hybrid Wall Treatment With Different Turbulent Prandtl Number Functions, ASME *Journal of Heat Transfer*, Vol. 118, pp. 562-568, 1996.
54.   G.K. Morris and S.V. Garimella, Orifice and Impingement Flow Fields in Confined Jet Impingement, ASME *Journal of Electronic Packaging*, Vol. 120, pp. 68-72, 1998.

Printed in the United States
by Baker & Taylor Publisher Services